A Summer Plague

A Summer Plague

Polio and its Survivors

Tony Gould

Yale University Press
New Haven & London · 1995

To Jen (again)
and in memory of
David Widgery (1947–92)
physician, friend and fellow polio

Set in Palatino by Best-set Typesetter Ltd, Hong Kong
Printed and bound in Great Britain by Biddles Ltd, Guildford and Kings Lynn

Library of Congress Cataloging-in-Publication Data

Gould, Tony.
A summer plague : polio and its survivors / by Tony Gould.
p. cm.
Includes bibliographical references and index.
ISBN 0-300-06292-3
1. Poliomyelitis–History. I. Title.
RC180.9.G68 1995
616.8'35'009—dc20 94-47253
CIP

A catalogue record for this book is available from the British Library.

For permission to reprint extracts from copyright material the publishers gratefully acknowledge the following: Heinemann for *Changing Patterns* by Sir Macfarlane Burnet; Hutchinson for *Virus Hunters* by Greer Williams; Farrar, Straus for *My Place to Stand* by Bentz Plagemann; William Morrow for *Patenting the Sun* by Jane S. Smith; MIT Press for *Tom Rivers* by Saul Benison; and *New Statesman and Society* for 'The Visit' and 'Down to Earth', articles by the author.

Contents

List of Illustrations vii

Acknowledgments ix

Introduction xi

Part I: *The Rise and Fall of Epidemic Poliomyelitis* 1

1 New York 1916 3

2 FDR 29

3 Warm Springs 41

4 Polio Crusaders 54

5 An Angel Abroad 85

6 A Planned Miracle 111

7 The Quick and the Dead 159

8 Born Too Soon 188

Part II: *Lives of the Polios* 227

In England 229

In America 276

A Civil Wound: an autobiographical coda 305

Appendix: Monkey Business 322

A Note on Sources 325

Notes 328

Select Bibliography 346

Index 354

Illustrations

Between pages 80 and 81

A placard nailed to a house, warning against the dangers of Infantile Paralysis, New York, 1916. (Range/Bettmann/UPI)

An Egyptian stele showing a man with a wasted leg (1580–1350 BC). (March of Dimes Birth Defects Foundation)

Loading a child into an ambulance in the New York epidemic, 1916. (Range/Bettmann/UPI)

Massage and electricity treatment for Poliomyelitis. From a *Manual of Infantile Paralysis*, 1914. (British Library, Department of Printed Books)

A child learning to walk again with leg braces and a walking frame. From Robert W. Lovett, *The Treatment of Infantile Paralysis*, 1916. (British Library, Department of Printed Books)

FDR and fellow polios in the swimming pool, Warm Springs, 1930. (Franklin D. Roosevelt Library, Hyde Park, New York)

FDR at a Thanksgiving dinner, Warm Springs, 23 November 1939. (Franklin D. Roosevelt Library, Hyde Park, New York)

FDR, Basil O'Connor and Mrs O'Connor, Warm Springs, *c.*1928. (Franklin D. Roosevelt Library, Hyde Park, New York)

FDR photographed in his wheelchair at Warm Springs. (Franklin D. Roosevelt Library, Hyde Park, New York)

FDR getting out of his car, Hollywood Bowl, 24 September 1932. (Franklin D. Roosevelt Library, Hyde Park, New York)

The first pool bus, Warm Springs. (March of Dimes Birth Defects Foundation)

The craft shop, Warm Springs. (March of Dimes Birth Defects Foundation)

The Warm Springs 'Poliopolitan Opera Company', Thanksgiving Day show, 1936. (Franklin D. Roosevelt Library, Hyde Park, New York)

Sun-lamp baths, Warm Springs. (Franklin D. Roosevelt Library, Hyde Park, New York)

Iron lung ward, Haynes Memorial Hospital, Boston, Massachusetts. (March of Dimes Birth Defects Foundation)

A six-week-old child in an iron lung, City-County Hospital, Houston, Texas. (March of Dimes Birth Defects Foundation)

Between pages 208 and 209

Still from the March of Dimes film, *The Daily Battle*. (March of Dimes Birth Defects Foundation)

March of Dimes propaganda in the Forties and Fifties. (Franklin D. Roosevelt Library, Hyde Park, New York)

March of Dimes poster child of 1954 with Danny Kaye and Bing Crosby. (March of Dimes Birth Defects Foundation)

March of Dimes poster designed by Erik Nitsche, 1949. (March of Dimes Birth Defects Foundation)

March of Dimes poster, 1949, with poster child Linda Brown. (March of Dimes Birth Defects Foundation)

March of Dimes poster, 1952, with poster child Larry Jim Gross. (March of Dimes Birth Defects Foundation)

The Barnet Ventilator. (British Polio Fellowship)

Administering the Sabin vaccine, Hull, 1961. (Hull Daily Mail Publications Ltd)

Children taking their vaccine impregnated sugarlumps in Scotland. (British Polio Fellowship)

The first British National Swimming Gala for polio survivors organised by the Infantile Paralysis Fellowship, Seymour Hall Baths, London. (Hulton Deutsch Collection)

Sister Kenny. (Minnesota Historical Society)

Cartoon of Sister Kenny. From the Washington *Times-Herald*, 31 March 1945.

Sister Kenny demonstrating treatment. (Sister Kenny Institute)

Jonas Salk. (March of Dimes Birth Defects Foundation)

Albert Sabin. (March of Dimes Birth Defects Foundation)

Schoolchildren participating in the field trial of the Salk vaccine, 1955. (March of Dimes Birth Defects Foundation)

Publicity photograph taken in a polio vaccine bottle company, 1955. (March of Dimes Birth Defects Foundation)

Acknowledgments

Several individuals and institutions have contributed to the making of this book. In the first place I would thank those who subsidised my research both in the United Kingdom and in the United States. I am particularly grateful to Mark Le Fanu and the Society of Authors for a generous grant from the Francis Head Fund; I am also indebted to the Wellcome Trust for a travel grant, and, in America, to the Franklin and Eleanor Roosevelt Institute for a Sidney J. Weinberg research fellowship.

My 'Note on Sources' (pp. 325–7) mentions the various libraries and medical institutions I visited. Here I would like to thank those individuals connected with them who went out of their way to facilitate my research. Professor Roy Porter, at the Wellcome Institute for the History of Medicine in London, was most helpful in getting me started on the project; several people befriended me in various ways at the Franklin D. Roosevelt Library in Hyde Park, none more than the librarian Sheryl Griffith; Joan Headley, at GINI in St Louis, was most generous with her time, allowing me to rummage through her office while she and her assistant, Joe Leone, were hard at work there; Andrew Kemp, of the British Polio Fellowship, was equally hospitable; and the editor of the BPF *Bulletin*, Sue Gearing, allowed me to use her pages to appeal for readers' stories.

So many polio survivors responded to my appeal that I was unable to get to see every one, and of the many that I did see only a small fraction could be included here. My thanks go to all whose stories appear in Part II of this book, but equally to those whose stories – many of them quite as interesting – I had to leave out, or just mention in passing in Part I. In England these included: Laura Andrew, Ernest Atkins, Cathy and Sid Battes, the late Wendy Boulton, Robin Buxton, Pamela Gavan, Mary Gillies, Pamela Goodman, Kathy Gordon, Evelyn Housman, Janet Keegan, Carole King (and her mother, Dora Lander), Ken Mercer, Rosalind Murray and Peter Preston. Among those who wrote to me more than once I would like to thank both Jean Saunders and Barbara Huelin (not herself a 'polio' but a mother of one); and I owe a special debt of gratitude to Grace Potts of the Wakefield branch of the BPF, a nurse who married a polio, for writing me wonderful letters and then organising a memorable tea party for me in Wakefield.

In America, too, I saw several polios whose stories could not be included in full. Foremost among them I should mention Jeanne Houghton, who took in a complete stranger and his wife and made us feel utterly at home during our week's stay in Washington DC; also the late 'Pidge' Cole, Margaret Ernest, Jack Genskow and his wife Lil, Adele Gorelick (whose gift of a book I would not otherwise have known about came in very handy), Henry Haverstock Jr, Todd Keepfer and his wife Pat, Stanley Lipshultz, Emelie Simha and Diane Rice Smith. Hugh Gregory Gallagher and Geoffrey Ward, both Roosevelt biographers as well as polios, were most friendly and helpful, as were Dr Richard Owen, now Emeritus Medical Director of the Sister Kenny Institute, and Jessica Scheer of the Washington DC Polio Society, an anthropologist who has made a special study of the polio-disabled. At Warm Springs, the Medical Librarian Michael Shadix gave me the run of the archive, and two retired physical therapists, Mary Hudson Veeder and Viva Erickson, the benefit of their experience and memories. Three medical historians, Saul Benison, Naomi Rogers and Daniel J. Wilson, all spoke to me either in person or on the phone about their work.

Various drafts or extracts from the book were read and commented upon by Victor Cohn, Professor George Dick, Dr Bernard Dixon, Frank Kelly, Joy Melville, Dr Geoffrey Spencer OBE and Ger Wackers. (They are not, of course, responsible for its faults.) Other useful information was gleaned from the late Duncan Guthrie and from Vivian Wyatt; and the orthopaedic surgeon Lancelot Walton (at the age of ninety) entertained me at his home on the island of Alderney with stories of Sister Kenny and games of table tennis.

Finally, a word of thanks to those (non-polio) friends in the United States who were so generous with their hospitality – Abby Collins and her daughter Kate, John Dyck and Beth Hardesty, Kathryn Dyck, Peter Metcalf, David Posnett and Nhu Thanh and their daughter Kim Thu, and Libby and Laurie Weisheit. Not forgetting the closest tie of all, my thanks to my wife Jenny, whose PhD. in history well qualified her for the role of joint researcher.

Introduction

Now, in the dying years of the twentieth century, mention the word *epidemic* and AIDS springs to mind. But for the greater part of the century, it was polio that spread panic and paralysis in unequal proportions through the populations of the Western world.

It was never a killer on anything like the scale of those hardy perennials, cancer and heart disease; it specialised more in maiming people and became known familiarly as 'the Crippler', after the opening sequence of a propaganda film put out in the United States by the National Foun-dation for Infantile Paralysis (or 'March of Dimes', as it was commonly called) after the Second World War. A figure leaning on a crutch – sinister in its invisibility – stalks the land like death itself, casting an elongated shadow over a farmer's boy who is innocently helping his father feed the horses, and over a pretty little girl sitting on the steps of her house. 'And I'm *especially* fond of children,' the voice-over intones with fiendish glee, as the boy wilts in the arms of his anxious father and the girl raises a hand to her fevered brow . . .

Diseases are no respecters of national boundaries, and polio first made its presence felt as an epidemic disease at the end of the nineteenth century in Scandinavia quite as much as in the United States. But the Americans made it their own – for two reasons initially. One was that the first *major* epidemic took place in New York and the surrounding states in 1916, and made an indelible impression on the American consciousness; and the other, that five years later the man who, from early in his career, was earmarked for the presidency, Franklin Delano Roosevelt, fell victim to what was then known as infantile paralysis – humiliating for a man of nearly forty and a father several times over to succumb to an *infant's* disease.

FDR's determination to walk again took him to all kinds of outlandish places, none further from his usual haunts than Warm Springs, Georgia, where the natural thermal waters were reputed to have healing powers. His creation there of a hydrotherapy centre for polio victims was the beginning of a 'crusade' which gathered momentum throughout his lifetime and culminated only after his death in the development of the vaccine forever associated with the name of Jonas Salk. The 'conquest of polio' was an American – perhaps *the* American – success story of the Fifties (the decade of conformism and the Cold War), and Salk, the bespectacled doctor in a white coat holding a test tube up to the light, became an instant celebrity.

An earlier celebrity was Sister Elizabeth Kenny, 'heroic daughter of the Australian bush', according to one citation – a kind of Dame Edna Everage *avant la lettre* in her self-presentation, at least. She took America by storm soon after her arrival there in 1940 and became such a legendary figure that she was made the subject of a Hollywood biopic starring Rosalind Russell. Her treatment for polio – based on the premise of getting patients up on their feet and moving at the earliest possible moment – was perhaps less revolutionary than she imagined, but still flew in the face of the prevailing orthodoxy, which prescribed long periods of immobility in plaster-casts as a precaution against developing deformities. Her theories of the nature of what she insisted on calling a 'new malady . . . misnamed "poliomyelitis"' were even more revolutionary than her treatment, but less soundly based. There was about her an extraordinary mixture of the dedicated healer and the charlatan, of the miracle-worker and the fraud. Yet the cultish aspects of her approach should not be allowed to detract from her exemplary achievement.

What Kenny was to treatment, Salk was to prevention. But that was not the end of the story. Within a few years of its triumphant launch – to the accompaniment of church bells ringing all over the United States – Salk's vaccine, which required a course of injections, had been superseded almost everywhere in the world, including his own country, by one that could be taken orally on a sugar-lump (until the dentists objected), developed by his great rival for polio honours, Albert Sabin. Between them these two medical scientists, backed by a host of others less well-known, effectively put paid to polio – in the West. But despite repeated predictions of its imminent elimination throughout the world by the World Health Organisation, and a commitment on the part of Rotary International to achieve this by the year 2005, paralytic polio is still a major scourge in some of the most populous parts of the globe – China and South-East Asia, India and Africa.

In the West – and, once again, particularly in the United States – the focus has shifted from disease control to the plight of the host of survivors of the massive polio epidemics of the late Forties and early Fifties, whose lives have been cast into jeopardy by alarming symptoms of fatigue and new muscular weakness suffered without warning thirty to forty years after the initial attack. Variously known as Post-Polio Syndrome and the Late Effects of Polio, the new ailments of polio survivors are still something of a medical mystery, but their effect on these survivors has been dramatic.

In the same way that Americans *en masse*, orchestrated by the populist and powerful National Foundation, set out to conquer epidemic poliomyelitis (and succeeded in doing just that), individual polio sufferers, following FDR's example, determined to triumph over it in their own lives. The great thing about polio, every victim of it soon learned, was that it was *not* a progressive, degenerative disease like muscular dystrophy or multiple sclerosis: what you gained, you held. This comforting belief enabled hordes of polio people, even quite severely disabled ones, to go about their lives as though they were not in any way disadvantaged. This was called 'mainstreaming'. It meant competing on equal terms with the able-bodied and 'denying' one's disability. It resulted, as with FDR, in some remarkable success stories; think of the great Israeli violinist Itzhak Perlman, for example, or the British pop singer Ian Dury (whose story appears below in Part II) or the Australian writer Alan Marshall, author of the classic of (polio) childhood, *I Can Jump Puddles*.

What the new weaknesses experienced by 'old polios' highlighted was the fragility of their hard-won position out there in the world of the 'normals'. With the loss of essential muscle power and the onset of debilitating fatigue, past heroics counted for nothing and the proud boast of independence acquired a hollow ring. For the first time, in many cases, polio people sought out one another and compared notes – not just about their current vulnerability, but about their past experiences too. People were both surprised and comforted to find that what had once seemed so unique was in reality part of a common pattern; it had just been occluded by the emphasis on mainstreaming. And the new-found solidarity was intensified by the sense of being left over from a previous era, like the survivors of some forgotten war.

The history of epidemic poliomyelitis has several aspects – scientific, clinical, political and experiential – and my aim, in the first half of this book, has been to weave together these various strands, which have in the past been treated separately, if at all.

Journalists and historians have inevitably focused on the scientific and political story of the development of the polio vaccines because that is such a dramatic story. But in the process the experience of the polio patient, the actual sufferer, often gets lost. Part II is therefore devoted entirely to the lives of individual polios; and in the historical part, too, I have tried to bring them centre-stage wherever possible.

It is not just a question of 'background', or of providing the context, but of insisting on the centrality of the polio experience, and of the recurring nightmare of increasingly severe epidemics over half a century, in the face of which the pressure to produce an effective prophylactic mounted. This led, in the Thirties in the United States, to premature human experimentation with vaccines that were either lethal or ineffective; and it accounts for the enormous investment of hope in the Salk vaccine two decades later and the correspondingly huge relief when it was pronounced 'safe, effective and potent' after the most extensive and thorough field trial in the history of medicine.

In Britain there was never the same urgency about polio; it was a comparatively rare disease, with only local and small-scale epidemics, until after the Second World War, and – partly for that reason – there was no pressure group comparable to the National Foundation for Infantile Paralysis. The NFIP was, indeed, an entirely new kind of voluntary organisation, aggressively propagandist in a manner that has since become commonplace, spending large sums of money in order to raise even larger sums; in many ways it was a precursor of today's telethons. The British Polio Fellowship (originally called the Infantile Paralysis Fellowship) is a completely different kind of outfit, a social organisation – as befits the word *fellowship* – intent on self-help and mutual support.

The nearest thing to a panic over polio in Britain was precipitated by the death of the England footballer Jeff Hall in April 1959 (the same month and year that I contracted the disease on the other side of the world in Hong Kong). By that time, the means of prevention – the Salk vaccine – was available; it simply wasn't being taken up on a mass scale. Hall's death changed that; the message finally got through to teenagers on the terraces, reinforced by broadcasts at half-time during football matches, and in the dance-halls where the more enterprising health authorities penetrated; emergency clinics were set up, and there was such a run on stocks that further supplies of vaccine had to be flown in from the United States.

Elsewhere in Europe, probably the most significant event was the Copenhagen epidemic of 1952, in which a shortage of tank respirators – or iron lungs, as they are more commonly (and colourfully) known – prompted the adaptation of respiratory techniques originally designed for use in the operating theatre. Several polio patients, mostly children, were kept alive by relays of medical and dental students ventilating them by hand, using a kind of bellows, in the absence of suitable machinery – a case, if ever there was one, of necessity proving to be the mother of invention. This response to an emergency led to a revolution in artificial respiration, with *positive*-pressure ventilation – air pumped in and out of the lungs, generally via a hole cut in the throat, or tracheotomy – by and large replacing (on this side of the Atlantic, at least) *negative*-pressure ventilation: for example, the iron lung, which works by *external* pressure, alternately contracting the chest by filling the tank with air and expanding it by creating a vacuum.

Nowadays, with polio so largely a thing of the past in the West, one must seek revolutionary techniques elsewhere in the world – as two English doctors have done. Dr Geoffrey Spencer, until his recent retirement Clinical Director of the Lane-Fox Respiratory Unit at St Thomas's Hospital in London (a veritable haven for polio survivors), and his colleague, the orthopaedic surgeon Mr Frederick Heatley, read an article submitted to the *Journal of Bone and Joint Surgery* in 1988 by a group of doctors at Military Hospital 208, at Changchun, in northern China, and saw at once that these Chinese surgeons had developed an original operation for straightening knee joints which were over-extending (bending backwards) as a result of polio. Spencer and Heatley were at first frustrated in their attempt to obtain permission to visit China and observe this operation, but after the Tiananmen Square massacre the official Chinese attitude changed overnight and their visit was encouraged. As a result of what they learned in the week they spent in Changchun, the St Thomas's doctors have incorporated the knee operation in their repertoire and performed it successfully over a dozen times, thus enabling polio survivors who might otherwise have been forced into wheelchairs to keep on walking.

When I set out to write this book I knew very little about paralytic polio, even though I'd had it myself. I knew that it was caused – like most things these days, it sometimes seems – by a virus; that the virus attacked the nerves in the spine, cutting off the impulse from the brain to the muscles and causing the affected muscle fibres to shrivel and die. I was aware that the unpredictable element came in through not knowing how much of the initial damage was temporary – the nerves, as it were, merely stunned, rather than taken out by the virus – and how much was permanent; and that the only way to find that out was through a rigorous programme of physiotherapy lasting anything up to two years. If you were fortunate, as I was, enough came back to enable you to lead an independent life and, so to speak, put it behind you.

I knew that Roosevelt had had polio, but not that he was effectively wheelchair-bound from then on. I had heard of the attractively named Warm Springs and knew that the place was connected with both Roosevelt and polio, but nothing more. I'd heard of Salk and the polio vaccine, but had no idea there was more than one vaccine and that Salk's was no longer in general use. I'd heard of Sister Kenny and may even have known that she was Australian, but I couldn't have said why I knew her name except that it related in some way to polio. I'd even heard rumours of something called Post-Polio Syndrome – and I was *quite* sure I didn't want to know any more about *that*.

For a writer, every book is some sort of voyage of discovery and self-discovery, and this one is no exception. My original intention was to preface the history of polio with my own story, thus moving out from the particular to the general. But the bigger drama took over and dictated a different pattern. I had always intended to present other people's stories, and I spoke at length – and sometimes more than once – to some twenty polio survivors in England and about the same number in the United States. It seemed to me that the testimony of witnesses would give resonance to the history, and vice versa. Then I came to realise that my own story belonged at the end rather than the beginning – among the other 'lives of the polios', many of them so much more remarkable than my own.

For what I discovered in the process of researching and writing this book was that I, too, had been so busy 'mainstreaming' and 'denying' my disability over the years that I had simultaneously both overvalued and undervalued it. I had overvalued it in the sense of regarding my experience of polio as somehow unique (hence my original intention of giving it pride of place here), and undervalued it by not allowing that it had made any substantial difference to my way of life and thinking. Certainly, I'd always regarded this as a pivotal episode in my life, one that had changed its direction; but I had also thought of it as over and done with.

I don't any longer. The knee on my 'good' leg may be cracking up a bit (providing me with a reason for being personally interested in the Chinese knee operation), but so far, mercifully, I have been spared the more alarming Late Effects of Polio. Yet I have learned that, however impressive a recovery you make, you don't 'conquer' or 'overcome' polio in any meaningful sense, you merely adapt to the limitations it imposes and – if you're fortunate – discover within yourself resources you might not otherwise have found.

Note: Any book dealing with both British and American experience comes up against the 'barrier of a common language', as Dylan Thomas called it. In addition to the usual lift/elevator, boot/trunk, ground floor/first floor, public school/private school dichotomies, a book about polio gives us caliper/brace and stick/cane. I have assumed that most people are conversant with the differences in Anglo-American usage and have therefore used whichever seemed appropriate in any given context.

Part I

The Rise and Fall of Epidemic Poliomyelitis

We can see polio more clearly than most diseases because its rise and fall took place within a single lifetime.

John Rowan Wilson, *Margin of Safety* (1963)

1
New York 1916

One of the most amazing things about poliomyelitis is that no epidemic of it was noted until seventy-one years ago. Large epidemics of other virus diseases, such as smallpox, yellow fever, influenza, and measles, are recorded much farther back in history.

Greer Williams, *Virus Hunters* (1960)

No doubt many scenes which occurred in London during the great plague of 1665 were reenacted in our Long Island and Westchester towns. Under the sway of panic people looked with skepticism and suspicion on government health officers. The selectmen of many villages, whose doctors were struggling with the impossible and failing to stop the epidemic or save the individual case from paralysis, resorted to home-made martial law. Deputy sheriffs, hastily appointed and armed with shot-guns, patrolled the roads leading in and out of towns, grimly turning back all vehicles in which were found children under sixteen years of age. Railways refused tickets to these selected youngsters, the innocent victims of ignorance and despair. Indeed, the notion was firmly held that below the magic age, called sweet at other times, there lurked the dread disease, whereas above it no menace existed either for the individual or the community.

George Draper, *Infantile Paralysis* (1935)

The epidemic of 1916 will go down in history as the high-water mark in attempts at enforcement of isolation and quarantine measures.

John R. Paul, *A History of Poliomyelitis* (1971)

In the summer of 1916, while European armies battered each other senseless, New Yorkers were engaged in a very different kind of struggle. An invisible enemy was killing and crippling children, in particular, quite as effectively as bullets and shrapnel killed and maimed the infantrymen stumbling through the wire and mire of the Somme. On 1 July – day one of the Somme offensive – New York City's health commissioner, Dr Haven Emerson (a great-nephew of Ralph Waldo Emerson), was quoted in the *New York Times* as saying: 'Our method of fighting the disease is this: whenever a case is reported in a block not previously affected, a house-to-house canvass of that block is made. In this way many unreported cases have been found. All cases are isolated at once . . .'[1]

Since the outbreak started in the Italian community, immigrants became the scapegoats. According to the *NY Times* that same day, 'There was a report yesterday that the disease had been brought to America by Italian immigrants,' though inquiries at the Quarantine Station at Ellis Island drew a blank: 'no cases had been noticed among immigrants and . . . Quarantine had no record of epidemics in any of the towns of Italy'. Yet the idea of a non-American source for such an unAmerican outbreak was too attractive for mere facts to be allowed to stand in its way. Haven Emerson himself favoured it. On 8 July, the *NY Times* reported him as saying that

> since May 15 . . . 90 immigrant Italians, including 24 children under the age of 10, had gone to live in Brooklyn, where the outbreak appeared, at about the date named. No symptoms of infantile paralysis were found among the Italians at Quarantine, but the Italian consul had been asked to ascertain whether the disease was present in any of the cities from which the immigrants came.[2]

The next day, the health commissioner was again on the offensive. 'Dr Emerson blamed the Brooklyn citizens themselves for whatever garbage there was in their streets. Brooklyn, he said, had not developed sufficient pride to keep its own streets clean.' Strict sanitary regulations were in force and reports of 'violations of the Sanitary Code', leading to court cases and fines, appeared in the press with monotonous regularity. Yet the same report that spoke of Emerson's condemnation of Brooklyn's insanitary habits contained an unusually frank summary of the Department of Health's predicament:

> In epidemics of typhoid fever and most other diseases the health authorities know exactly what to do. But fighting infantile paralysis consists largely in doing everything that seems effective in the hope that some of the measures taken will be effective.[3]

This statement encapsulates Emerson and his colleagues' dilemma. The emphasis on hygiene was, in a sense, all that they knew: cleaning up New York could do no harm, and might well do good; whether it would radically alter the number of children, or indeed adults (since by no means all the victims were minors), stricken with poliomyelitis was a question to which they did not have the answer. But they could not afford to doubt;

like the Scottish preacher who underlined a part of his sermon and wrote in the margin, 'Weak point here, shout like hell', the authorities were relentless in enforcing sanitary measures.

In response to a recommendation by an advisory committee of doctors, they ordered 'the placarding of premises for poliomyelitis, a practice previously confined to smallpox, scarlet fever, diphtheria and measles'.[4] And on 4 July the *NY Times* reported, 'Commissioner Emerson and his lieutenants in the fight discussed yesterday a proposal for the police to compel every child in the city under 16 years of age to remain at home continuously for two weeks.'[5] Fortunately for parents, this proposal was not adopted and playgrounds remained open, though an order to exclude children under sixteen from 'moving picture shows' came into effect the next day.[6]

Even before the end of June, the Board of Health had passed resolutions increasing the period of isolation of patients from six to eight weeks, and demanding immediate hospitalisation of all those who could not satisfy stringent quarantine requirements in the home.[7] This meant that the children of the comfortably-off could be nursed at home, while children from poor and overcrowded tenements had to be removed to hospital, despite the protests of their parents – not to mention their doctors. (The Department of Health recognised that 'the administrative regulation of the disease implied not only interference with the personal liberty of the members of many households, but a sacrifice of important professional opportunities and income by the physicians to the poor, whose poliomyelitis patients were removed from insufficiently equipped homes to the hospitals'.[8]) But keeping a child at home was not always an easy option, since quarantined premises were visited every other day by patrolmen of the so-called Sanitary Squad, who ensured the regulations were upheld.

Mrs Anna Henry, a nurse at a babies' clinic in Brooklyn, was so zealous in reporting cases of infantile paralysis and violations of the sanitary regulations in the Italian colony known as 'Pigtown' that she received a 'Black Hand' letter threatening her life and had to be escorted to and from work by police.[9]

Many mothers felt that sending their children to hospital was tantamount to condemning them to death or crippledom. They were hardly reassured by the information that suspected cases were kept separate from confirmed ones, and they protested against the forcible removal of their children. One woman in Brooklyn was reported in the *New York Journal* to be receiving medical attention as a direct result of a police raid on her house to remove her two-year-old nephew. Three policemen with drawn revolvers apparently burst into her room, tore the infant from her arms and passed him out of the window to a waiting surgeon.[10]

It was probably in response to such stories that Dr Emerson substituted nurses for policemen on ambulance duty. He told reporters that 'trained women were more successful in persuading parents to let their children go to hospital'.[11] But nothing could remove parents' fears. A social worker in a poor district reported to the social services committee of the Volunteer Hospital:

> The mothers are so afraid that most of them will not even let the children enter the streets, and some will not even have a window open. In one house the only window was not only shut, but the cracks were stuffed with rags so that 'the disease' could not come in. The babies had no clothes on, and were so wet and hot they looked as though they had been dipped in oil, and the flies were sticking all over them. I had to tell the mother I would get the Board of Health after her to make her open the window, and now if any of the children do get infantile paralysis, she will feel that I killed them. I do not wonder they are afraid. I went to see one family about 4 p.m. Friday. The baby was not well and the doctor was coming. When I returned Monday morning there were three little hearses before the door; all her children had been swept away in that short time. The mothers are hiding their children rather than give them up . . .[12]

One woman took the Department of Health to court for forcibly removing her child to hospital. The judge was sympathetic. In his summing-up, he said:

> This good lady, the mother, tender, affectionate, cleaving to her child, wanting to have it at home was uncertain what she ought to do. . . . She was evidently confused by the conflicting opinions of the medical gentlemen with whom she . . . came into contact . . . The safeguards which are recognized as necessary to be thrown around a case of this particular disease, were taken down and removed by her, largely upon the opinion of . . . Dr Smith and Dr Flynn, and naturally enough, too, because their opinions were in the line of her motherly affection and her motherly instinct, and in the line also of her personal desires as to herself, her children and others, that she should be free to receive anybody into the house and go into other houses and out upon the street, as though there were no communicable disease affecting the child.[13]

Nevertheless, Mr Justice Garretson ruled in favour of the Board of Health, dismissing the suit.

The sad story of 'Paul Hughes, 5 years old, who lived with his parents at Great Kills, SI [Staten Island]', indicates the high level of collective hysteria engendered by the epidemic. Paul had fallen ill one night and his condition had so far deteriorated by dawn that his father sought medical attention:

> Unable to obtain a physician, he put the boy into an automobile and drove to the Smith Infirmary at New Brighton, but the child died on the way and

> the doctors at the hospital would not receive the body, nor would they permit it to go in the hospital morgue for fear that infantile paralysis had caused death and might spread from the morgue to the hospital. Mr Hughes was unable to reach any Health Department official in the early morning, and he drove around Staten Island with the boy's body for hours looking for some one who would receive it. Thomas McGinley, an undertaker at Stapleton, finally communicated with the Department of Health in Manhattan and obtained permission for the father to leave the body at the department's disinfecting plant at Four Corners. Coroner Vail will hold an autopsy today to determine whether infantile paralysis caused death.[14]

Another indication of the panic precipitated by this latter-day plague was the wholesale exodus from the city of the children of the well-to-do. On 5 July, the *NY Times* 'conservatively' estimated that 50,000 of them had been sent out of New York 'to places considered safe by their parents'; and on 7 July, 'Reports of persons fleeing from town continue to come in.' Equally panicky was the response of several neighbouring states and communities, which took defensive action against the unwanted intruders. 'Hoboken [New Jersey] led the way by isolating itself from the world, so far as new residents were concerned', the *NY Times* reported on 14 July.

> Policemen were stationed at every entrance to the city—tube, train, ferry, road, and cowpath—with instructions to turn back every van, car, cart, and person laden with furniture and to instruct all comers that they would not be permitted under any circumstances to take up their residence in the city.

(Despite these precautions, the first case of infantile paralysis in Hoboken was recorded twelve days later.) On Tuesday 18 July, under the headline 'QUARANTINE GUARDS ON ROAD', the *NY Times* informed its readers that 'from noon Saturday until noon yesterday 150 families who attempted to enter Hastings-on-Hudson by train and automobile from New York were turned back by policemen stationed at every entrance to the village'.[15]

The Department of Health responded to pleas from the railroads and other transport organisations by issuing 'health certificates, or more properly travellers' identification cards . . . to those who presented themselves . . . for medical inspection and could prove residence at an address from which no case of poliomyelitis had as yet been reported'.[16] Dr Emerson attempted unsuccessfully to calm the public by repeating that there was 'no cause for alarm' and that the epidemic appeared to be 'on the wane'. (This was not true: the climax was not reached until the week beginning 5 August, during which 1,151 cases and 301 deaths from polio were recorded, though the worst single day was 3 August – the day on which

Sir Roger Casement was executed for treason in England – with 217 new cases reported.) On Wednesday 19 July, 1,031 health certificates were issued to anxious parents and on the Thursday the number rose to 2,079. By then the US Public Health Service had taken over responsibility for the supervision of interstate travel, though the chaos continued. 'Scenes in railroad and steamboat terminals of the city yesterday', the *NY Times* reported on Monday 24 July, 'were indicative of the numbers going away and of the numbers returning against their will.'[17] Emerson's policy of publishing in the newspapers' daily lists of the names and addresses of confirmed cases of polio can have done little to reassure an already jittery public.

In the continuing search for scapegoats, black Americans were exonerated since 'it has been found by the Department of Health that negro children are less susceptible to the disease than white children.' This was announced in early July. A week or so later it was reported that 'Dr Ager [assistant health commissioner] has concluded that negro children are more or less immune and that the virus attaches itself more often to blondes than brunettes.'[18] By 26 August, however, Haven Emerson was writing to the director of the Rockefeller Institute for Medical Research, Dr Simon Flexner:

> Since the conference on Poliomyelitis on August 3rd and 4th, there has been such an increase in the number of cases of polio among colored people that the impression which I had at the time of our meeting and which I conveyed to you, I believe has been entirely dispelled, and it is apparent that the race incidence of the disease among colored people is not essentially different from its incidence among other racial groups in this city.[19]

But the myth that black people did not get polio persisted.

Cats and dogs were suspected of being carriers of the disease; strays were rounded up, and pets put down. In early July animals were being destroyed at a rate of '300 to 450 a day'. A headline in the *NY Times* for 26 July proclaimed '72,000 CATS KILLED IN PARALYSIS FEAR' and in Brooklyn, according to another account:

> Much assistance was rendered the police by small boys, due to the circulation of a report that a bounty was to be paid on all the animals turned over to the authorities. Some stations were literally flooded with cats and small boys, the former demanding their liberty and the latter their money.[20]

Unfortunately for the latter, the rumour of a payment of ten cents per head turned out to be false. (Perhaps it was put about by the police themselves to facilitate their task.)

In any event, the wholesale slaughter of inoffensive animals was misguided. 'The epidemiological evidence collected during the progress of this investigation', C.H. Lavinder, A.W. Freeman, and W.H. Frost of the

US Public Health Service conclude, 'lends no support to the idea of any relationship between paralytic disorders of animals and the occurrence of poliomyelitis in human beings.'[21]

Dr W.H. Frost's was one of the few voices of sanity to be heard during the epidemic. The *NY Times* reported him as saying that 'the incidence of poliomyelitis is proportionately about the same among persons living under good conditions as among persons living under poor hygienic conditions'. He went on:

> This practically eliminates from consideration as of great importance in the causation of the disease such factors as are intimately associated with poor hygienic conditions, insufficient and improper food, overcrowding, personal uncleanliness, and association with verminous insects.[22]

Dr Charles Caverly, the health officer who had made a study of the first significant outbreak of infantile paralysis in America, which took place in Vermont in 1894, had made precisely the same observation in the *Journal of the American Medical Association* in 1896:

> That the general sanitary surroundings and methods of living were in anywise responsible for the outbreak is . . . more than doubtful, since the disease showed no partiality to that class of the population whose habits and surroundings are the most unsanitary. The so-called laboring classes were oftenest affected, but not out of proportion to their numbers . . . general sanitary conditions did not seem to have any influence on the epidemic.[23]

But three days after printing Dr Frost's words of wisdom, the *NY Times* reported, '"Clean-up" is being more persistently shouted at tenement house dwellers every day by public and private institutions.'[24]

Two American doctors who wrote a textbook on 'Infantile Paralysis' in the wake of the 1916 epidemic described it as 'a disease of comparatively recent origin'.[25] It was not the disease itself, but outbreaks of epidemic proportions that were of recent origin. An Egyptian stele, dating from the period 1580–1350 BC and depicting a young man with a withered leg leaning on a long staff, suggests that polio has been endemic since ancient times.

The term *poliomyelitis* derives from the Greek words, *polios*, meaning 'grey', and *myelos*, 'matter', and refers to the grey matter of the spinal cord. The disease was called by many names in the nineteenth and early twentieth centuries, including: Dental Paralysis, Infantile Spinal Paralysis, Teething Paralysis, Essential Paralysis of Children, Regressive Paralysis, Myelitis of the Anterior Horns, Tephromyelitis (from the Greek *tephros*, meaning 'ash-grey') and, most poetically, Paralysis of the Morning – after the way in which a child goes to bed apparently healthy, wakes feverish in the night and then is unable to get up in the morning.[26] The number of names – and there were several others – reflects the confusion over the nature of the disease.

Perhaps the earliest recorded case is that of Sir Walter Scott (born in Edinburgh in 1771), who was led to believe that he 'showed every sign of health and strength' till he was about eighteen months old. Then:

> One night, I have been told, I showed great reluctance to be caught and put to bed, and after being chased about the room, was apprehended and consigned to my dormitory with some difficulty. It was the last time I was to show much personal agility. In the morning I was discovered to be affected with the fever which often accompanies the cutting of large teeth. It held me three days. On the fourth, when they went to bathe me as usual, they discovered that I had lost the power of my right leg. . . . There appeared to be no dislocation or sprain; blisters and other topical remedies were applied in vain . . .
>
> The impatience of a child soon inclined me to struggle with my infirmity, and I began by degrees to stand, to walk, and to run. Although the limb affected was much shrunk and contracted, my general health, which was of more importance, was much strengthened by [my] being frequently in the open air, and, in a word, I who in a city had probably been condemned to helpless and hopeless decrepitude, was now a healthy, high-spirited, and, my lameness apart, a sturdy child . . .[27]

The first attempt at a clinical description of the disease was made by the English physician, Michael Underwood, in the second edition of his treatise on the *Diseases of Children*, published in 1789. He calls it 'Debility of the Lower Extremities' and writes: 'It is not a common disorder, I believe, and seems to occur seldomer in *London* than in some parts. Nor am I enough acquainted with it to be fully satisfied, either, in regard to the true cause or seat of the disease, either from my own observation, or that of others.' Nevertheless, he is inclined to attribute it to 'teething and foul bowels'. Where both lower extremities 'have been paralytic, nothing has seemed to do any good but irons to the legs, for the support of the limbs, and enabling the patient to walk'.[28] (A later editor of Underwood's treatise, obviously unfamiliar with the flaccidity of the paralysis resulting from polio, comments parenthetically: 'If the limbs are paralytic, how are irons to the legs to enable the patient to walk?')[29]

The first account of an outbreak of the disease was written by a young doctor called John Badham, the son of a distinguished professor of medicine at the University of Glasgow. It took place in 1835 in Worksop, in north Nottinghamshire. There were four cases, described by Badham in meticulous detail. He comments on the 'extraordinary youth' of all four patients; on the 'cerebral symptoms', such as drowsiness or abnormality of the pupils of the eye; on the 'remarkable [fact] that in no one instance has the health of the child been in any degree impaired'; and on the strabismus (squinting) apparent in one case, leading him 'to suspect a cerebral complication, rather than a spinal one'.[30] Unfortunately, the thirty-two-year-old John Badham died of consumption in 1840, the very year in which the first systematic investigation of poliomyelitis, written partly in response to his account of the Worksop outbreak, was published in Germany.

Like Badham, Jacob von Heine draws attention both to the extreme youth of patients (six months to three years) and to their good general health (though he is referring to their health *preceding* the attack). But where Badham sees only 'drowsiness', Heine recognises fever and pain in children during the pre-paralytic phase of the illness, which makes him think that the disease may be contagious, and far from suspecting 'a cerebral complication, rather than a spinal one', he finds no cerebral involvement and concludes that all the symptoms 'point to an affection of the central nervous system, namely the spinal cord'.[31]

In the matter of treatment (which had baffled Badham), Heine steered clear of fashionable nostrums such as purges, emetics, blisters and bleedings, and recommended 'exercise, baths, and various simple surgical procedures, followed by the application of braces and apparatus'.[32] As the historian of the disease and a leading participant in its 'conquest', Dr John R. Paul of Yale University points out, 'Considering the degree to which the handling of a given disease is wont to change over a period of 125 years, Heine's treatment of paralyzed limbs and the resulting deformities and disabilities of children has undergone remarkably little alteration.'[33]

In 1907 the Swedish paediatrician, Ivar Wickman, named the disease 'Heine-Medin disease' after both the great German orthopaedist and the Swedish pioneer, Karl Oskar Medin, whose pupil Wickman was. Medin's involvement in an unprecedented outbreak of forty-four cases in Stockholm in 1887 led him to categorise various types of the disease – spinal, bulbar, ataxic, encephalitic and polyneuritic – as well as, crucially, to conclude that its acute phase consisted of two separate fevers, sometimes with a fever-free remission (what the American doctor George Draper mis-labelled the 'dromedary form' – it is the Bactrian camel, not the dromedary, that has two humps). The initial fever was no more than a general malaise; it was the second attack that did the damage to the central nervous system.[34]

The significance of this finding was not lost on the young Ivar Wickman when he came to investigate the infinitely more serious Scandinavian epidemic of 1905, in which there were more than 1,000 cases. The questions that concerned him were: was the disease contagious and, if so, how was it spread – by direct contact with infected children, or by carriers of the virus who themselves showed no sign of infection? Others, including Dr Charles Caverly in his notes on the 1894 Vermont outbreak, had observed that there could be 'abortive' or non-paralytic cases. Wickman's originality lay in suggesting that non-paralytic cases were both far more widespread than anyone had supposed and instrumental in spreading the disease. His experience of the 1905 epidemic convinced him that Heine-Medin disease was highly contagious, and that apparently healthy or only mildly affected persons played a key role in spreading it.

Although he committed suicide at the age of forty-two, just two years before the New York epidemic, Wickman's historic monograph of 1907 earned him a place in the Polio Hall of Fame erected in 1958 at Warm Springs, Georgia, to celebrate the twentieth anniversary of the National Foundation for Infantile Paralysis (where fifteen doctors and scientists are honoured, along with the founders of the NFIP, Franklin D. Roosevelt and Basil O'Connor).

The end of the nineteenth and the beginning of the twentieth century was once considered a medical golden age, but has recently been dubbed 'the Childhood of Scientific Medicine, a period of great stimulus and rapid growth, filled with the excitement of learning new things – and also filled with childish certainties'.[35] Following Robert Koch's painstaking and brilliant work on anthrax, tuberculosis and cholera, and Pasteur's discovery of a rabies vaccine (a century after Jenner had started the whole business of vaccination by deliberately infecting people with cowpox as a preventive of smallpox), 'Everybody, everywhere, tried to hunt microbes, see them, grow them, identify them, explain them, escape them. Their primary activity . . . was finding and naming the causes of infectious diseases. It was the Day of Diagnosis.'[36] The microbes being so assiduously hunted were largely bacteria, and the essential tackle needed for their capture included the microscope, dyes that highlighted the micro-organisms, culture dishes, test tubes and some unfortunate laboratory animal to act as involuntary host. Bacteriology, or microbiology, was the name of this sport; virology had yet to be born.

As far as microbe hunters were concerned, the only difference between bacteria and viruses was one of size. Bacteria were the organisms which would not pass through a porcelain filter; viruses the ones which would. If the filtration process succeeded in sterilising cultures of organisms, then they were bacteria; if not – and laboratory animals could be reinfected *after* filtration of the culture – it was called a virus. Bacteria became visible under the microscope if stained with certain types of dye; viruses, however, were too small to be seen, even through an optical microscope, and it was practically impossible to study them visually before the invention of the electron microscope in 1937.[37]

Despite the virtual invisibility of viruses, immunisation was first discovered in relation to a virus – smallpox – and the first two human vaccines (the second being rabies) were virus vaccines. But as John Rowan Wilson points out, 'almost all the really productive developments in this field after the death of Pasteur until 1930 were in connection with bacteria. The reason for this lay in the technical difficulties of culturing the organisms'.[38] In Vienna in 1908, Drs Karl Landsteiner and Erwin Popper discovered that the infectious agent for poliomyelitis was not a bacterium, but a 'filterable virus' – as any micro-organism that passed through the porcelain filter came to be called. When Landsteiner and Popper injected filtered fluid taken from the spinal cord of a polio victim into the brains of two monkeys, both animals went down with the disease. Through their experiment, the two scientists not only established the cause of polio, but also set the pattern for future research, with monkeys as polio's primary 'guinea pigs'.[39]

The importance of their discovery was widely recognised, not least by Simon Flexner in New York. Flexner had been appointed director of the newly established Rockefeller Institute for Medical Research in 1903 and had already successfully developed an antiserum (serum is the watery fluid left when blood coagulates and contains proteins called globulins which comprise antibodies) for cerebrospinal meningitis. In doing this, he had worked out one of 'only about four big ideas [needed] in order to prevent human diseases via vaccination'.

> The first big idea, at least two thousand years old, is that people who recover from certain infectious diseases are safe from a second attack. The second is that a scientist can find a suitable animal host, susceptible to the infection, that will manufacture virus for him in quantity. The third idea is that in such a host, or through some laboratory manoeuvre, the scientist can find a way of taming, stunning, or killing the virus so that it will still produce disease resistance but not the disease. The fourth idea is the use of antibodies in an immune serum – or antiserum, as it is also called – as a quantitative index for virus presence. This is called a virus-neutralization test.[40]

The neutralisation test was originated by George M. Sternberg, a military doctor who was promoted to Surgeon General of the US army during Cleveland's presidency. It was already known that blood serum contained antitoxic properties in relation to bacteria, but Sternberg was the first to show that it was true for viruses, too. Following Sternberg's – and Landsteiner's – lead, Flexner 'demonstrated in 1910 that the serum of monkeys convalescent from experimental [i.e. artificially induced] poliomyelitis contained antibodies, spoken of as "germicidal substances" – a finding that was made almost simultaneously by Landsteiner and others'.

> Shortly after this, Netter and Levaditi [in Paris] and others also found these neutralizing substances in the blood of humans recovering from

> poliomyelitis. This demonstration of antibodies in convalescent patients was to prove another landmark in the therapeutic history of the disease. Its significance ranked almost on a par with the discovery of the virus, a fact unappreciated until some years later.[41]

One of several sons of Jewish immigrant parents, Simon Flexner had had a distinctly unpromising childhood in Louisville, Kentucky. When he was ten, his father, Morris, had taken him along to the local jail as a tacit warning of what lay ahead of him if he did not mend his ways. But this visit filled the young Flexner with excitement rather than dread and he surprised his older brothers, who had gathered round to gloat over his humiliation, by saying that he had 'had a swell time'.[42] He left school at fourteen and was apprenticed to a plumber, according to one historian, who writes: 'At the end of a week the plumber returned him to his father with the blunt evaluation that he was too dumb to be a plumber.'[43] It is a good story, perhaps too good, since Flexner's son tells it rather differently:

> Simon received a curt command to follow his father and was led into a plumber's shop. Morris pushed the boy forward and offered him to the plumber as an apprentice. The plumber said he did not need an apprentice. Morris went out and walked off, leaving the boy standing on the street.[44]

Whichever version one accepts, there can be no doubt that during his adolescence Flexner was, in his own words, 'in and out of wretched jobs leading nowhere'.[45] It was not until he was apprenticed to a pharmacist that he found a sense of direction. After that, his rise was meteoric, through Johns Hopkins Medical School in its heyday and a professorship of pathology at the University of Pennsylvania to the Rockefeller Institute, whose first director he was – a post he held for more than thirty years. Considering the – perhaps disproportionate – influence he would have on medical research in the United States over three decades, it is worth noting that his initial medical education at the University of Louisville Medical School was a farce. Flexner recalls, 'I did not learn to practice medicine, indeed, I cannot say that I was particularly helped by the school. What it did for me was give me the MD degree.'[46] Appropriately enough, his own younger brother, Abraham, was to put an end to such anomalies by exposing them in his 1910 'Flexner Report' on *Medical Education in the United States and Canada*.[47]

Simon Flexner's lack of clinical competence and experience, combined with his belief that 'medicine derived from such basic sciences as pathology, physiology, chemistry, and bacteriology', meant that his initial staff appointments to the Rockefeller Institute 'were not of physicians interested in pursuing problems in clinical medicine, but rather of investigators skilled in the basic sciences who sought to cast light on medical problems through experimental research'.[48] The success of the institute's experiments with monkeys in developing the antiserum for cerebrospinal meningitis, reducing the mortality rate from three in four to one in four, convinced Flexner of the validity of his methods and determined his approach to polio research, encouraging a degree of confidence which was scarcely justified by subsequent events.

Initially, however, he succeeded in taking Landsteiner's work a crucial stage further by transferring poliovirus not just from humans to monkeys, but from monkey to monkey. (Others, including Landsteiner himself, also achieved this, but Flexner did it first.) When Flexner published his report, he omitted the word *experimental* from the title – a very significant omission, according to Dr John Paul:

> Indeed this was a major mistake that was to dog Flexner's footsteps throughout his entire professional life – his failure to distinguish between certain aspects of experimental poliomyelitis in the monkey and the disease in man . . . It was an error with unfortunate implications that were to influence thought at the Rockefeller Institute for a generation.[49]

Paul compares Flexner's role as 'laboratory doctor' unfavourably with the clinical investigations of a contemporary Swedish team headed by Carl Kling. Sweden had suffered another epidemic in 1911 (at nearly 4,000 cases the largest to date anywhere in the world), and Kling and his colleagues succeeded in isolating poliovirus from living patients – not just from those who had been paralysed, but from abortive cases as well, thus confirming Wickman's theories about the way the disease spread. From autopsies they also made important discoveries of the sites in the body favoured by the virus other than the central nervous system, where the damage was done; as they expected, they found it in the throat, but they were surprised to find it in the intestinal wall as well. This caused them to ponder such key questions as how the virus entered the body and how, once there, it penetrated the central nervous system. They did not come up with all the answers but, by combining clinical and laboratory techniques, at least they were asking the right questions.[50]

A news item in the *New York Times* of 9 March 1911 suggests that Simon Flexner was so confident of the rectitude of his approach that he was not looking to the Swedes or anyone else for help in solving the mysteries of polio:

> The Rockefeller Institute in this city believes that its search for a cure for infantile paralysis is about to be rewarded. Within six months, according to Dr Simon Flexner, definite announcement of a specific remedy may be expected.
>
> 'We have already discovered how to prevent the disease,' says Dr Flexner in a statement published here today, 'and the achievement of a cure, I may conservatively say, is not now far distant . . .'[51]

No cure for polio has ever been achieved and more than forty years would elapse before a safe and reliable method of prevention was developed.

In North America and in Britain, as in Scandinavia, there were alarming outbreaks of infantile paralysis in the early years of the twentieth century. On 4 November 1911, Dr A. Bertram Soltan writes in the *British Medical Journal* of an outbreak in Plymouth, Stonehouse and Devonport:

> It is hardly an exaggeration to say that a new terror has been added to the approach of the summer months . . . In our Three Towns . . . with a population approximating to a quarter of a million, the disease has this year for the first time assumed an epidemic form: the number of cases, so far as I have been able to collect them, was seventy-three . . .
>
> If the cases have been more aggregated [in one place] than elsewhere it has been in Stoke, which is one of the best situated localities, with wide roads and no slums. Some of the worst parts of Plymouth, as the Barbican and the neighbourhood of King Street, where overcrowding and poverty are at their worst, have escaped almost entirely, while families living elsewhere in every possible comfort and luxury have been attacked . . .[52]

Two weeks later, in the correspondence columns of the *BMJ*, a medical officer of health in London's East End raises the spectre that was to haunt the popular imagination on both sides of the Atlantic in relation to polio for half a century:

> I have for some considerable time interested myself in [bathing-water purification] at Poplar, where I have endeavoured to give every bather a clean and sterile bath. In September, 1909, in two somewhat lengthy reports . . . I pointed out to the Baths and Washhouses Committee of Poplar Borough Council the horrible dangers of the public swimming baths, *inter alia* mentioning how quickly swimming-bath water changes its pristine sweetness even after being used only by a few bathers . . . and becomes after use by a number of bathers nothing more nor less than diluted sewage, and this condition exists often before the first day's use is finished . . .
>
> As it is during the months of July, August, and September that swimming baths are mostly used . . . it would probably be of considerable interest to bacteriologists to take into consideration the possible connection of polluted swimming-bath water . . . and the disease and possible determination of one of the causes of poliomyelitis.[53]

A measure of the increasing seriousness with which the authorities in Britain regarded poliomyelitis is that the disease became officially notifiable in 1911 – a year or two later than in the United States and six or seven years later than in Norway and Sweden. But just how many cases does it take to transform *endemic* into *epidemic*? Seventy-odd cases in

Plymouth and the surrounding district were enough to convince the physician to the South Devon and East Cornwall Hospital that he had an epidemic on his hands; many times that number filling the Rockefeller Hospital and elsewhere in New York failed to convince Simon Flexner, who wrote to his wife Helen in July 1911:

> The poliomyelitis situation at the Hospital is entering its tragic stage. The isolation pavilion is full and other quarters will probably have to be found. But the tragedy is the severe and fatal cases. One child (an infant) died this morning and I have just come from the side of a lovely boy of five who is dying. My heart has been torn into shreds. The little fellow has extensive paralysis that has affected the nerves of the diaphragm. He is a little pale-haired almost red-haired fellow obviously the idol of his grief-stricken parents – two simple, dear American people. The Hospital staff has suddenly wakened up to the importance and seriousness of the disease – it is a tragedy. And yet there is no epidemic yet; but many cases scattered through this immense population.[54]

In 1916, however – and as early as 29 June – Flexner had no doubt that New York was in the throes of 'a severe epidemic', and he wrote to Helen: 'We have known of the beginnings for a month and have been astonished at the scant attention given by the Department of Health to this matter. But this is for your eyes alone!'[55]

Even in 1911, he and the director of the hospital he had been instrumental in creating as part and parcel of the Rockefeller Institute for Medical Research, Dr Rufus Cole, had been sufficiently concerned about the situation – whether or not it justified the epithet *epidemic* – to initiate a clinical study of polio. The three doctors who made up the investigating team were all on the hospital staff: Francis Peabody, George Draper and Alphonse Dochez.

Flexner's lack of clinical training and his focus on two fundamental objectives in research – a cure and a means of prevention – meant that he was ill-equipped to provide answers to doctors who sought his advice on how to treat an ever-increasing number of polio patients. Peabody, Draper and Dochez's study – whatever its shortcomings in comparison with the work of contemporary Swedish epidemiologists – is packed with clear information, case histories and suggestions for making prognoses and treatment. The following lengthy extract, in which the sheer brilliance of the clinical observation is matched by the clarity of the prose, was written by Francis Peabody:

> With the onset of respiratory difficulty, it seems almost as if the children were suddenly awakened and made to realize the struggle before them. Little children seem to age in a few hours. One sees a heedless, careless, sleepy child become all at once wide-awake, high-strung, alert to the matter

> in hand, and this is, breathing. The whole mind and body appear to be concentrated on respiration. Respiration becomes an active, voluntary process, and every breath represents hard work. The child gives the impression of one who has a fight on his hands, and who knows perfectly how to manage it. All he wants is to be left alone, not to be interfered with, to be allowed to carry out his fight on his own lines. Instinctively he husbands his strength, refuses food, and speaks, when speech is necessary, quietly and with few words. One little child of four, so helplessly paralyzed that she was unable to move, but with a mind that seemed to take in the whole situation, said to the nurse clearly but rather abruptly between her hard-taken breaths, 'My arm hurts'; 'Turn me over'; 'Scratch my nostril'; and then when the doctor approached, 'Let me alone, doctor!' 'Don't touch my chest.' Pressure on the chest, tight neck bands, anything that obstructs easy respiration is immediately resented. The child demands constant attention, is irritated unless everything is done exactly as he wishes it, and often shows an instinctive appreciation for some especially efficient nurse. He is nervous, fearful, and dreads being left alone. The mouth becomes filled with frothy saliva which the child is unable to swallow, so he collects it between his lips and waits for the nurse to wipe it away. He likes to have his lips wet with cold water, but rarely attempts to take it into his mouth, for he knows he cannot swallow it. During the whole course it is remarkable that cyanosis is absent. There is a little bluish tingeing of the lips and tongue, but much more distinctive is the pallor, which is sometimes striking. Sweating is profuse. Then, as respiration gets weaker, the mind becomes dull, and with the occasional return of a lucid interval, he gradually drifts into unconsciousness. An hour or more later respiration ceases. This peculiarly alert, keen mental state has been much less noticeable in small babies. They tend to be dull and drowsy most of the time; but in the older children this alertness has been such a characteristic feature of the fatal cases, that it is preferable to find a child in a stuporous condition, rather than with a mind whose nervous acuity seems due to a perception of impending danger.[56]

The only word in the entire passage likely to give trouble to the layman is *cyanosis* and that is explained in the reference to 'bluish tingeing' in the next sentence. Unsurprisingly, Peabody, Draper and Dochez's monograph 'quickly became the bible on polio throughout the United States' and was greatly in demand in 1916.[57]

In August of that year, Dr George Draper (co-author of this polio 'bible' and brother of Ruth, who became famous for her one-woman stage shows) was put in charge of the New York State Department of Health's work on Long Island. In that capacity he experienced what has been described as a 'citizens' revolt' in the Oyster Bay township. Even before the polio epidemic focused attention on sanitation, Oyster Bay had a problem: the decline of the shellfishery which gave the village its name due to the amount of untreated sewage emptying directly into the bay and contaminating the oyster and clam beds. But the effect of this decline was largely offset by a boom in tourism throughout Nassau County and particularly at Oyster Bay, whose most famous – summer – resident was ex-President Theodore Roosevelt. The village might need a proper sewage system, but the year-round population was reluctant to raise taxes for such a purpose.

When the health officer reported sixteen cases of polio in the township on 21 July, the town council hurriedly approved panic measures. Three days later Roosevelt himself intervened, along with a large number of other prominent residents, and pressured the town council into establishing a 'Committee of 21' to conduct a clean-up campaign. While this well-heeled committee went about its self-appointed task with a will, another local citizens' group, comprising Italian and Polish immigrants, formed a 'Committee for the Suppression of Infantile Paralysis' designed to disseminate information and help to those in need (as in New York itself, the Oyster Bay outbreak began in the Italian community). The unfortunate health officer found himself – in the words of a modern historian – 'in a no-win situation': wealthy summer residents accused him of negligence, while poor immigrants threatened or abused him for removing their children into quarantine or, as they saw it, into danger.[58]

Draper was drawn into the dispute when yet another local group attacked not just public health officials but the medical profession in general for creating a state of 'frenzy and terror' and being motivated more by 'greed for gold' than by a desire to help the sick. This group drafted a resolution which declared that 'both profane and modern history are replete with the medico-politico barbarism which we are now experiencing, as anyone who reads the history of quarantine must understand and acknowledge'. It was committed to exposing the 'hysterical inconsistencies of the present quarantine' and was at one with the immigrants in wishing to put an end to the 'brutal removal' of children from their parents without the latter's consent.[59] In marked contrast was the group's attitude to the family doctor, who was seen as kindly and supportive in a difficult situation; and there is evidence that these practitioners shared the group's hostility towards officious health bureaucrats. On 25 August, Draper wrote to Simon Flexner, 'Strangely enough, during the past 48 hrs, there has been an almost complete cessation of the cases in the whole of Nassau County,' only to add a handwritten note saying, 'This turns out to be due to antagonism of physicians to the State Dept. They aren't reporting to us.'[60]

At an Oyster Bay town council meeting in early September Draper found himself under attack. If the germ that gave you infantile paralysis, it was argued, was something you breathed in (the orthodox theory of the time, propounded by Simon Flexner, was that the main 'portal of entry' for the virus was the nose), then doctors themselves were surely guilty of spreading the disease when they went from house to house, examining patients. Draper's defence, that doctors averted their faces when their patients coughed or talked, is less than convincing.[61]

Theories abounded as to the genesis of this modern plague, and many laypersons were exercised by it. On 29 August, Kathleen Doherty of New Haven wrote to John D. Rockefeller, complaining that she had received no answer to an earlier letter and offering 'my reasons for thinking that sharks are the cause of Infantile Paralysis':

> The poisonous gases used in the war in Europe and the germs from the dead bodies there are breathed by the sharks. The sharks carry these germs to America and breathe them into the air. The air is inhaled by people who are affected by the germs and gases. Children are the chief victims of this disease because they are the weakest.[62]

This was just one of 230 suggestions received by the authorities. Doctors at the Department of Health hardly knew 'whether to laugh at the fantasies or weep over the ignorance and superstition exhibited'.[63] They analysed the letters they received and found that the largest number (eighty) blamed foodstuff such as ice cream, soft drinks, 'candy' and summer fruits. Other suspects included sewers, rubbish bags, automobile exhaust, animals (wild and domestic). Also atmospheric conditions, 'moisty air laden with coal gas and gasoline', 'a condition similar to "Bermuda High" permeates the young child through the pores, reducing blood pressure'; insects, 'even including the tarantula, which was supposed to inject the virus into bananas shipped to New York City', and tobacco.[64]

The Department of Health's monograph on the epidemic offers 'a few gems' worthy of 'special mention', including a communication from a 'high Priest of Iris' in Londonderry to the effect that 'fine hydrogen from the earth is incarnated into the body, the paralysis resulting from the hydronizing of the parts affected'. A resident of San Francisco blamed the electrical companies: 'The increasing amount of radio calls and wireless electricity in the air gives off vibrations which pass through children's bodies, act on delicate tissues and minute capillaries which press on nerves preventing nourishment, thereby causing paralysis.'[65] Other culprits that cropped up in the correspondence noted by the department included 'pollen of plants, subluxated vertebrae, maggots in the colon, [and] tickling of children'.[66] To these might be added: feather pillows, cemeteries, ground dampness, mould and lack of personal hygiene – 'Is it not possible that poliomyelitis has a sexual origin?' one writer asked

Simon Flexner. 'May not unclean males of *all* ages (who are uncircumcised) be the infection carriers?'[67]

Popular notions of the causes of the 1916 epidemic might be fantastic or absurd, or both, but to some extent they were simply filling the vacuum left by the lack of a credible official explanation.

Work on other diseases, such as malaria and yellow fever, in conjunction with the notion that polio was a summer disease, had prompted the thought that insects might indeed be carriers; and an announcement by two Harvard professors that they had done experiments which proved that poliomyelitis was transmitted by the common stable fly had electrified an international congress on hygiene and demography held in Washington in 1912.[68] Further research soon cast this theory into doubt, but such was its plausibility that one of the Harvard entomologists was employed by the New York City Department of Health in 1916 to make an investigation. Not the least of this theory's attractions was that it helped to explain the apparently 'random' spread of the disease, its penetration not just of slums but of clean-living middle-class homes as well; and no less an authority than Dr Flexner 'warned against the presence of flies, insects, and especially the stable fly, as possible carriers of the germ which causes infantile paralysis'.[69] The epidemiologist W.H. Frost had originally supported this theory, but subsequent research of his own had convinced him it was false. After the 1916 epidemic he and his colleagues from the US Public Health Service reported that entomological surveys 'led to no results of consequence'.[70]

Frost and his fellow-epidemiologists freely acknowledged in their 1918 bulletin that of the essential cause of poliomyelitis, 'nothing is known'. But they went on to point out that 'the origin of most epidemics of endemic contagious diseases, such as scarlet fever, measles, diphtheria, cerebrospinal meningitis and others, is equally mysterious'.[71]

The prevention and treatment of infantile paralysis raised as many questions as its causation. In the middle of July, Dr S.J. Meltzer of the Rockefeller Institute urged that injections of adrenalin into the spine be used to preclude paralysis; he had had promising results treating monkeys in this way. Haven Emerson was unimpressed, and he 'cautioned those discussing the treatment against proclaiming it as a cure'.[72] When Dr E.J. Bermingham, chief surgeon of the New York Throat, Nose and Lung Hospital, did precisely that in a report he published a week later, 'neither Dr Emerson nor Dr Louis C. Ager . . . considered that Dr Bermingham's tests were sufficient evidence of the value of adrenalin'.[73]

More promising was a new immune serum developed from Dr Simon Flexner's experiments with monkeys, which was being used at the Willard Parker and Kingston Avenue Hospitals:

> To make it, blood taken from a person who has recovered from infantile paralysis is placed in a test tube and whirled rapidly in a machine until a white fluid remains. This, it is believed, is an infantile paralysis antitoxin, the same as that which, remaining in the recovered patient, gives immunity from the disease.[74]

The treatment became so popular that the Willard Parker Hospital was soon appealing for donors, though faith in serum therapy was by no means universal. George Draper had been sceptical when he had heard that it was being tried out in France the previous year.[75] But he changed his mind in 1916 and wrote to Simon Flexner on 7 August that none of the four cases he had treated with immune serum had developed paralysis, while three others 'which had similar high cell counts on diagnostic [lumbar] puncture are all paralyzed or dead'. 'I cannot tell you how impressed I am with this method', he went on, 'but it all depends upon earliest possible puncture, and, of course, it is very early to be sure'.[76] He became so enthusiastic a proponent of it that he could not bear not to use it even in the interests of scientific investigation, apologising to Flexner for not letting him have the 'fluid and blood which you suggested should be collected, but . . . I have not the lack of conscience necessary to withhold serum, when we have it, from any early case'.[77]

Draper's colleague and co-author of the Rockefeller clinical study of poliomyelitis, Francis Peabody – though he too treated a number of cases with it in Boston during 1916 – was more cautious:

> At the present time we have only an imperfect idea as to what proportion of persons affected with the disease become paralyzed even if no treatment is instituted. Nevertheless there is apparently general agreement among those who have used immune serum as to its harmlessness, and as to the fact that in certain, possibly in numerous instances its administration is beneficial.[78]

The debate over the use of immune serum would not be resolved until the mid-Thirties, when a properly controlled experiment found no evidence to support its therapeutic value. Nevertheless, the idea behind it was valid and its time would come in the early Fifties when a concentrate of that portion of serum known as gamma globulin was used in combating the large-scale epidemics which ravaged the United States just before Jonas Salk perfected his vaccine.[79]

In the absence of proven treatments, the field was open to all comers. On 23 July, the *NY Times* reported that Dr Oscar M. Leiser, Assistant Director of Public Education at the Department of Health, had warned the public against 'the numerous quack remedies being sold' and had told how 'a number of ignorant persons, believing that a bath in ox blood would protect them from the disease, had gone to slaughter houses with buckets to obtain the blood'.[80] A week later, the same paper carried a news item to the effect that a certain Joseph Frooks had been charged with selling bags of 'Infantile Disease Protector', which were found to contain 'one-half ounce of a mixture of wood shavings saturated with some preparation having a naphthaline odor'.[81]

The public itself was not slow to come up with helpful suggestions in the matter of remedies, 'cures' and preventives. Some of these – such as a serum made from 'blood of frogs' and intraspinal injections of fresh human saliva, not to mention 'earthworm oil' – reminded the New York City Department of Health of the witches in *Macbeth*.[82] But the prize suggestion came from abroad:

> Place hydrogen conductors at soles of feet and hands, and cause attraction for this fine hydrogen by neg. electricity or neg. applications. Apply cantharides and mustard plasters. Diet must be high in fine oxygen, such as rice, bread and oxygen waters. Give oxygen through lower extremities, by positive electricity. Frequent baths using almond meal, or oxidising the water. Applications of poultices of Roman chamomile, slippery elm, arnica, mustard, cantharis, amygdalae dulcis oil, and of special merit, spikenard oil and Xanthoxolinum. Internally use caffeine, Fl. Kola, dry muriate of quinine, elixir of cinchona, radium water, chloride of gold, liquor calcis and wine of pepsin.

'One's only embarrassment', the department spokesman comments, 'would be which one to use first'.[83]

At the height of the epidemic, Haven Emerson invited members of the press to accompany him to the Willard Parker Hospital and see for themselves the condition of the children with infantile paralysis there. The *NY Times* correspondent wrote:

> Two striking impressions received from the visit were in the indefatigable optimism of the little ones and in their prevailing physical and mental excellence. The statement of epidemiologists that infantile paralysis has a preference for healthy children has been emphatically borne out in the present epidemic . . .
>
> As the line of white-robed newspapermen . . . passed between the cribs, the little girls and boys looked in wide-opened eyes, as many as were not prevented by paralysis sitting or standing in their cribs. Many of them smiled, others laughed outright at the 'funny looking' visitors, and others shouted playfully . . .[84]

But the time was approaching, as Emerson recognised, 'when patients will begin to leave hospitals, and the most important consideration will then be the after-care of them'.[85] The Rockefeller Foundation, which was already funding a special investigation under the medical direction of Dr Alvah H. Doty into the means by which polio was transmitted, now turned its attention to this problem. A committee prepared a rough estimate of the cost of caring for 2,000 crippled children for a year, which worked out at $75,000, and concluded that:

> In view of the fact that the eventual ability of the paralyzed child to earn its living is largely dependent on the extent to which its muscular functions are destroyed, the *per capita* expenditure would seem to be amply justified as an investment.[86]

By the time the secretary of the Rockefeller Foundation, Jerome D. Greene, wrote to Simon Flexner nine days later, the budget had been whittled down to $16,500–$22,000:

> You will notice that it covers only the cost of administration and does not touch the matter of salaries for orthopedic nurses and extra orthopedic surgeons. As the care of crippled children is one of the easiest things for which to raise money, under ordinary circumstances, I would suggest that we make no commitments at present in regard to the salaries of doctors and nurses and that any help we may give in this direction be subject to the formulation of plans which we can consider whenever they are presented.[87]

Greene was aware that aftercare involved more than orthopaedic nursing and exercises. Families would get a preliminary visit from a nurse, the purpose of which was to provide 'helpful information and advice of a very general character':

> For example, some parents will be so overjoyed at getting their children back, apparently well, that they will need to be cautioned, or perhaps a little frightened, as to the importance of care lest impaired muscles be overworked during the tender period or deformities caused. Other parents, whose children are brought back with considerable paralysis, will be much depressed and will need to be cheered by the prospect of doing something to mitigate the paralysis, provided due care is taken and the right treatment secured.[88]

As to the quality of treatment available to the children, Greene was not too sanguine:

> I am afraid that to a very large extent they will have to suffer the consequences of ignorance and unprogressiveness in the medical profession and that only the minority can expect to obtain treatment comparable with what Dr Lovett could give.[89]

Robert W. Lovett of Boston was an orthopaedic surgeon and a Harvard professor; he was *the* authority when it came to the treatment of infantile paralysis and had written a book on the subject which, by serendipity, was published in 1916.[90] 'The therapeutic measures at our disposal in fairly early cases', he taught, 'are massage, electricity and muscle training.'[91] Of the three (in another account he listed four, the fourth being 'heat'),[92] only muscle training was unreservedly recommended by him. In an article in the *Journal of the American Medical Association*, he provides a useful analogy:

> If a railway wreck occurs in the main line and the track is blocked, it is often possible to send trains by means of a branch line around the obstruction, so that service between the terminals is maintained. In the same way, after a wreck of certain nerve centers, it may be possible by a modified route to send a motor impulse from brain to muscle. On this principle of establishing new connections and opening new paths rests most of the claim of muscle training.[93]

Dr Lovett divided the course of the disease into three stages: acute, convalescent and chronic. During the acute stage of the illness ('a matter generally of from four weeks to three months'), his advice was to leave well alone: the prime requirement then was rest, 'in order that nature may be given a chance to repair the damage so far as possible by absorption'. But once the period of convalescence ('a matter of about two years') had begun, Lovett's enlightened approach was to get patients on their feet at the earliest possible moment 'in order to call forth the instinctive muscular actions induced by the efforts and balance'. While he warned against overfatigue and overuse of weakened muscles, he deplored 'the old practice . . . of allowing patients to sit or lie around for months or years with no attempt to walk until many of them have developed serious flexion deformities'.[94] Deformities might be corrected, or reduced, by operations during the third – chronic – phase of the disease, but it was better that they should not have been allowed to develop in the first place.

Drs Lovett and Draper, both authorities on infantile paralysis, would pool their resources five years later in the treatment of a single case, one that was to influence most profoundly the history of both the disease and the United States. But in the summer of 1916, Franklin Delano Roosevelt was thinking more of his children than of himself when he wrote to his wife from Washington (where he had been obliged to return early from the family holiday island of Campobello due to the exigencies of his war work as Assistant Secretary of the Navy in Woodrow Wilson's administration): 'The infantile paralysis in NY and vicinity is appalling. *Please* kill all the flies I left. I think it really important'.[95]

The continuing virulence of the epidemic obliged Haven Emerson to announce, on 9 August, the postponement of the opening of New York's public schools from 11 to 25 September, at the earliest.[96] Private schools and colleges were outside his jurisdiction but, when some of those imposed on New Yorkers what he regarded as excessive quarantine requirements, he protested in a letter to Simon Flexner, who – he understood – had been consulted by them. Flexner did not deny his advisory role, but was quite unmoved. The irony is that Emerson, who had introduced such fiercely protective quarantine measures in New York itself, was indignant when they were imposed on New Yorkers from outside, and there is little indication that this pioneer of public health, whose lanky figure and ascetic good looks earned him the sobriquet, 'The last of the great Puritans',[97] saw any inconsistency in this. But a report in the *NY Times* of 6 August does suggest that he was not entirely unaware of the glaring contradiction between his obsession with sanitation and the mounting evidence of its ineffectiveness in preventing the spread of polio:

> Dr Emerson repeated yesterday that, while unsanitary conditions were bad for the health on general principles, it had not been established that rubbish and refuse carried the disease. He pointed out that there had not yet been a case on Barren Island, where hundreds of children live and go to school. On the other hand, he said, a person had been attacked in an isolated dwelling on a peninsula in Lake Michigan, 18 miles from a town, where the only visitor to the house had been a rural mail carrier. He said a large number of the cases in this city and other cities had been in immaculately kept apartments in clean neighbourhoods.[98]

If ever there were a hellhole designed to foster an outbreak of infectious disease, the aptly named Barren Island was surely it:

> There were 350 children, under 16 years of age, on Barren Island in Jamaica Bay, Borough of Brooklyn. To this island all the city garbage and offal is taken for reduction in the large rendering plants. Flies and mosquitoes are abundant. Rats are numerous. There is no public water supply and there are many shallow surface water wells. There is no sewage system. There are few, if any, cellars. There is no garbage collection. There are no public highways. The population of about 1,300 people represents the lower grade of unskilled labor, Poles, Italians and Negroes predominating. The standard of living is low . . .[99]

Yet, 'No cases of poliomyelitis developed on Barren Island'. Emerson compares this with the situation on Governor's Island, a military post, where eighty or ninety children were living 'under as nearly ideal sanitary conditions as may well be ordained'. The children were not permitted to leave the island during the epidemic, and no visitors under sixteen years of age were allowed on to it. There were no cases on Governor's Island either. So what was the explanation? 'It is probably correct to say that social and geographical isolation of this Barren Island group accomplished by accident what was enforced by regulations at Governor's

Island, namely, group isolation which was effective *regardless of environment* [my italics].'[100] Despite this rather grudging admission, Emerson continued to beat the drum for his clean-up campaign, arguing that: 'Notwithstanding the prevalence of the epidemic of poliomyelitis . . . both the infant mortality and the general death rates of the city were as low or lower, throughout the summer of 1916, than they were during the same period of 1915, when there was no epidemic to contend with.'[101] (In other words, the sanitary measures were effective in controlling every infant malady except for the one they were designed to curb.) By such sleight of hand did Emerson seek to transform defeat into victory. He concludes his monograph on the epidemic:

> Unfortunate as the recent epidemic undoubtedly was, and in some respects unproductive from an epidemiological point of view, this disastrous visitation may yet turn out to be a blessing in disguise, if it fixes indelibly in our minds one obvious and incontestable truth – that the control not only of poliomyelitis, but of all preventable diseases, does not depend upon the mysterious power of any supernatural agency, but that the remedy lies largely within ourselves.[102]

As a rousing finale to a lay sermon preached by the last of the great Puritans, this may pass muster; but as a summary of the lessons learned in 1916 it does not compare with the tentative and down-to-earth conclusions reached by Dr W.H. Frost and his fellow epidemiologists from the US Public Health Service:

1. That poliomyelitis is, in nature, exclusively a human infection, transmitted from person to person without the necessary intervention of a lower animal or insect host, the precise mechanism of transmission and avenues of infection being undetermined.
2. That the infection is far more prevalent than is apparent from the incidence of clinically recognized cases, since a large majority of persons infected become 'carriers' without clinical manifestations. It is probable that during an epidemic such as that in New York City a very considerable proportion of the population become infected, adults as well as children.
3. That the most important agencies in disseminating the infection are the unrecognized carriers and perhaps mild abortive cases ordinarily escaping recognition. It is fairly certain that the frank, paralytic cases are a relatively minor factor in the spread of infection.
4. That an epidemic of one to three recognized cases per thousand, or even less, immunizes the general population to such an extent that the epidemic declines spontaneously, due to the exhaustion or thinning out of infectible material. Apparently an endemic incidence relatively small in comparison to that prevailing in an epidemic may produce a population

> immunity sufficient to definitely limit the incidence rate in a subsequent epidemic.[103]

It was much easier to impose stringent public health measures than to admit that, in John Paul's words, 'the infection was simply too widespread, too hidden for any such measures to be effective'.[104]

It had taken an epidemic in which there had been 9,000 cases in New York City alone (a leap from a previous maximum of under 8 per 100,000 population to a staggering 28.5) and 2,343 deaths – 27,000 cases nationwide, yielding 6,000 deaths – to arrive at the realisation of how ineffective quarantine restrictions were. But they were not immediately abandoned; far from it. Emerson and his cohorts had done such a thorough job of public education that people continued to expect them and would have felt bereft without them.

2

FDR

> As in other diseases, the personality affects the disease more than the disease affects the personality. The disease may unmask a person but I doubt if it really changes him. People almost generally need obstacles to bring out their best qualities. It isn't the condition itself but the view one takes of it that is the real crux of the situation. The real personality comes into clearer focus, but there are no essential changes, at least in those older than a few years when they become sick.
>
> Merritt B. Low, MD, 'Poliomyelitis with Residual Paralysis' from *When Doctors Are Patients* (1952)

The story has been told so many times that it has become encrusted with myth; and its subject was not averse to a little embellishment of his own. Franklin Delano Roosevelt, the laughing cavalier who 'conquered' polio with all the grace and style expected of a scion of an aristocratic Hudson River valley family straight out of the pages of an Edith Wharton novel, so that it was little more than a hiccup in his triumphal progress from Assistant Secretary of the Navy and Democratic Vice-Presidential candidate to Governor of New York and President of the United States; FDR, the prince who, like Shakespeare's Henry V, put behind him his playboy youth in order to take the responsibilities of the world onto his athlete's ample shoulders; FDR, the 'happy warrior' (as he dubbed another in a memorable speech written by a third party) whose most famous one-liner – 'the only thing we have to fear is fear itself' – came to epitomise the man . . . His eldest son James recalls it differently:

> It's easy to look back now and say that he contracted polio and learned to live with it; that with the encouragement of Louis Howe, Eleanor Roosevelt paved the path for Franklin Roosevelt's return to politics over the opposition of [his mother] Sara Roosevelt; that he swiftly ascended to the

> governor's chair and finally to the presidency. But that isn't fair to father. It was more than three years before he hobbled briefly back into the spotlight, eight years before he became governor, and twelve years before he became president. The suffering and struggling of all those days, weeks, months and years are not to be taken lightly. I was thirteen when he was stricken and twenty-five when he became president. I grew up with his suffering and struggling.[1]

When he joined his family on the island of Campobello, off the coast of Maine and New Brunswick, in August 1921, Franklin Roosevelt was at a low point in his political career. In his fortieth year, he was not only out of office after the massive Democratic defeat in the previous year's presidential election, but he had also been singled out for attack by Republican senators investigating a scandal at the naval base in Newport, Rhode Island, during his time as Assistant Secretary of the Navy. He was held personally responsible for allowing naval personnel to be used as *agents provocateurs* to entrap homosexuals at the base. Despite his efforts to get an early copy of the report indicting him, he was prevented from seeing it until the day of publication, which scarcely gave him time even to scan it before replying to its damaging allegations. It was his attempt to clear his name in Washington that had kept him away from his family on Campobello during July.

His personal life, too, was at a low ebb. The only child of an adoring mother, he had wanted to have several children himself, and now he had five (a sixth had died). But his marriage had virtually ended three years earlier when Eleanor, who was his cousin as well as his wife (a Roosevelt by birth, and a niece rather than a mere cousin to the great Theodore), found evidence to confirm her fears that her husband was having an affair. To make matters worse, the object of his affections was none other than her social secretary, a well-connected but impecunious beauty called Lucy Mercer. For Eleanor, betrayal was no light matter. In one sense she was used to it – both her father and her brother were destroyed by drink. But whereas she could forgive a feckless father or brother, she could not forgive a feckless husband. From the day she married her handsome prince, and was transformed from an ugly duckling into a swan, she had subordinated her wishes not just to Franklin's but to his mother's as well; effectively she had no life, or even home, of her own. At her bedside after her death in 1962 was found a poem, the first verse of which runs: 'The soul that has believed/And is deceived/Thinks nothing for a while/All thoughts are vile.' Across the top of the faded cutting on which it was printed she had written, '1918'.[2]

Eleanor apparently offered Franklin his 'freedom' and he seems to have taken his time in deciding whether or not to accept it. Various factors came into consideration: there were the children, of course, and there was Eleanor herself, whom he almost certainly continued to love despite his infatuation with Lucy Mercer; and there was the question of Lucy's Catholicism. What evidence there is suggests that Lucy would have been more than willing to marry a divorced Franklin, had she been given the opportunity; but she was told – probably by Franklin himself – that Eleanor was not prepared to divorce him. These personal considerations, however, were outweighed by two others. The first was Sara Roosevelt's threat to cut off the money supply which guaranteed her son's independence and enabled him to pursue his political career, if he divorced; and the second was the irretrievable damage such a step would do to that career, as his *éminence grise*, Louis Howe, was quick to point out. In the end, his love for Lucy was sacrificed on the altar of his political ambition.

The marriage survived, but it was transformed into something else, an apparently sexless partnership in which Eleanor played an increasingly independent role. She found an unlikely ally in her struggle for self-expression in Louis Howe. Initially Eleanor had shared her mother-in-law's (and her children's) snobbish distaste for this grubby, chain-smoking little ex-newspaperman who monopolised Franklin's attention. Sara never got over her dislike of Howe, but Eleanor came to love him and value his company as much as and, latterly, even more than her husband did.

The reason for this change of heart was the attention that Louis started to pay Eleanor during the 1920 campaign in which her position as camp-follower in her husband's nationwide vote-catching tour was making her feel superfluous. Louis saw that she was ignored by Franklin, who preferred to play poker with his aides than talk to his wife in his off-duty moments, and this may have alerted Howe to the danger of disaffection which beset his dream of taking the 'boss' to the White House. So Louis talked to Eleanor about politics, gave her speeches to read and consulted her over their contents. From there, it was but a step to encouraging her to participate, and Louis coached her in the art of public speaking and managed to triumph over her timidity. If husband and wife were no longer lovers, they could still be genuine partners, with Eleanor actively involved in promoting Franklin's career – at least that was how Louis Howe saw it. In Joseph Lash's words, 'Long ago he had set his hand to making a king; now he began to make a queen.'[3]

So while Franklin picked up the threads of his non-political activities, in law and business, after the failure of the Cox/Roosevelt Democratic ticket in 1920, Eleanor joined the board of the League of Women Voters and made the first of many friendships among politically-minded women. Her mother-in-law did not approve of the changes in her lifestyle, but Eleanor was no longer prepared to be ordered around by Sara, though she continued to humour her when she could do so without compromising her new-found independence.

That was the situation when Roosevelt made his way from Washington to Campobello by way of Bear Mountain, where he attended a boy scout rally – being president of the Boy Scout Foundation of Greater New York was one of the ways in which he kept himself in the public eye during this politically lean period. He arrived on the island worn out by his exertions, but instead of resting immediately flung himself into such holiday activities as sailing and swimming.

When he was out sailing with Eleanor and the children on the afternoon of Wednesday 10 August, they spotted a forest fire on a nearby island. Attacking it with brushwood, they fought it down to the last ember before Franklin and the boys plunged into the water and then raced one another home. The mail had arrived during their absence, and Franklin sat down to go through it without bothering to change out of his wet bathing suit – an omission to which his mother (who was in Europe at the time) forever after attributed his subsequent illness, rather than to the virus which he had almost certainly picked up at the boy scout camp.[4]

In a letter written in response to a query from a doctor in North Carolina more than three years later, Roosevelt described the onset of his illness with quasi-professional detachment:

> I first had a chill in the evening which lasted practically all night. The following morning the muscles of the right knee appeared weak and by afternoon I was unable to support my weight on my right leg. That evening the left knee began to weaken also and by the following morning I was unable to stand up. This was accompanied by a continuing temperature of about 102 and I felt thoroughly achy all over. By the end of the third day practically all muscles from the chest down were involved. Above the chest the only symptom was a weakening of the two large thumb muscles making it impossible to write. There was no special pain along the spine and no rigidity of the neck.[5]

This was written with hindsight as to the nature of the malady. At the time, neither the local family physician nor an eminent Philadelphian surgeon holidaying nearby diagnosed it correctly. Considerable harm may have been done to his chances of eventual recovery by too early and too heavy massage of his legs, ordered by the eighty-four-year-old Dr W.W. Keen – who was so unabashed by his mishandling of the case that he stung the Roosevelts $600 initially ($1,000 in all) for the privilege of consulting him.[6] It was not until Franklin's uncle, Fred Delano, consulted the Harvard Infantile Paralysis Commission, set up during the 1916 epidemic, that a proper diagnosis was made – by proxy. The young Dr

Samuel Levine recommended that a lumbar puncture be done within twenty-four hours, which would both confirm the diagnosis and relieve pressure in the spine. For some reason this was not carried out and it was only when Dr Robert Lovett, who was on holiday at the time, was contacted and persuaded to go to Campobello that the damaging massage treatment was stopped.

Franklin's helplessness imposed a renewed intimacy with Eleanor, who was called upon to nurse him and see to his most personal needs, like giving him enemas to empty his bowels and catheterising him to draw off his urine. Her dedication earned her praise from Dr Keen, who wrote, 'You have been a rare wife and have borne your burden most bravely . . . [W]hen the catheter has to be used your sleep must be broken probably at least once a night.'[7] It cannot have been easy for Franklin to have his wife stick a tube up his urethra, or for Eleanor to do it in such a way as to avoid both unnecessary pain and the risk of infection. But Eleanor came into her own in such a situation. She once wrote, 'In all our contacts it is probably the sense of being really needed which gives the greatest satisfaction and creates the most lasting bond.'[8] The enduring strength of the Roosevelt marital partnership grew out of the agony of those early days and weeks of his illness, which were an ordeal for both partners. Years later FDR would confide to his Secretary of Labor, Frances Perkins, that

> for the first few days he had been in utter despair, feeling that God had abandoned him. Then his buoyancy and strong religious faith reasserted themselves; he felt he must have been shattered and spared for a purpose beyond his knowledge.[9]

Another stalwart was Louis Howe, who was as ready as ever to put the needs of the Roosevelts ahead of his own family's. Even at this crisis in his patron-*cum*-protégé's life, he never for an instant lost sight of the ultimate goal, using all his considerable wiles, first, to keep the news of the boss's illness out of the papers and then, when that was no longer feasible, to minimise its impact.

From the outset, according to Roosevelt's biographer Frank Friedel, 'Howe helped create the illusion – so important if Roosevelt were to continue active in politics – that there was nothing vitally wrong.'[10] Howe was a master of illusion – and deception. When Roosevelt, still utterly helpless, had to be moved from his Campobello holiday home to a New York hospital, Howe stage-managed the whole manoeuvre in such a way that reporters were deflected from witnessing the tricky transference of the stretcher from boat to train and only arrived in time to see FDR propped up comfortably in a private railway carriage, with a lighted cigarette in his mouth, smiling out of the window (which had had to be removed to get him into the carriage).

Sara paid for the carriage. Eleanor had arranged that Franklin's half-brother Rosy meet her on her return from Europe and give her a letter in which the news of her son's illness was broken gently. Sara took it well, and she had gamely participated in the little charade put on for her benefit when she arrived at Campobello. She wrote to her brother Fred that she 'came up to a brave, smiling, and beautiful son, who said: "Well, I'm glad you are back, Mummy, and I got up this party for you!" He had shaved himself and seems very bright and *keen* . . .'[11] To her sister, she reported:

> Below his waist he cannot move at all. His legs (that I have always been proud of) have to be moved often as they ache when long in one position. He and Eleanor decided at once to be cheerful and the atmosphere of the house is all happiness, so I have fallen in and follow their glorious example . . . [The family doctor] just came and said, 'This boy is going to get all right.' They went into his room and I can hear them all laughing, Eleanor in the lead.[12]

The harmony between mother and daughter-in-law did not survive their return to New York, where they lived in adjacent and interconnecting houses. His mother, who had never wholly approved of Franklin's political activity, now wanted him to follow in his long-deceased father's footsteps and lead the retired life of a Hudson Valley squire at their country estate in Hyde Park, near Poughkeepsie. She later denied that she had ever wanted to 'make an invalid' of him,[13] but that was how it seemed to Eleanor both at the time and retrospectively.[14]

Eleanor had an altogether less indulgent approach to illness. In her autobiography, she claims to have always been 'a particularly healthy person' except during the first three months of each pregnancy. Her sickness on those occasions made her 'a little more understanding and sympathetic of the general illnesses human beings are subject to'. But, she adds, 'I always think we can do something to conquer our physical ailments.'[15]

When Sara found she could no longer influence Eleanor directly, she tried to achieve her aims in a roundabout way. Louis Howe, 'that ugly, dirty little man', as she called him, became her target, and her disaffected adolescent granddaughter Anna the means by which she hoped to dislodge him from the bosom of the family.[16] She had often undermined Eleanor's – and even Franklin's – authority over the children by giving them expensive presents their parents disapproved of, or, in some instances, had withheld as a punishment. Now she persuaded Anna that she should really have the large, sunny room in the front of the house that Louis occupied instead of her small back room. But when Anna, thus primed, took her complaint to her mother, Eleanor gave her extremely short shrift. Without Howe's support, Eleanor might well have collapsed under the strain. It was, she recalled, the 'most trying winter' of her life.[17]

Roosevelt himself, when he came out of hospital in November, was subject to 'the intense and devastating influence of the interplay of these high-voltage personalities', as George Draper put it.[18] He was fortunate in having Draper as his New York physician. Not only was Draper an infantile paralysis specialist second only to the Boston panjandrum, Dr Robert Lovett, but 'Dan', as he was known to his Groton and Harvard contemporaries (of whom FDR was one), was also a family friend. From the moment he took Franklin into the Presbyterian Hospital, Draper was as sensitive to his patient's state of mind as to his bodily ills. Lovett had been inclined to think that Roosevelt's would turn out to be a mild case of infantile paralysis. Draper was not so sure; he worried about the slow rate of recovery, the persistence of pain and even the cheery optimism of the patient, in which he detected signs of self-delusion. He wrote to Lovett:

> I feel so strongly after watching him now for over a week that the psychological factor in his management is paramount. He has such courage, such ambition, and yet at the same time such an extraordinarily sensitive emotional mechanism that it will take all the skill which we can muster to lead him successfully to a recognition of what he really faces without utterly crushing him.[19]

Perhaps Roosevelt himself knew instinctively that he would be crushed by such a recognition; at any rate neither then nor later did he accept that the loss of the use of his legs was permanent. He tried all sorts of therapy and always believed – or professed to believe – that he was on the brink of a breakthrough. In the letter he wrote in 1924 to the doctor in North Carolina, he claimed that 'the muscles have increased in power to a remarkable extent and the improvement in the past 6 months has been even more rapid than at any previous time'.[20] Yet if there was any muscular improvement at all in his legs at that time, it was absolutely minuscule.

Draper agreed with Eleanor and Louis Howe that, in Eleanor's words, 'it was better for Franklin to make an effort to take an active part in life again and lead, as far as possible, a normal life'.[21] Unlike Franklin, Eleanor was under no illusion that he was going to recover full use of his legs; nor did she share Howe's belief that he could still be President; but she did think that – according to Frances Perkins – 'he would die spiritually, die intellectually, and die in his personality, if he didn't have political hope'.[22] So she encouraged him in his daily exercises with the visiting physiotherapist, Mrs Kathleen Lake, and welcomed visits from politicians and businessmen, which his mother deplored, believing they tired him out unnecessarily. His nurse, Miss Edna Rockey, also thought there was 'a good deal too much pushing ahead being done by the wife and Mr H', as Mrs Lake reported to Dr Lovett.[23]

At the beginning of February 1922 Roosevelt suffered a setback. His hamstrings contracted so that he found it increasingly difficult to straighten his legs. The doctors decided to put the legs in plaster, and he had to endure the painful stretching of his hamstrings by means of wedges driven in behind his knees. Before he could stand, or attempt to walk, he was fitted with full-length braces – calipers – with a pelvic support. Yet by the end of March he was 'walking quite successfully', as Draper reported to Lovett in a cautious assessment of his progress. Indeed, he now wanted 'to rid himself of his nurses and have his wife take over the joint duties of Mrs Lake and Miss Rockey'. Obviously, Draper went on,

> this is an impossible plan. In the first place, Mrs R is pretty much at the end of her tether with the long hard strain she has been through, and I feel that if she had to take on this activity, that the whole situation would collapse.[24]

Despite this accurate assessment of Eleanor's state, it may have been she who was pressing to resume nursing her husband. Now that Franklin was surrounded by professional helpers, she was in danger of being pushed to one side just at the point when they had succeeded in re-establishing a degree of intimacy. Insecure in her marital relationship, at odds with her mother-in-law and her daughter, and yet unable to pursue the independent life for which she had begun to acquire a taste, Eleanor had reason enough to be miserable. And she recalls that she did have a moment of collapse:

> . . . one afternoon that spring, when I was trying to read to the youngest boys, I suddenly found myself sobbing as I read. I could not think why I was sobbing, nor could I stop. Elliott [her second son] came in from school, dashed in to look at me and fled. Mr Howe came in and tried to find out what was the matter with me, but he gave it up as a bad job. The two little boys went off to bed and I sat on the sofa in the sitting room and sobbed and sobbed. I could not go to dinner in this condition. Finally, I found an empty room in my mother-in-law's house . . . I locked the door and poured cold water on a towel and mopped my face. I eventually pulled myself together, for it requires an audience, as a rule, to keep on these emotional jags. That is the one and only time I ever remember in my entire life going to pieces in this particular manner.[25]

In the event, Franklin retained the services of his nurse, who accompanied him to his mother's house in Hyde Park, where Mrs Lake also stayed several days a week to supervise his physiotherapy.

So, although Eleanor won the war over Franklin's future, Sara won at least some of the early battles, in that Roosevelt took very little part in public life for a period of seven years. When he was not in the warmer south during winter seeking treatment, he was more often in Hyde Park than in New York, pursuing his interests in stamps, birds, fishing and model boats, as well as his physical therapy, and hobnobbing with some of the grander Hyde Park neighbours who formed his mother's social circle. According to one commentator, 'Less than 25 per cent of his time during those years did he live in the Roosevelt Manhattan townhouse where he was able to attend to business and politics in person.'[26]

As his presidency would demonstrate, FDR was a pastmaster of the art of 'divide and rule'; he would habitually set up two individuals or departments in opposition to one another, leading both to believe that they had his personal backing. People often made the mistake of imagining that he agreed with them when he was merely being agreeable and keeping his real thoughts to himself. It was a characteristic he had developed early in dealing with his mother, who had to be kept at arm's length without being alienated. It became second nature to him and served him well in this time of crisis; though he was as determined as Eleanor and Louis Howe that he should lead a full life, he did not take sides in the dispute between his wife and his mother; he used both as he needed them and resisted the domination of either. Eleanor's assessment of her function in his life is penetrating, even if it is less than the whole truth:

> He might have been happier with a wife who was completely uncritical. That I was never able to be, and he had to find it in other people. Nevertheless, I think I sometimes acted as a spur, even though the spurring was not always wanted or welcome. I was one of those who served his purposes.[27]

For uncritical support and devotion he turned, not to his wife or to Louis Howe, or even to his mother, but to the young woman who had only reluctantly agreed to become his secretary (because she feared that the work, in law and business, might be too boring) shortly before he fell ill: Marguerite – or 'Missy' – LeHand. Did Missy, or did she not, become FDR's mistress? This is one of the most hotly debated issues in Roosevelt biography. Elliott Roosevelt says yes, she did; James and other members of the family – and most biographers – say no. Certainly there was great intimacy between FD, as she – and she alone – called him, and Missy, and equally certainly she adored him and dedicated her life to him in such a way as to preclude the possibility of her marrying anyone else. On the other hand, Eleanor, who was prone to jealousy, was never jealous of her *in that way*, and remained well-disposed towards her for more than twenty years. Anna Roosevelt perhaps gets closest to the real nature of the relationship when she describes Missy as 'the *office* wife, quote, unquote'.[28]

One thing can be stated categorically: polio does not impair sexual functioning; the motor nerves are affected, not the sensory ones. FDR's son James may write, 'From my observation, it would have been difficult for him to function sexually after he became crippled from the waist down with polio.'[29] But it does not alter the fact that Roosevelt was perfectly capable of having sexual intercourse, though obviously his disability limited the positional options at his disposal and might have inhibited him psychologically. That is unlikely, however: he was a handsome man who became, if anything, better-looking after his illness, and his self-confidence was boundless. If he did not have sex with Missy, or anyone else for that matter, it was because he (or she) chose not to. Unlike Eleanor, he enjoyed flirting, and he charmed women – as he charmed men – effortlessly. Even if he had to be lifted in and out of cars and carried up and down steps, he personified power; and power is a great aphrodisiac.

Missy's health was always delicate and she had at least one serious breakdown before suffering the stroke that ended both her career and her life prematurely during the Second World War. Whether that breakdown was the consequence of an affair that was never going to lead to marriage and children, or of an unrequited – or partially requited – love, or something entirely different can only be a matter of conjecture. The fact is that for twenty years FDR lived in much greater intimacy with Missy than with Eleanor, and his disability made it inevitable that there was physical closeness of a more or less innocent kind.

In the early years his mother disapproved, but there was never a hint of scandal about his relationship with Missy. FDR was a reserved man and with increasing power came increasing loneliness; he could relax with Missy as he never could with Eleanor. If Eleanor resented Missy, she tried to hide her resentment behind expressions of affection and motherly concern, but her children felt no such need. Franklin Jr once reduced Missy almost to tears by saying, 'Are you always agreeable? Don't you ever get mad and flare up? Do you always smile?'[30] But Missy was not a cipher, she had a mind of her own; FDR's political advisor, Sam Rosenman, called her 'the frankest of the President's associates, never hesitating to tell him unpleasant truths'.[31] The difference between Missy and Eleanor was that Missy's style was not confrontational.

FDR, with his chameleonic qualities, his capacity to be 'all things to all men', is the inscrutable one. There can be no doubt of his warmth and closeness to Missy, or of his dependence upon her; it is hardly coincidental that he should resume seeing his old love, Lucy (who had married a much older man, the wealthy widower Winthrop Rutherfurd, who was now dying), only after Missy's stroke had removed her from his daily life. But it may be that a formal working relationship – however much more to it there was than that – suited him better than a surrogate marriage, with all the complications that that would have entailed. At any rate, the relationship endured and, though in the event Missy predeceased him, FDR had made a provision in his will that the interest on up to half of his estate should go towards her medical costs. He told James, when he made him executor of his will, 'If it embarrasses mother, I'm sorry. It shouldn't, but it may.'[32]

Another question that exercises Roosevelt's biographers and memorialists is: what difference did polio make to his life? At one extreme there is Frances Perkins, who had known and not greatly admired the brash young politician trying to make his mark in the New York State Senate. She recalled that Roosevelt then 'had an unfortunate habit – so natural that he was unaware of it – of throwing his head up'.

> This, combined with his pince-nez and great height, gave him the appearance of looking down his nose at most people.
>
> It is interesting that this habit of throwing his head up, which when he was young and unchastened gave him a slightly supercilious appearance, later had a completely different effect. By 1933, and for rest of his life, it was a gesture of courage and hope . . .[33]

Frances Perkins had no doubt that polio changed his life. Franklin Roosevelt, she writes, 'underwent a spiritual transformation during the years of his illness'.

> I noticed when he came back that the years of pain and suffering had purged the slightly arrogant attitude he had displayed on occasion before he was stricken. The man emerged completely warmhearted, with humility of spirit and with a deeper philosophy. Having been to the depths of trouble, he understood the problems of people in trouble. Although he rarely, almost never, spoke of his illness in later years, he showed that he had developed faith in the capacity of troubled people to respond to help and encouragement . . .
>
> I saw Roosevelt only once between 1921 and 1924, and I was instantly struck by his growth. He was young, he was crippled, he was physically weak, but he had a firmer grip on life and on himself than ever before. He was serious, not playing now. Politics had become important to him as a means to a good life. He had become conscious of other people, of weak people, of human frailty . . . His viability – his power to grow in response to experience – was beginning to show.[34]

Elliott Roosevelt stands at the opposite extreme from Frances Perkins. He castigates her for romanticising his parents, and takes his mother to task, too, for contributing to the legend by saying that FDR's illness 'proved a blessing in disguise; for it gave him strength and courage he had not had before. He had to think out the fundamentals of living and learn the greatest of all lessons – infinite patience and never-ending persistence.'[35] Elliott dismisses this as platitudinous piffle, saying that his father 'grew steadily in stature, resourcefulness and zest for power from the time his schooldays ended'; polio made only one difference and that was physical: 'After 1921 he could not walk.'[36]

James Roosevelt is inclined to agree with his brother about this aspect of their father's life, if not about the role Missy played in it. 'There are those who say that having polio helped father in a strange sort of way,' he writes.

> Possibly it did. It decreased his distractions. He could not get around and do as others could. It made him more human in the eyes of the public. On the other hand, if he had not been able to convince the public that he could handle his handicap, it would have curtailed his career. Possibly it increased his feelings for the troubled people of the world, but I believe that he was at heart a humanitarian who did not have to be crippled to care about others.
>
> In short, I do not think, as has been suggested, that the ordeal of being crippled built father's character. I believe he had the basic strength of character to overcome his handicap. It was not easy, but many others have overcome their handicaps . . .[37]

This is surely correct. As Louis Howe, in particular, had recognised very early on, FDR had always had qualities of leadership that could take him to the top. From that point of view, his illness was little more than an interruption – albeit a lengthy one – in his brilliant career. In terms of character, at the age of thirty-nine he was unlikely to undergo a sea-change.

Yet the impact of infantile paralysis was not negligible. The polio virus could neither be charmed nor conquered; it tempered his ambition with intimations of mortality; without in any way diminishing his sense of fun, it sobered him; this 'infantile' ailment knocked the last vestiges of childish conceit out of him. It did not make him a great man, but it gave his inherent greatness a pathos it might otherwise have lacked.

3

Warm Springs

> Polio is a contraction of the jaw-breaking Anterior Poliomyelitis, the scientific term for the disease of Infantile Paralysis, which no lay person is sure about spelling or pronouncing correctly. There is something jolly about the brief Polio. It sounds like a game, and it is played like a game at Warm Springs . . .
>
> Reinette Lovewell Donnelly, 'Playing Polio at Warm Springs' (*Macon Telegraph*, 16 Aug 1931)

From pre-Columbian times it was a place of healing. Legend has it that Indians of all tribes were given safe passage to 'the land where the waters are warm', the only thermal springs in Georgia. The wounded both drank the water – rich in minerals – and bathed in the mud pools. These springs at the foot of Pine Mountain gushed out some 1,800 gallons of water per minute at a temperature of 88–90°F., and it has been estimated that 'it would take 20 tons of coal every 24 hours to duplicate nature's heating plant'.[1] With the arrival of European settlers in the nineteenth century, Warm Springs was developed as a spa.

By mid-century it was being advertised as a 'fashionable watering place, and delightful Summer Retreat . . . All that a visitor can desire shall be furnished with pleasure' – a claim disputed in an early ledger entry: 'Landlord tight as two ticks, but doesn't know it.'[2] For some years after the Civil War, the Springs were not much visited. The building of the railroad brought the place renewed popularity, but by the end of the 1880s it had again gone into a decline. (A contemporary traveller encountered an aspect of life at Warm Springs that was to trouble Eleanor Roosevelt when she stayed there on one of her fleeting visits half a century later. 'There was a sense of abundance', he wrote, 'in the sight of fowls tiptoeing about the verandas, and to meet a chicken in the parlor was a sort of guarantee that we should meet him later on in the dining-room.')[3]

The building of the Meriwether Inn in the gingerbread style of the Victorian era, following the destruction of an earlier hotel by fire, ushered in the Gay Nineties, the fashionable years at Warm Springs. But the development of the automobile in the first two decades of the twentieth century made the Atlantic coast more accessible, and Warm Springs, once again, was bypassed. By 1923, when Thomas W. Loyless, formerly editor of the *Augusta Chronicle*, took over the management of the inn and surrounding cottages, the future of the resort looked distinctly unpromising. But Loyless persuaded George Foster Peabody, a Wall Street financier with a winter home in Columbus, Georgia, to invest in the place; and the canny Peabody, in turn, drew the attention of Franklin Delano Roosevelt to the health-giving properties of the waters there.

At the 1924 Democratic Convention, in which Roosevelt made his first post-polio public appearance at a major political event and created a dramatic impact as much with his brave attempt at walking with braces and crutches as with his 'Happy Warrior' speech nominating Al Smith for the presidency, Peabody took him aside and told him about Warm Springs. He followed this up by sending Roosevelt an account by a young man of the astonishing recovery he had made there from infantile paralysis. After persistent exercising in the pool over three summers, Louis Joseph had discarded his two crutches, braces and back support, and was walking with a cane, and no aids at all indoors. Roosevelt, whose political comeback, as he saw it, depended on his acquiring the ability to walk unencumbered with braces and crutches, was prepared to try anything to achieve this goal.

The view that further gains could not realistically be expected after the two-year convalescent period of infantile paralysis was unacceptable to FDR. So when orthodox methods failed he looked for alternatives.

On 23 July 1923, he had written from Hyde Park to his physiotherapist, Mrs Lake, for her 'personal and confidential information' – by which he meant he did not want her to report what he was telling her to Dr Lovett – that he had consulted two osteopaths from Kansas City, whose 'whole method of examination and theory' appealed to him. While Drs Starr and Barrett approved of his exercise treatment, they maintained that it was too long drawn out and it would speed up the process if he were to spend an hour every morning under 'a simple light machine giving the physical – not the chemical – sun rays'. This would go to 'the seat of the trouble – i.e., the nerve cells'. FDR added, a little defensively, 'Frankly, I can see no harm in doing this', and assured Mrs Lake that 'I have not given up Lovett, do not intend to, and want to be able to discuss the case with Lovett at any time'.[4] In fact, he had already seen Lovett for the last time – the doctor died suddenly during a visit to England a year later.

In October, a fellow polio survivor in neighbouring Poughkeepsie, a young First World War veteran by the name of Paul Hasbrouck, wrote to Roosevelt asking him whether he had found any worthwhile treatment for his paralysis other than 'exercise, massage and swimming'. FDR replied:

> I have . . . found for myself one interesting fact which I believe to be a real discovery, and that is that my muscles have improved with great rapidity when I could give them sunlight. Last winter I went to Florida and was much in the open air under the direct rays of the sun with very few clothes on, and there is no doubt that the leg muscles responded more quickly at that time than when I am at home when I am, of necessity, more in the house. This summer also I have made a real effort to sit in the sun for several hours every day, and the improvement has undoubtedly been much more rapid. I am about to try out a new artificial light for use this winter. This light is said to give 4 times the strength of the sun rays with the red or burning rays of the sun extracted. It is like a lamp on a high stand which can be attached to any electrical fixture. If it seems to me to be good I will let you know about it. My theory is that by exercise we can only develop the muscles up to a certain point, i.e. that it is necessary to build up the nerve centers of the lower spine in order to make more exercise possible . . .[5]

When Roosevelt next wrote to Hasbrouck, in January 1925, he had been to Warm Springs and he encouraged the young man to go there, too. He retailed the story of Louis Joseph, whom he had found 'practically entirely cured'. His own month – three weeks, in fact – down there had done him 'more good than any thing I had done before'.[6] He went on to explain that he planned to rent a cottage there for a month in the spring before the hotel opened and that if Hasbrouck was interested in trying out the place he should write to Loyless, who would arrange accommodation for him. (Hasbrouck did go to Warm Springs, but not for another two years, by which time the place had been transformed.) It seems that Roosevelt was already taking a proprietorial interest in Warm Springs, though he had been there only once and stayed for less than a month.

During this first visit he had given an interview to a reporter from the *Atlanta Journal* called Cleburne Gregory. The article, entitled 'Swimming Back to Health', had been syndicated all over the United States. Full of FDR's usual exaggeratedly optimistic claims about both his own progress and the possibilities of the place, the piece acted as a magnet to less fortunate 'polios' from all over North America. They wrote letters and the braver – or better-off – among them made their way there, arriving even before the hotel opened the following spring and finding a place which, though inadequately equipped to care for them, nevertheless made them welcome.

Roosevelt himself, meanwhile, was basking in the sunshine off the coast of Florida on board the houseboat *Larooco*, which he had purchased jointly with a friend. These annual houseboat trips were ostensibly occasions of great jollity, but there were days – according to what Missy, who was ever in attendance, later told Frances Perkins – 'when it was noon before [FDR] could pull himself out of depression and greet his guests wearing his lighthearted façade'.[7] Following his discovery of Warm Springs, Roosevelt would abandon houseboat trips, but not immediately. His last trip was in 1926, after he and his partner had tried unsuccessfully to sell the boat (which was then destroyed by a hurricane).

At about the time he first went to Warm Springs, Roosevelt teamed up with Basil O'Connor, a smart corporate lawyer ten years his junior – indeed, O'Connor may have visited him at Warm Springs. But the official announcement of the new legal partnership was not made until New Year 1925. An Irish Catholic 'one generation removed from servitude' – as he was fond of pointing out by way of contrast with his later ostentatious wealth – O'Connor supported himself through Dartmouth College by playing the violin in a dance band he had organised.[8] From there he went to Harvard Law School. By the time Roosevelt met him, 'Doc' O'Connor was married and had his own law firm in New York.

The authorised version of how the two partners met is the stuff of legend. Monday 9 October 1922 was Roosevelt's first day back at the office of the finance company of which he was a vice-president after an absence of fifteen months. His chauffeur lifted him out of the car and helped him on to his crutches while people passing by stopped out of curiosity to watch. He successfully negotiated the short distance into the lobby of the office building, but once inside his crutch slipped on the polished floor and bystanders looked on helplessly as the big man crashed to the ground. O'Connor, whose office was in the same building, was one of the men who helped Roosevelt to his feet; and he was impressed by the cheerily casual way in which Roosevelt handled the incident (though FDR was obviously shaken by it, since he did not attempt to go back to the office for another two months). Turnley Walker, a polio sufferer of a later generation who went to Warm Springs, takes up the story:

> Basil O'Connor . . . stood near the place Roosevelt had fallen. With a strange intentness, he watched the struggle towards the nearest elevator. He looked down at the scratches which the steel braces had made on the gleaming floor. He was a man who understood the anguish of physical helplessness: for a time in his own life he had been blind [at Harvard Law School, as a result of excessive study].[9]

Cue, perhaps, for that violin O'Connor had once played. In fact, the Roosevelt-O'Connor partnership had little to do with sentiment – though obviously there was an affinity between the two men which developed into a lifelong friendship – and a great deal to do with mutual self-interest. Roosevelt got $10,000 a year for doing almost nothing, and his name had pride of place in the firm of 'Roosevelt & O'Connor'; and O'Connor gained respectability and prestige from the association.

'Dear Franklin,' he wrote nearly two years after the formation of their partnership:

> In connection with our law practice a thought has occurred to me which at first may not impress you. I am sure, however, that you will be glad to consider it seriously. The thought is this: – If when you are in NY you could spend a couple of days at our office (or even more if such a thing were possible) it would impress much more forcibly all those who call on you that you really have a law firm and are active in it . . .[10]

The curious mixture of belligerence and subservience that characterised O'Connor's relationship with FDR – a determination to assert his independence even as he hastened to do his partner's bidding – is manifest in the postscript he felt obliged to add to this letter: 'Don't gather from this that I'm dissatisfied in any way. When I am I'll tell you! I can make money but I want you to make a lot of money.'[11] This was the purpose of the partnership: to make 'a lot of money' for both partners – Roosevelt, whose rehabilitation did not come cheap, no less than O'Connor. The difference was that while O'Connor knew he had to work for it, Roosevelt wanted it to work for him.

FDR's purchase of Warm Springs pitted the idealist against the realist. Roosevelt, sanguine as ever, had a vision of developing a resort in which guests would go out and play golf or happily swim alongside polio sufferers intent on regaining the function of their limbs. O'Connor told him he would have to make a choice between a spa for the wealthy and a polio rehabilitation centre if either were to have any chance of success – and he was proved right, though Roosevelt was reluctant to abandon his dream.

O'Connor later recalled the acquisition of Warm Springs in a studiedly off-hand manner:

> I thought he was crazy to want that big goddam four-story firetrap with the squirrels running in and out of the holes in the roof. I couldn't have been less interested in the project. But in 1926 he bought it and made a nonprofit foundation of it and in 1928 he ups and becomes Governor of New York and nonchalantly says to me, 'Take over Warm Springs, old fella: you're in.' I tell you, I had no desire to be 'in'. I was never a public do-gooder and had

no aspirations of that kind. But I started enjoying it. Like Andrew Jackson at the battle of New Orleans, I found myself up to my rump in blood and liked it.[12]

At the time, he merely remarked, 'Something tells me Peabody's doing all right,' referring to the fact that George Foster Peabody charged FDR double what he himself had paid for the property three years earlier.[13] Roosevelt is generally reckoned to have put two thirds of his entire fortune, a sum of around $200,000, into purchasing Warm Springs, though one recent writer suggests that, given 'the intricacies of the laws regarding corporate liabilities . . . this probably is an overstatement'.[14] But it was still the biggest personal financial undertaking of his life.

In 1925, when the first 'polios' (as I shall continue to call them for convenience) were drawn to Warm Springs, regular patrons of the Meriwether Inn resented their presence in the public pool. They complained to the unfortunate manager, Tom Loyless, that it exposed them to the disease (regardless of the fact that these were 'old' polios, long past the infectious stage). As a result, Roosevelt had another pool constructed nearby, so that 'he and his "gang"' – in the words of Fred Botts, an early arrival – 'could continue their unsupervised, unregulated, groping efforts to reduce their afflictions in the warm water of the springs'. But that concession was not enough to satisfy the resort guests. Botts recalls that Loyless was forced 'to bar even the use of the regular dining room to the "pariahs"', who were obliged to eat their meals in a cheerless basement.[15]

Fred Botts, almost as much as FDR himself (whose absences grew ever longer once he took on the responsibilities of political office), came to epitomise the 'spirit of Warm Springs'. An aspiring opera singer, he was a victim of the 1916 epidemic, since when he had been living – or half-living in the manner of a hopeless cripple – in his father's house in Elizabethville, Pennsylvania, one of the innumerable polios (somewhere in the region of 125,000) shut away in back rooms all over the United States. As Botts tells it, he was slumped in his wheelchair by the fireside one December evening when his father casually asked him if he had ever heard of a place called Warm Springs, Georgia. Fred replied 'wearily' that he had not, whereupon his father 'proceeded to read me one of the most beautiful stories I have ever heard'.[16]

Galvanised by the newspaper account of the Hon. Franklin D. Roosevelt swimming his way back to health, the following April Botts had himself conveyed in the baggage car of a train – along with 'three sacks of mail, three chicken crates, one very expensive-looking casket and various other and sundry things not worthy of mention' – to Warm Springs, where he was greeted by 'a kindly and inquisitive crowd of villagers'.

These villagers were shocked by his skeletal appearance, and he heard one old lady remark, 'He's so pale he's like to die' (though the Shakespearean English sounds more like Botts than the anonymous lady). Another person tried to cheer him by saying he'd soon put on weight at Warm Springs and in a few days would 'look like the tanned side of a buffalo hide'. Yet others reckoned that 'a boll weevil or two' in his soup would fatten him up.[17]

The Meriwether Inn was not yet open, so Fred was taken to a nearby cottage where, after being served a large meal by a 'colored girl', he was visited by Tom Loyless and his wife. Looking around him, Fred could not help noticing the contrast between the grandeur of Loyless's vision of the future (which Loyless, who was already dying of cancer, would not live to see) and the 'sad deficiencies' evident in his new surroundings: 'The screening was torn and rent asunder, and through the roof the night threatened . . .' But he forgot about this the following morning when he joined two other polios who had reached Warm Springs a day earlier than he, and his 'ignominiously white and angular and thin' body was lowered into the pool:

> It was great sport just paddling and floating around. Soon Mr Roosevelt's cheery 'Good Morning!' sounded on the air and in a few minutes he had joined us in the pool. At once we were told and shown a series of exercises he had worked out. It was 'Catch hold of the bar this way – now – swing – in and out – Hard! *harder!* that's it – that's fine! Now – again, this way –,' and so through the entire regime of things he had worked out that morning. It was all along the line of lending to the affected limb and muscles the normal actions as near as possible.[18]

When Roosevelt, who also loved to reminisce, recalled these early days, he linked Fred's case with that of a woman whom out of delicacy he forbore to identify except as 'a lady from St Louis who weighed about 200 pounds'. In a speech he made at Warm Springs in March 1937, he said:

> Old Dr Roosevelt finally persuaded Fred to see if he could put his legs down to the bottom of the pool. That was easy, because they did not have any flesh on them anyway. But when it came to the lady, that was different. Old Dr Roosevelt put the lady alongside the edge of the pool – there was a handrail there – and I said, 'Just concentrate. Use your mind. Just think about getting that leg down to the bottom of the pool.' Well, she would get about halfway down and I would take hold of her right leg and push a little and push a little and finally got the whole leg down to the bottom of the pool. And finally I said, 'Concentrate and hold it there.' And she would say, 'I have it there!' And then, gently, I would move over to get the left leg down and as I moved the left leg down, up came the right leg.[19]

He went on to claim, above the laughter of his polio audience and with pardonable exaggeration, that 'these girls who think they are physiotherapists don't know anything about it. I invented it first.'[20]

In fact, hydrotherapy was not new. As Dr LeRoy W. Hubbard, whom Roosevelt brought to Warm Springs in 1926, writes:

> Since the epidemic of 1916, hot salt baths at a temperature of 95 to 100 degrees F. have been used quite generally. These baths are enjoyed by the patients; they stimulate the circulation, and generally relieve the pain and tenderness. It was soon discovered that the children, who had room in the bath tub, could move their legs and arms in the water when they could not do so in bed or on a table, and often the first movement of some of the muscle groups would occur in the bath. It was also observed that those patients who knew how to swim, and went to the sea shore in the summer made considerable gain[s] in muscle power.[21]

Yet Roosevelt had a struggle to convince the American Orthopedic Association (AOA), which happened to be meeting at Atlanta, Georgia, in 1926, that the 'Hydrotherapeutic Center' he was setting up at Warm Springs deserved their endorsement – though that may have had more to do with his effrontery as a layman in trespassing on medical preserves than with the nature of the treatment on offer. But FDR was not a politician for nothing: when the surgeons refused to invite him to speak at their annual convention, he turned up anyway, wheeling up and down the corridors of the convention hotel and lobbying them until he succeeded in obtaining unofficial approval for his venture and an agreement that three surgeons would monitor the treatment of the next batch of patients.

Then he persuaded Dr Hubbard (who brought with him the formidable Miss Helena Mahoney, a physiotherapist with experience of the 1916 epidemic) to come from New York. Dr Hubbard's progress reports on the twenty-three patients who came to Warm Springs that summer convinced the AOA that the hydrotherapeutic centre was worthy of their endorsement, which they gave in January 1927. As soon as he had secured that, FDR pushed ahead with his plans for a nonprofit foundation and appointed Hubbard surgeon-in-chief.

Although he himself was not a doctor, Roosevelt had read all about the disease he was fighting and his 'Old Dr Roosevelt' line was not entirely an act. Pidge Cole contracted polio in 1927 when she was two and a half; she was taken to Warm Springs by her parents and for a while she actually lived in 'Uncle Franklin's' cottage. (FDR was a kind of universal uncle to the polio children at Warm Springs; others remember him as 'Uncle Rosey'.[22]) His valet, Irwin McDuffie, would take Pidge into Roosevelt's room every morning, 'and Uncle Franklin and I would eat breakfast on his bed'. She remembered that one morning, as he was bouncing her up and down, he told her mother that Pidge should be wearing a corset as he could see signs of incipient scoliosis (lateral curvature of the spine). Pidge hated the corset and would not speak to Uncle Franklin for a while; but in retrospect she was grateful for his quickness in spotting the potential danger. Besides, it was impossible to be angry with him for long, he radiated such *joy*: 'The joy of living – it just exuded from him.'[23]

Old Dr Roosevelt's most valuable legacy, however, was the product of his psychological, rather than medical, insight. From the beginning, though he recognised the need for qualified medical supervision – and authorised the recruitment of young physical education graduates from Peabody College in Nashville to help the overworked Miss Mahoney – he made sure that there was no attempt to turn the place into a hospital. Warm Springs inevitably became an institution, but regimentation was kept to a minimum: everybody was called by their first names and no one wore a uniform. Just how radical an innovation this was only becomes apparent when it is seen in the context of its time. Hugh Gregory Gallagher (a Warm Springs alumnus of a later generation as well as a Roosevelt biographer) writes:

> In the 1920s, to be handicapped in some visible way carried with it social opprobrium. The handicapped were kept at home, out of sight, in back bedrooms, by families who felt a mixture of embarrassment and shame about their presence. The well-to-do were able to afford custodial nursing care for their handicapped family members, and the loving family was able to care at home for its crippled loved ones. Many of the handicapped, however, were simply ignored by their family and society.[24]

In the Victorian charitable world, which survived in the early part of our own enlightened century, the urge to 'improve' was sometimes indistinguishable from the need to punish – a point Gallagher reinforces with a quotation from a medical textbook, on *The Care of Invalid and Crippled Children in School*, published in London in 1911: 'A failure in the moral training of a cripple means the evolution of an individual detestable in character, a menace and burden to the community, who is only too apt to graduate into the mendicant and criminal classes.'[25] The novelty of Warm Springs, in terms of charity, was that the gap between donor and recipient, usually so vast, was eliminated: the philanthropist who created it was one of the objects of his own philanthropy. This meant that he had a vested interest in making the place as pleasant as possible to live in. Even Eleanor, who was originally inclined to dismiss the venture as one of Franklin's fads, came to see that this was different: ' he feels' – she wrote to a friend – 'that he's trying to do a big thing which may be a financial success & a medical and philanthropic opportunity for infantile & that all of us have raised our eyebrows & thrown cold water on it . . .'[26] FDR might pursue other more or less seductive treatments – the sun and sea-water of Florida, or the contrastingly draconian regime of the Massachussetts neurologist, Dr William McDonald, who tried to get him walking without braces – but he returned to Warm Springs for comfort and renewal for the rest of his life.

Mary Hudson Veeder, one of several slender and attractive Peabody graduates (they were once photographed lined up on the poolside like bathing beauties as a publicity stunt for the resort), originally came to Warm Springs to teach swimming but ended up, as all those who stayed did, doing physiotherapy as well. She learned it on the job, having anatomy lessons in the afternoons and studying in the evenings. According to her, Miss Mahoney was the guiding spirit of Warm Springs, 'the foundation of the whole thing'. Even when Mahoney was ill, she controlled things from her bed.[27]

By the time Mary arrived the routines were well established. After breakfast in the Meriwether Inn (now re-colonised by the polios, who had effectively ousted the able-bodied guests and established themselves as the new élite), a bus took patients down to the pool. This was divided in two with a little bridge – not unlike the one depicted on willow-pattern china – spanning the channel which connected them. In the first pool there were underwater treatment tables where individual polios were put through their paces by the physios, after which they were free to swim and relax in the second, or play, pool. There was a colander-like wooden chair – the holes in the seat allowed the water to drain away – for lifting the more seriously disabled in and out of the pool, and wheelchair polios were pushed around by young white southern high-school boys specially employed for that purpose and, for that reason, called 'push boys'. In the sexually charged atmosphere of Warm Springs, many a romance developed between push boys and patients, and some ended in marriage (just as some of the physios, like Mary herself, married polios).[28]

The bus took the patients back to the inn for lunch, after which they rested; the physios massaged selected individuals during this 'quiet hour'. Later in the afternoon, the polios were taken outside again for walking practice – on ramps with parallel bars, on steps with rails, on crutches with whatever apparatus they required. When he was at Warm Springs, FDR followed the same routine as the rest, though he lived apart in one of the cottages (till he built his own, which is now preserved as the Little White House) and did not have his meals in the Meriwether Inn.

In 1927, under Mahoney's guidance, FDR began to practise walking with a cane in one hand and a supporting arm for the other. A cane was far less off-putting than the crutch he had used at the 1924 Democratic Convention, when he had appeared on the arm of his son James; and in the election year of 1928 he set about perfecting this 'cosmetic' mode of walking. 'He walked on sheer determination,' his physiotherapist, Mary Hudson Veeder, recalls. She provided an arm for him then, and now she remembers how the other 'companions' – as they called themselves – observed his efforts:

> I used to walk him every day . . . You talk about sweating it out; he did. He was trying with everything he had . . . So what were the rest of them doing? They weren't saying anything, but they all knew what was going on. You're dealing with people that have *very* sharp minds . . .[29]

So what *was* going on? Was all this walking practice simply to enable FDR to make another pretty speech on behalf of Al Smith at the Houston Convention, or was he preparing to re-enter the fray and offer himself as a potential office-holder?

Prompted by Louis Howe – who regarded 1928 as likely to be another disastrous year for the Democrats – Roosevelt publicly took the line that he was not yet ready for a return to politics: he needed more time to build up his physical strength, as well as to get the Warm Springs Foundation properly established.

Meanwhile, Democratic Party managers were telling Smith that Roosevelt was the man to succeed him as Governor of New York – as a Protestant and a patrician, he was the perfect foil to the urban and Catholic presidential candidate. But some Smith supporters saw FDR as a dangerous rival to their man and attempted to sideline him during the early part of campaign. Smith himself, whose poor Irish Catholic background (like Basil O'Connor's) was at the opposite end of the social spectrum from Roosevelt's, was not particularly fond of FDR despite his public expressions of admiration; he also fatally underestimated him, mistaking the playboy exterior for the man. FDR was far less malleable than Smith imagined.

Roosevelt took little part in the early stages of the campaign, escaping to Warm Springs and leaving Eleanor and Howe to represent his interests. The story of how, at the beginning of October, he played a kind of hide-and-seek with Al Smith, who now desperately needed him to run for Governor, is a familiar one: FDR deliberately made himself inaccessible by going out for a picnic the day before the nominations were due, and then took advantage of bad telephonic connections to tease Smith (and Eleanor, who was used as a decoy to get hold of him when he might not otherwise have come to the phone) and gain time for himself before finally acceding to the pressure. What persuaded FDR to change his mind and allow himself to be nominated for the governorship is still an open question. One theory is that he was won over by a promise from the millionaire John

J. Raskob to finance Warm Springs, and it is certainly true that Raskob became one of its leading benefactors, contributing over $100,000 in three years.[30] But after the fateful phone call, Missy sulked, Eleanor was lukewarm and Howe furious – 'Mess is no name for it', he wired. 'For once, I have no advice to give.'[31]

As in the purchase of Warm Springs, Roosevelt risked alienating all those who were closest to him, yet he went ahead. He is alleged to have said to Egbert Curtis, a friend of Missy's whom he'd appointed manager of the Meriwether Inn after Loyless's death, 'Curt, when you're in politics, you've got to play the game.'[32] Or, to put it another way:

> There is a tide in the affairs of men,
> Which, taken at the flood, leads on to fortune;
> Omitted, all the voyage of their life
> Is bound in shallows and in miseries.[33]

In a lightning calculation, FDR must have weighed up the pros and cons – first and foremost, the likelihood of his winning the governorship of New York, the prerequisite for a shot at the presidency; then the potential gains and losses for Warm Springs, which was proving a heavy financial burden; and finally the question, not of his general health, which was good, but of his chances of continuing improvement once he was in office, not to mention his ability, as a cripple, to 'pass' in the able-bodied world. In the end, his decision probably had less to do with any premonition of victory (he shared Howe's view of it being a Republican year) than with a gambler's instinct that he had to try: 'It Was Time', as one biographer heads the chapter in which he tells the story of these events.[34]

The rest – as the saying goes – is history: how for most of election night New York seemed to be going Republican like the rest of the country; how FDR and his downcast aides gave up and went to bed, one by one; how Frances Perkins refused to concede defeat and sat the night out along with Sara Roosevelt, whose disapproval of her son's return to active politics did not prevent her willing him to win; how Ed Flynn, boss of the Bronx Democrats, excitedly woke up FDR to tell him he might carry the state after all, only to be told he was 'crazy' for having woken him, and wrong into the bargain; how Smith was defeated nationally and Roosevelt victorious, finally, by the narrowest of margins, in New York State.[35]

For Warm Springs, the future was assured by Roosevelt's personal triumph. If the polios lost their chief companion – for FDR had no intention of being an absentee governor, whatever Smith might think, and would therefore spend considerably less time with them – they gained a hero, an inspiration, a role model. Polio had not prevented FDR from being elected Governor; neither would it stop him becoming President of the United States.

The companions celebrated their founder's victory with a parade. With the help of the push boys, they dragged an old stagecoach out of a museum and hoisted Fred Botts – magnificent in top hat and tails – into the driving seat. Other vehicles lined up behind the coach, and their passengers – also in fancy dress – waved banners as they progressed in triumph through the Georgian countryside.[36]

When FDR himself finally arrived, exhausted by the campaigning he had done and preoccupied with the future, his attention was monopolised by a host of visiting New York politicians. But he still found time to join the companions on Thanksgiving Day and carve the turkey at dinner, after which Fred Botts – this time dressed up 'in an outrageous minstrel costume' – sang humorous songs.[37] Thanksgiving became the official Founders' Day, and for the rest of his life, if he possibly could, Roosevelt tried to be in Warm Springs on that day.

4
Polio Crusaders

> Can we, who have felt the spirit of Warm Springs, do less than our best in a crusade launched against the vast human loss inflicted by infantile paralysis? We all have glimpses, however vague, of the whole picture; we understand at least a little of the human values involved in the stupendous problem presented by hundreds of thousands of polio victims. We have a gospel to preach. We need to make America 'polio conscious' to the end that the inexcusable case of positive neglect will be entirely eliminated . . .
>
> Franklin D. Roosevelt, President-Elect of the United States, (Warm Springs) *Polio Chronicle*, Vol. 2 No. 5, Dec. 1932

Warm Springs was still trying to combine the attractions of a resort with the benefits of a therapeutic centre, but despite the golf course and riding facilities the guests were now very much secondary to the patients. Arthur Carpenter came to Warm Springs as a patient in 1928 and so impressed Roosevelt that he was appointed Business Manager of the foundation when FDR went off to be Governor of New York. Carpenter (who had been Advertising Manager of *Parents' Magazine* before he got polio) had a close friend called Keith Morgan, who was a successful insurance salesman in New York. Morgan visited 'Carp' and his wife at Warm Springs, met Roosevelt and ended up as the foundation's chief fund-raiser. Until 1931, when he was appointed Chairman of the Finance Committee, his 'association with the Foundation was very definitely on the basis of an avocation and not a vocation'.[1] According to the Warm Springs historian Turnley Walker, 'Morgan "sold" the Foundation everywhere his busy footsteps took him.'[2]

In 1929, for instance, he tried to interest the publisher of the *American Magazine* in the idea of running a story about Dr Hubbard – 'At 75 years of age he is one of the most active and keen minded men that it has ever been our privilege to meet.'[3] In 1930, after another visit to Warm Springs, he reported to Governor Roosevelt that it had 'made tremendous progress in public definition' – on their way home from Florida he and his mother had stopped at a number of hotels and gas stations in the neighbourhood to find out how familiar people were with the foundation and what it stood for: 'The common knowledge was amazingly on the plus side.' He was also impressed with his friend's work: 'Carp has certainly done a remarkable job in a year's time.'

Keith Morgan also came up with an idea that benefited both FDR and Warm Springs. He arranged to have Roosevelt's life insured by twenty-two companies for over half a million dollars in the name of the foundation. This not only provided security for Warm Springs; it also silenced those critics who questioned FDR's fitness to govern. The physicians who examined him gave him a clean bill of health, one of them writing to him:

> Frankly, I have never before observed such a complete degree of recovery in organic function and such a remarkable degree of recovery of muscles and limbs in an individual who had passed through an attack of infantile paralysis such as yours.[4]

When doubts were voiced again during the 1932 presidential campaign, FDR would use the same tactic to quell them.

Despite Morgan's initiatives, donations to the foundation dropped off dramatically during the early Depression years. In July 1931, the first issue of the *Polio Chronicle*, a magazine put out by the patients and staff at Warm Springs, appeared with a front-page call to arms above the signature of FDR himself. 'WANTED', it proclaimed, 'Enlistments for a Crusade'. Although Warm Springs was primarily concerned with after-treatment, the Polio Crusader Number One argued,

> we cannot divorce ourselves from participation in all phases of the problems presented by this disease. These problems may be briefly summarized as follows:
>
> 1. Research for prevention.
> 2. Diagnosis and treatment during acute stage.
> 3. After-treatment.
> 4. Rehabilitation – Adjustment of lives of handicapped people to happy productivity.

The article continued: 'Let us try, individually and collectively, to provide stimulus and leadership to the ends of coordinating all of these forces into one vast national crusade against infantile paralysis.'[5] Bold and – as it must have seemed in 1931 – fantastical words; yet, as the existence of Warm Springs itself demonstrated, FDR had the happy knack of turning Utopian dreams into reality.

The crusade began, modestly enough, with the creation of a new National Patients Committee of Warm Springs alumni, with the slogan 'Every Patient, a Polio Crusader'. Fred Botts, 'The Dean' of Warm Springs as he was now known, reported that 'the formation of this committee fulfils a long felt need in providing an outlet for the enthusiasm of those who have received treatment at the Foundation and who feel that they should help to carry on its work'. To that end, the alumni should individually pledge themselves to:

1. Help the Foundation and all others who are Crusading against Polio.
2. Aid the growth of the Patients' Aid Fund of the Georgia Warm Springs Foundation and contribute to this fund, annually, in ratio to my means and appreciation.
3. Spread the gospel that physical handicap of polio does not make one a 'cripple'.[6]

One Roosevelt biographer, or perhaps 'psychobiographer' – he trained as a psychologist of the handicapped – writes that FDR's 'return to high office coincided with a new stage of his psychological rehabilitation. . . . His commitment to the improvement of the Warm Springs Foundation for all patients became a vicarious gratification for his own rehabilitation.'[7] Whatever the motivation, there can no doubting the commitment. As FDR grew in stature and moved from the Governor's Mansion in Albany to the White House in Washington DC, he oversaw the development of the foundation 'from a local treatment center to a national and international center for rehabilitation'.[8] Its horizons widened along with his.

Despite the Wall Street crash and the Depression, a series of new buildings went up at Warm Springs. Already in 1928, an enclosed pool had been built out of funds donated by Edsel Ford; in 1930, an infirmary was built in memory of a young polio from money raised directly by patients and their friends; in 1933, Georgia Hall, a classical-style building with pillars and porticoes modelled on Jefferson's University of Virginia and funded by the people of Georgia, was erected as the centrepiece and administrative hub of the foundation; other constructions – dormitories, brace shop, chapel, school, occupational therapy building and, finally, hospital – went up before the end of the Thirties. And the old Meriwether Inn came down.

There were changes in personnel, too. Helena Mahoney, sadly incapacitated with arthritis, was the first to go, and she was replaced in 1930 by Alice Lou Plastridge, who was less of a martinet but a more highly qualified physiotherapist and just as influential over the next two or more decades. (She was still alive and well in Florida at the time of writing, though well over 100 years old.) Dr Hubbard, an elderly man who found it difficult to accept the loss of Miss Mahoney, was the next to be ousted; in his place came the bristling, bustling orthopaedic surgeon, Dr Michael – 'Mike' – Hoke:

> His grey hair was cropped to bristles and his eyes glared out through glasses often askew on his bony, thrusting nose, giving him the appearance of a slightly comic eagle. He sometimes thought that he was dying simply because he could not get air enough into his lungs. Close sleeping quarters made him rage, and once he smashed out a train window with his shoe, because the berth's green curtains and the pressing shadows were more than he could bear. But when he came into a room where a crippled person waited, a great gentleness was in him, and children found only comfort in his big, bony hands. He had blazed new trails in orthopaedics, an entire system of hospitals had been built on the foundation of his personal reputation.[9]

Arthur Carpenter handled the negotiations with Mike Hoke. When Hoke decided he would have to give up his surgical practice in Atlanta if he was to take on the responsibility of Warm Springs, Carp wrote to Basil O'Connor: 'There isn't a thing about him to indicate a man who wants to settle down to a dull and easy routine. He is a driver and a bundle of energy, if ever there was one . . .'[10] Over the next few years. Carp would have several confrontations with Hoke, but each retained the respect of the other. 'It was at this time', according to Keith Morgan, 'that the idea became clearly crystallized that Warm Springs was a laboratory, specializing in the after-treatment of Infantile Paralysis, to which patients, doctors, physiotherapists, and nurses could come for specialized information.'[11]

The rapid growth of Warm Springs did not make it any easier to balance the books; patients were charged $42 a week initially, though this was later reduced to $39, and there was a patients' aid fund for the needy. In Mary Veeder's rather jaundiced view, you could always tell who were the non-paying patients: they were the ones who 'bellyached about the food in the dining room and went out and bought things at the gift shop'.[12]

FDR insisted on signing the patients' aid fund cheques himself, and when Basil O'Connor complained that if they continued to dole out so much money they would never balance the books, he is alleged to have replied, 'If we did, we'd be a failure.' O'Connor learned the lesson and was later wont to say, when asked about the organisation of philanthropic institutions: 'This kind of thing we're talking about is a red-ink operation. Don't let that scare you. You've got to be more than a little broke or you're no good.'[13]

When FDR visited Warm Springs towards the end of the 1932 presidential campaign that took him to the White House and, in the process, gave substance to the dream of a 'vast national crusade against infantile paralysis', the foundation put on a show, coming to meet him with three cars, representing three phases of his public life: 'A Ford of 1914 vintage carried his double as "Assistant Secretary of the Navy". A 1928 model carried the "Governor of New York", and in a snappy 1932 model was the "Next President".'[14]

That year 250 people sat down to Founder's Day dinner on Thanksgiving Day, but that still represented only a fraction of 'The Warm Springs Family'. The companions were ecstatic:

> Our Polio Crusader Number One has become Citizen Number One . . . Forgetting the election, he has won a great triumph as a man. He has licked polio. None can realize as his fellow patients at Warm Springs do, just how much that really means.[15]

'Licking polio' was, of course, a relative achievement; to lick it completely he would have had to have made a full recovery. But what FDR had accomplished was more remarkable than that: by the same combination of will power and guile that was to make him a great president he made it *cease to matter*. In this he was fortunate in having the willing co-operation of the media – an unimaginable thing today. James Roosevelt once asked the photographer Sammy Schulman why he and his colleagues had obeyed the 'unwritten rule' that FDR was never to be photographed below the waist or sitting in his wheelchair. Schulman replied:

> First of all, he treated us well, so we treated him well. There never was a public figure who was so accessible to the press, who was so responsive to them and easy with them, who treated them as equals and would joke with them. He was a decent human being, we genuinely liked him, and we didn't want to embarrass him. Then, you have to remember that he arrived during Depression days and remained into the war. These were hard times, and he was our hope. To have done anything to tear him down in the eyes of the public would have been unthinkable.[16]

While taking nothing away from the sincerity of this answer, one may wonder whether Schulman would have put it in quite the same way to someone other than a Roosevelt – because, in addition to the rapport that undoubtedly existed between FDR and journalists, there was an element of coercion. If someone attempted to photograph the President locking his braces or being lifted out of his car, a staff aide or secret service agent would move in smartly and give the order that no picture be taken; and should that be ignored, they were not above removing the offending film from the camera. Hugh Gallagher makes this anomaly the focus of his study of Roosevelt:

> Although there are over thirty-five thousand still photographs of FDR at the Presidential Library, there are only two of the man seated in his wheelchair. No newsreels show him being lifted, carried, or pushed in his chair. Among the thousands of political cartoons and caricatures of FDR, not one shows the man as physically impaired. In fact, many of them have him as a man of action – running, jumping, doing things. Roosevelt dominated his times from a wheelchair, yet he was simply not perceived as being in any major sense disabled.
>
> This was not by accident. It was the result of a careful strategy of the President. The strategy served to minimize the extent of his handicap, to make it unnoticed when possible and palatable when it was noticed. The strategy was eminently successful, but it required substantial physical effort, ingenuity, and bravado.
>
> This was FDR's splendid deception.[17]

Deception perhaps, but not self-deception. Those who knew FDR, particularly those who knew him at Warm Springs, where he could relax in the bosom of his 'family', insist on his acceptance of his disability. Pidge Cole felt strongly about this:

> Yes, it was a deception for the general public. But it was *not* a deception for the people that he was around. We were people who happened to have polio – like they're now saying, you know, you're not a disabled person, you're a person with a disability. That started with Roosevelt . . . My parents brought me to St Louis when I was six. Mother said, 'You were walking down a street' – she used to tell this a lot, that I looked up at her and said, 'Mother, don't let it bother you if people stare at me. Everybody's got something wrong with them, only with some of us it shows.' Now that's pretty astute for six years old. But where did I get it? I got it from Roosevelt. He wasn't fooling people. He *wasn't* fooling people. As for the public . . . that was a planned thing.[18]

Cissy Lord, a Warm Springs polio whose father was a Groton and Harvard contemporary of FDR, told one Roosevelt biographer that FDR was 'not self-conscious about his disability', but after Cissy had learned to walk again he still preferred it if she wheeled in to dinner at the White House alongside him. He 'enjoyed the company of beautiful women because he then didn't seem so handicapped'.[19]

The *necessary* deception, as one might call it, was in no sense a 'denial', as the current jargon has it. FDR might not admit that he would never get any more use out of his legs – in a newspaper interview in 1928, he was still challenging the idea of a two- or three-year limit to the period of recovery and was quoted as saying, 'In my own case I have received more improvement during my sixth year than during all the previous five'[20] – but it is impossible to distinguish what he actually felt from what he found it politic to say.

Fulton Oursler, author of a bestselling life of Christ, *The Greatest Story Ever Told*, and editor of *Liberty* magazine, gives a highly coloured account of Governor Roosevelt vetting an article on himself that Oursler planned to publish:

> 'Mr Oursler', he said, in what I can describe only as the voice of a fireside chat, 'there is only one statement in this article that I want corrected. The author says in this line here that I have "never entirely recovered from infantile paralysis".' He banged his great fist on the desk and shouted: *'Never recovered what?'* Now he leaned forward and pointed a finger at me as he added in a mild and mincing tone: 'I have never recovered the complete use of my knees. Will you fix that?'[21]

FDR's sensitivity over how his disability was reported did not prevent him from publicly identifying with – even personifying – the struggle against polio. In 1933, Keith Morgan recruited the public relations wizard Carl Byoir and a client of his, 'a crusty old utilities tycoon and grinder of political axes' by the name of Henry L. Doherty, to the cause.[22] Colonel Doherty was surprised to be approached, since he was in trouble with the Federal Trade Commission and had been called by one congressman 'the biggest tax evader in the country'.[23] When he asked Morgan why he should help, Morgan is alleged to have replied, 'Because it might get an old pirate like you into heaven.'[24]

The group that Morgan got together under Doherty's leadership came up with the idea of giving a national birthday party for the President to raise money for Warm Springs. Arthur Carpenter, who was present at the meeting, could not afterwards remember which of them actually made the suggestion: 'We were all so pleased by the idea we paid little attention to who made it.'[25] FDR's agreement was sought, and word came back from the White House: 'If my birthday will be of any help, take it.'[26] The first of the President's Birthday Balls took place on 30 January 1934 with people all over the United States 'dancing' – in the words of the slogan – 'so that others might walk'. Roosevelt's initial popularity as President ensured success, but nobody could have predicted the extent of it: the first Birthday Ball raised over a million dollars for the Warm Springs Foundation – almost an embarrassment of riches.

There was one aspect of this extraordinary fund-raising exercise that would have repercussions. Keith Morgan was being somewhat economical with the truth when he wrote to FDR, 'As we emphatically point out to some people there is no political connection whatsoever between the White House, the Democratic Party, and Warm Springs and certainly Warm Springs is not an adjunct of any political leader.' Morgan himself had been responsible for using postmasters, who were political appointees, to organise and promote the charity balls across the nation – and in his next sentence he seeks to justify what he has just denied:

> But there is a distinct connection between all those who under your leadership can unite in setting up effective machinery throughout the country which will do actual good and which will arouse and enlighten public opinion to the seriousness and cost of infantile paralysis.[27]

Nowadays the use of civil servants for such charitable ends would be illegal, but in 1933 the need for the swift establishment of a regional network was paramount and ready access through the US postal service was too inviting a shortcut to be eschewed on ethical grounds. The honeymoon period following FDR's election to the presidency could not last, however, and the identification of the polio crusade with the Democratic Party would backfire when people – people with money, at least – started reviling Roosevelt as 'that man in the White House'.

As far as Warm Springs was concerned, the success of the appeal and the notoriety it brought were not unmixed blessings. Dr Hoke was scathing about all the shenanigans and wrote to Basil O'Connor that the publicity was better suited to 'a box factory or a foundry' than a polio treatment centre, and its perpetrators 'fit in with a medical outfit about as accurately as a jackass would with a symphony orchestra'.[28] He was also alarmed at the prospect of thousands of polios labouring under the illusion that they had only to make their way to Warm Springs in order to be cured, and complained to Missy LeHand that he was getting an 'enormous number of letters that no sane person could answer from all sorts of poor derelicts all over the United States who couldn't be helped here'.[29] When Hoke later erupted over people pulling strings by writing to various senators, and to the President or Mrs Roosevelt, in the belief that it was the only way to get into Warm Springs – and over the fact that their letters were then sent from department to department when all that was needed was to refer all enquiries directly to the foundation – Missy was contrite. '. . . from now on we all promise to be good', she wrote.[30]

Even before FDR became President, Mrs Roosevelt had received one such 'begging letter' from a young woman in Seattle, Washington, which is remarkable for its dignity, anguish and articulateness. 'I know how such letters are received by people in your position,' this girl writes. 'I therefore despise the necessity of writing it.' But she goes on to say that she has read an article in *Reader's Digest* about Governor Roosevelt and the Warm Springs Foundation and feels that she owes it to her mother and father, if not to herself, to try everything she can:

> I am a girl, 19 years of age. Shortly after graduation from high school in 1928 an epidemic of infantile paralysis swept through Washington and I was one of the victims. Since that time I have been practically helpless. I

> walk a very little with the aid of crutches and a brace on my left leg. Once a week I am able to get to a swimming pool for half an hour's exercise. This is such a little, but it is all we can afford . . .
>
> With the stock [market] crash a number of investment houses went into bankruptcy and with them my mother and father's life savings. Fortunately we were able to get enough out of it to pay the hospital bills. But that's all . . . Mother is frail and it's just a matter of time before she breaks under the strain. Before I became ill we were ordinary people in ordinary comfortable circumstances. Now we have nothing. No financial security. Nothing ahead but more uncertainty and worry. It is my lot in life to realize all this and know that I can't help in any way . . . I want my mother and father to have a life of their own without me always to wait on and take care of.
>
> If I could only go away from them and fight it out alone, and lift entirely the burden of my care from them. And then to come back well and strong and with all the vigor and energy I lost 2 years ago . . . I should be so much more useful a citizen if I were an able one. I'm not at all a good invalid. Just that little ray of hope is enough to keep me from doing anything that would definitely establish my invalidism. I can't settle to anything definite unless I know whether I am to be a normal person ever again . . .
>
> This is just another hard luck story but it's important to me. I don't expect you to realize how much it is costing me to write this . . . It is more revealing than I am in the habit of being . . . But won't you help me if you can?[31]

Eleanor scribbled a note to Missy, soliciting FDR's support, and one can only hope the young woman got to Warm Springs in the end.

The difficulty was that such cases were commonplace. Epidemic poliomyelitis was worldwide in its distribution, but over half the reported cases occurred in North America, mainly in the United States. According to statistics provided by the Health Section of the League of Nations, '54.3% of the 178,328 cases reported from 1919 to 1934 inclusive, originated in the United States and Canada'.[32]

In the summer of 1934, there was what Keith Morgan described as 'a rip snorting epidemic in California'.[33] Dr John Paul, who went to Los Angeles as part of a three-man commission to investigate this, writes: 'A striking feature of the outbreak was that although most of the cases were mild, an atmosphere of stark dread prevailed, particularly among hospital personnel who had the task of caring for patients . . . It was as if a plague had invaded the city, and the place where cases were assembled and cared for was to be shunned as a veritable pest house.' But Los Angeles, he concludes wryly, 'is by no means a typical American city'.[34]

The importance of poliomyelitis, epidemiologists argued, was 'not so much in its incidence, since at its worst, it is an infrequent disease', but in the fact that *'it is one of the major causes of crippling'*; and because a majority of people who contracted it were under the age of twenty-one there was great physical and mental stress and even greater economic loss – 'inasmuch as such persons are handicapped at the onset of the useful and productive period of their lives'.[35]

The scale of the problem had been foreseen by FDR when he launched his crusade at the beginning of the decade, and it had now become obvious that the focus of the campaign had to be shifted away from the over-publicised Warm Springs. Republican opponents were already insinuating that, as a worried friend wrote to FDR, '"knowingly or unknowingly" you would . . . benefit through your ownership of property at Warm Springs Ga. by the expansion and extension of the wonderful sanitarium you have so splendidly and generously advanced there'[36] – to which FDR replied that he was 'not in the least bit surprised at the snarling curs who would gladly crucify their own mothers if it would help their own pocketbooks'.[37] But even some Democrats cast slurs. The racist ex-Governor of Georgia, Eugene Talmadge, 'the red-galluses demagogue' – so called because he would snap his braces (in the English sense of the word) for emphasis while he was speaking – went around saying that the Warm Springs Foundation was 'a racket, being disguised under the name of charity, by the President of the United States'.[38]

When Roosevelt agreed to let his birthday be used for a second time, it was on the understanding that seventy per cent of all the money raised should remain within the community to be used 'to combat its own local Infantile Paralysis problem', and the other thirty per cent would go to the newly formed President's Birthday Ball Commission for Infantile Paralysis Research (PBBC); not a cent was to go to Warm Springs.[39] (Though according to one document the foundation did get $256.42; perhaps this represented the sum raised locally in Georgia.[40])

A prime mover of research was the bacteriologist-turned-writer, Paul de Kruif, who is chiefly remembered for his scientific bestseller, *Microbe Hunters*, published in 1926 (the book that inspired the young Albert Sabin to change his course of study from dentistry to medicine, thus launching a distinguished career in research that culminated in the development of a live vaccine for polio[41]). After the first Birthday Ball, de Kruif asked his friend Arthur Carpenter at Warm Springs, 'Why do you use all that dough to dip cripples in warm water? That doesn't cure them any more than it cured you or the President. Why don't you ask the President to devote a part of that big dough to research on polio *prevention*? Nobody knows a thing about that.' Word of this got back to the great man via Keith Morgan, and de Kruif suddenly found himself appointed Secretary of the PBBC 'with the job of spotting virologists, neurologists, and epidemiologists who might be interested – now that at last there was a bit of money – in research on polio prevention'.[42]

A big man both physically and in terms of his appetites – he 'drank hard, talked loud, worked late, wrote long, praised to excess, and damned to oblivion, depending on where his current passions lay'[43] – de Kruif was an excellent propagandist but too opinionated and excitable, perhaps, to be the ideal arbiter of scientific projects. He had trained as a microbiologist at the University of Michigan and then worked at the Rockefeller Institute in the early Twenties. He was regarded as a very promising scientist, a young genius even, until he fell foul of Simon Flexner, who dismissed him when he discovered that de Kruif was the author of an anonymous attack on the institute in *Century* magazine. A colleague, Dr Peter Olitsky, bumped into de Kruif on the day of his dismissal and found him in a fine frenzy, saying that he had a 'waspish pen' and threatening to get even with Flexner. Olitsky tried to dissuade him from any such attempt.[44] But de Kruif used his waspish pen to considerable effect when he collaborated with Sinclair Lewis of *Babbitt* fame in writing a novel about a scientist, the eponymous Arrowsmith, in which the Rockefeller Institute and many of its members, including Flexner, are satirised.

The feud with Flexner has been given as a reason for de Kruif's surprising neglect of the Rockefeller Institute when he was distributing PBBC money for research.[45] But Flexner's successor as Director of the institute, Dr Thomas Rivers, told the historian Saul Benison he doubted that the institute would have taken money even if it had been offered: 'first, because Mr John D. Rockefeller, Jr, did not allow anyone from outside – either government or private agencies – at that time to contribute to the support of the Institute research projects, and, second, because Dr Flexner did not think it was possible to make a practical vaccine against polio at that time.'[46] (Rivers had reason to remember that the Rockefeller Institute refused outside funding, since he had not been allowed to supplement his budget with PBBC money when – encouraged by de Kruif – he had wanted to apply the technique he had pioneered of 'cultivating viruses outside the bodies of living animals . . . to the study of infantile paralysis'.[47]

Simon Flexner's pessimism over the practicability of making a vaccine against polio may seem strange, given that he had been so positive about the prevention and cure of Infantile Paralysis back in 1911. But there had been remarkably little scientific progress in understanding the polio virus and its behaviour in the intervening years, despite ever more frequent outbreaks of the disease. Part of the problem was the invisibility of the virus; it was difficult to study something you could not see even through an optical microscope (and the electron microscope was still unavailable). Then there was the expensiveness of monkeys, the only experimental animals yet to be used in polio research, which limited the extent of possible research.

Added to these difficulties were the many misapprehensions about the virus and the way it worked. Flexner himself was largely responsible – so influential was he, even after his retirement as director of the Rockefeller Institute in 1935 – for giving these currency. In *Breakthrough: The Saga of Jonas Salt*, Richard Carter summarises the situation succinctly:

> As of 1935, when the Birthday Ball Commission began scattering money around, virologists tended to agree that the disease was caused by only one type of virus (an error), that the virus grew only in living nerve cells (an error), and that it entered the body through the nose (an error), traveling thence to the brain and spine by way of nervous tissue (an error).[48]

It was Flexner who – according to John Paul – 'had convinced himself and his colleagues that the virus was a strictly neurotropic one that entered the body by the nasal route and proceeded directly to the central nervous system'.[49] He simply ignored or discounted any evidence that challenged his prejudices. When the Australian doctors, Frank Macfarlane Burnet and Jean Macnamara, produced evidence in 1931 that there was more than one type of poliovirus, American microbiologists, led by Flexner, responded sceptically – 'Burnet was a comparatively unknown figure . . . and Australia was remote; anything, it was argued, might come out of that far-off continent, including an exotic variety of poliovirus' (which was not allowed to be poliovirus at all, but labelled dismissively 'Australian X disease').[50] Following the 'experimental path', as Dr John Paul puts it, 'only led [Flexner] further and further away from the human disease and deeper into the woods'.[51]

Here, it might be thought, was another opportunity for de Kruif to revenge himself on Flexner and, at the same time, benefit humanity by challenging his old chief's dogmatic views. Unfortunately, de Kruif too was an enthusiastic proponent of 'the new concept of the way the virus travels from the outside world to the spinal cord . . . which reveals how closely, and maybe exclusively, the virus of poliomyelitis prefers the Central Nervous System . . .'[52] The question – 'the fundamental and moot question now vexing all engaged in attempts at immunization against infantile paralysis' – was: 'If there is evidence of immunity in the *blood* of a person vaccinated, does that mean that he will resist the disease as he is exposed to it in nature?'

> The difficulty of answering this question will be apparent when you remember that the infecting virus of infantile paralysis *does not* reach the spinal cord of the victim by way of the blood, but travels so far as has been determined experimentally by way of the nose, up the olfactory nerves, across the brain, and so down into the nerve cells of the spinal cord. So that, even if the victim's blood did have immunity, that would not protect him

> from the disease, if the vaccine had not also immunized the mucous membranes of his nose, or the nervous tissues of the nerves of smell, or the brain and spinal cord tissues, or all of them.[53]

Flexner himself could not have put the case against the feasibility of a vaccine better – or got the facts more wrong.

Fortunately, perhaps, de Kruif saw his own role as Secretary of the PBBC as no more than an enabling one; the commission had an advisory medical committee – originally consisting of three doctors, but expanded to four when they were joined by the new Director of the Rockefeller Institute, Tom Rivers, at the end of 1935 – that decided which projects to support. De Kruif regarded this 'co-operative activity of those members of the Commission who have been truly interested, of the advisory medical committee, and of the various grantees' as a welcome 'new development in medical science'.[54] Privately he wrote, 'My only part in this whole affair has been some small contribution to getting these fellows to work together suppressing their creative "egos".'[55] But he was being unduly modest about his own considerable influence on the direction of research.

In September 1935 there was a severe outbreak of polio in Raleigh, North Carolina. Keith Morgan proudly reported that it was through the efforts of Dr McCoy, Chairman of the advisory committee of the President's Birthday Ball Commission, that the health authorities had immediately sent for a supply of a new vaccine developed by Drs Park and Brodie of the New York Health Department laboratories:

> The fact that Dr Park had this vaccine available was made possible through the earlier grants which the Foundation made to his work in the amount of $21,000 so that he could prepare the necessary laboratory equipment and facilities with which to make this vaccine from the spinal cords of monkeys.[56]

Despite the reservations – to put it no more strongly – of de Kruif and a number of microbiologists about the practicality of a vaccine, further PBBC grants, amounting to $64,000 in all, were given to Park and Brodie. (It probably did Park's case no harm in de Kruif's eyes that there was no love lost between him and Flexner.[57]) Park was a highly respected figure in virus research, but he was getting on a bit; and the vaccine was the brainchild of 'a bumptious young research worker named Maurice Brodie', who had been born in Britain and trained at McGill University in Canada.[58] The idea was that an injection of the virus, inactivated or 'killed' by formalin, would provide immunity to the disease by producing antibodies in the blood; and to that end, 9,000 children were injected after Park and Brodie had tried it out on themselves without ill-effects, and

Park, who – according to Dr John Paul – 'was never one to let grass grow under his feet',[59] had pronounced it safe.

The breathtaking speed and casualness – by latter-day standards – were partly due to the fact that there was a competitor in the field, Dr John Kolmer of Temple University, Philadelphia. This 'quiet, unassuming, earnest, little man' was a quite prodigious worker.[60] He, like many others then and later, did not believe in the possibility of an effective killed vaccine and he attacked Brodie's concoction, saying that he had a superior one, a 'live' vaccine – in other words, one in which the virus had not been inactivated but merely weakened or *attenuated* (a key word that he was the first to use in this context) by repeated passages through monkeys. The race was on, prefiguring another, far more significant one that was to occur two decades later. As John Paul puts it, 'Competition in this matter should not have been a factor which played a vital role; nevertheless it usually is and certainly was in this particular instance.'[61]

Judged by later standards, Kolmer's vaccine – which was *not* funded by the PBBC – was 'a veritable witch's brew',[62] and his method of production 'hair-raisingly amateurish, the therapeutic equivalent of bath-tub gin'.[63] But like Park and Brodie, Kolmer indulged in 'the usual preliminary heroics' of trying out his vaccine on himself and 'remaining vertical'.[64] Unfortunately, he went on to try it out on 12,000 children.

Sceptics among medical scientists were already busy testing Brodie's and Kolmer's vaccines in their laboratories, as de Kruif reported:

> In regard to the opinion held of the work of Dr Park and Dr Brodie by other investigators, it can be stated [that] the production of *blood immunity* by the injection of the formaldehyde vaccine into monkeys has been confirmed by Professor E.W. Schultz, of Stanford University, and by Dr P.K. Olitsky, of the Rockefeller Institute for Medical Research. Both of these investigators, however, do not believe that the vaccine has been proven to develop immunity against direct inoculation into the brains of monkeys.[65]

As for the Kolmer vaccine, Schultz had not completed his tests when de Kruif wrote his progress report in mid-November 1935. 'He states, however, that, even if this vaccine did show immunity, he would not recommend it for injection into human beings. It is too dangerous.' (This last sentence is deleted in the report.)[66] In other words, these vaccines were considered by other researchers to be either ineffectual or lethal.

In the summer of 1935 there had been a dozen cases of paralytic polio, half of them fatal, in areas where there had been no epidemic. The vaccines, particularly Kolmer's, were implicated. It was a moot point whether or not Brodie's had killed anyone (there were reports from California that at least one child had died as a result of his vaccine[67]); but then it was equally questionable whether it had immunised anyone. Things came to a head at a meeting of the southern branch of the American Public

Health Association held in St Louis on 19 November 1935, where Dr Rivers (of the Rockefeller Institute) and Dr Leake (of the US Public Health Service) accused Kolmer, Park and Brodie of disregarding safety in their unseemly rush to produce a viable vaccine. According to Rivers, James Leake was 'hot under the collar': 'He presented the clinical evidence to the effect that the Kolmer live-virus vaccine caused several deaths in children and then point-blank accused Kolmer of being a murderer.'[68] This accusation was watered down in the printed record of the event in the *American Journal of Public Health*, which has Leake saying – in the manner of a nineteenth-century gentleman called upon to defend his daughter's honour – 'I beg you, Dr Kolmer, to desist from the human use of this vaccine.'[69]

Rivers himself – 'a thickset man with a round, ruddy face, incipient dewlaps, a strident voice compared by some to the moose's and by others to the bullfrog's, and an accent that would have revealed his Cracker origins if his social prejudices had not'[70] – was no stranger to strong language and, as he put it, 'didn't mind jumping on Dr Brodie and Dr Kolmer.[71] (Though he had earlier given it as his opinion that Brodie's vaccine was probably safe, if not necessarily effective, and should be given an extended trial.[72]) He recalled that, after being attacked from two sides,

> Dr Brodie got up and said, 'It looks as though, according to Dr Rivers, my vaccine is no good, and, according to Dr Leake, Dr Kolmer's is dangerous.' He sat down and Dr Kolmer got up. He didn't refer to me at all. He just said, 'Gentlemen, this is one time I wish the floor would open up and swallow me.' He then sat down.[73]

That was the end of it: 'The vaccines were dead and so were careers.'[74] In fact, only Brodie's career was destroyed; Park was allowed to retire with honour (he died in 1939), and Kolmer returned to Temple University, where he continued as a Professor of Medicine, working hard and producing textbooks and innumerable articles until his retirement in 1957. Brodie, unlucky to be tarred by the same brush as Kolmer and condemned by the influential Rivers, has had a bad press – at least until John Paul partially rehabilitated him in his history of polio: 'After all, he had almost succeeded in a good cause and was at least on the right track in choosing formalin to inactivate the virus'[75] (as a young medical student by the name of Jonas Salk, who was at New York University Medical School when Brodie lost his job there, would eventually and triumphantly prove). Paul gives currency to the rumour that Brodie committed suicide, but this is not borne out by Brodie's death certificate, which gives the cause of death (in his early thirties) as coronary thrombosis. 'But we can say, figuratively', the author of *Virus Hunters*, Greer Williams, comments, 'that this young scientist died of a broken heart'.[76]

Dr Rivers was invited by de Kruif (the man he would eventually replace as the main arbiter of polio research projects) to join the medical advisory committee of the President's Birthday Ball Commission as a direct result of the vaccine fiasco. He later boasted that they needed a 'rough-neck' like him, and that after he came on the commission Dr Park 'never got another cent'.[77]

The twin misconceptions about the neurotropism of the polio virus and the nasal passage as its so-called 'portal of entry' led researchers to attempt to block children's noses against it with such chemicals as picric acid and zinc sulphate. When tried out on monkeys, this proved effective. Dr Edwin Schultz of Stanford University and Dr Charles Armstrong of the US Public Health Service were the first to apply the principles of this experimental work to human infection. A number of microbiologists, including Rivers, favoured the use of preventive nasal sprays in humans. 'If I had a child in an area where poliomyelitis appeared', Rivers was quoted in *Time Magazine* as saying, 'I would take my child to a good otolaryngologist and ask him to apply the spray in the manner set forth by [the neurosurgeon] Dr Peet'.[78]

Armstrong's picric acid and alum nasal spray was used in an epidemic in Alabama in 1936 but in too haphazard a fashion to provide convincing evidence either way. The next year, however, an outbreak of polio in Toronto gave an opportunity to test the zinc sulphate spray known as the 'Schultz-Peet prophylactic' with proper controls. The trial proved a dismal failure; the spray was useless as a preventative and, in addition, had the undesirable side-effect of causing many children to lose their sense of smell, some permanently. Chemical prevention of polio, hailed by de Kruif in his characteristically exaggerated fashion as 'the greatest advance in the fight against infantile paralysis since it was found that the disease could be given to monkeys', turned out to be even more of a fiasco than the Park/Brodie vaccine. Thirty years later, Tom Rivers summed up 'the research accomplishments of the President's Birthday Ball Commission' thus:

> Minus. If you take the good things that they did, and subtract the bad things, you get a minus. It doesn't mean that everything they did was rotten or useless. It means that when you add and subtract you get a minus. That's all.[79]

This negative judgment is challenged by the more temperate John Paul (whom Rivers labels 'conventional'[80]). Paul points to grants made to a number of leading researchers – including himself and his colleagues at the Yale Poliomyelitis Unit, though he is too modest to dwell on their pioneering work aimed at establishing whether or not there was more than one type of polio virus – and concludes: 'It was at least another few years before the National Foundation for Infantile Paralysis, successor to the PBBC, could claim to have supported investigations of any such importance.'[81]

It was not just virologists such as Brodie and Kolmer, working with inadequate data, who made costly and dangerous mistakes; clinicians too could pursue promising leads into blind alleys, as the experience of George Draper illustrates. Since 1916, Draper had been obsessed with the idea – not in itself a disreputable one – that a certain constitutional type of person was more susceptible to paralytic polio than others. He developed his 'anthropological' theory at great length in his 1935 textbook on *Infantile Paralysis*. His method was a combination of observation and 'mensuration' – or 'measuring the distance between certain important points on the surface of the body'. He writes:

> Ever since Heine's time many physicians have referred to the large size and healthy appearance of their poliomyelitis patients, and even lay visitors to hospital wards set apart for those cases have often been heard to remark upon the magnificent specimens which filled the beds.[82]

So far so good. But, he goes on, this picture of a fine constitution may be deceptive: 'somewhere in that constitution is a vulnerable point, like the heel of Achilles, which fails to resist the attack of the poliomyelitic virus'. The search for this Achilles' heel is conducted in a quasi-scientific manner – 'Indeed, the training and experience necessary to make adequate observations of a human being's constitutional characteristics must be at least equal to those of a good bacteriologist or zoologist.'

But Draper's observations are, to say the least, bizarre. The susceptible child, according to his far from flattering picture, has a broad face, wide-set, often mongoloid, eyes, widely spaced teeth (in susceptible young adults, by contrast, 'there was a tendency to crowding of the teeth'), moles and pigmentation of the skin, little or no 'moons' at the base of thumb or finger nails, broad – 'feministic' – pelvis and, in the case of a boy, under-developed genitalia. The dodgy nature of this kind of approach to disease susceptibility becomes apparent when physical characteristics are transformed into mental defects, as with 'the mongoloid eyes', reinforcing – albeit subliminally – the already rich confusion of physical and mental handicap:

> There were three mongoloid idiots with poliomyelitis in Willard Parker Hospital during the [1931] epidemic. This is a rather high percentage of cases of a rare condition to appear among a comparatively small group of similarly diseased people.[83]

This from the doctor whose most famous poliomyelitis patient was currently President of the United States.

Other doctors, including the influential Lloyd Aycock of the Harvard Infantile Paralysis Commission, took the question of a genetic predisposition to paralytic polio seriously, as de Kruif reported: 'It may be said that Aycock is a partisan of the idea that some still unknown physiological kink in people is at the bottom of the mystery of susceptibility or of resistance to polio.'[84] At a meeting of the New York Academy of Medicine in the early Thirties to discuss how immunity was acquired in the general population 'the "new and exciting" physiological explanation' – as John Paul had cause to remember – was greeted with far more enthusiasm than his own and a colleague's endorsement of the Swedish view that the prevalence of mild abortive cases and healthy carriers accounted for the large number of people who were naturally immune to the disease, not some inherited faculty. Paul recognised that Draper's views 'had some truth in them, but the notion that built-in susceptibility could be recognized by physical characteristics was at least problematic'.[85]

Paul also thought that Draper – particularly towards the end of his life – 'rode his views too hard'.[86] Rivers thought so, too. In 1935, when Flexner asked him if he would support Draper's application for a grant from the PBBC, he said he would, as he then regarded the relation of constitution to susceptibility to polio as a respectable research topic, but he advised against giving him 'the whopping sum he asked for'. Draper later got a small grant from the National Foundation for Infantile Paralysis, too – though not everybody on the virus research committee was persuaded of the virtues of his approach. (In fact, as the minutes of the 6 July 1938 meeting record, 'Rivers occupied somewhat lone wolf position here, believing that this inquiry should be prosecuted'.[87])

> Well, he set to work, to compare the buffy coats [the thin layer of leucocytes which form on top of a mass of red cells when blood is centrifuged] of bloods taken from patients who had polio with those of patients who had other diseases and set up control groups with persons of various sizes and shapes and racial extraction. He sent in progress reports with interesting pictures of various kinds of blood cells, but for the life of me I didn't know what it all meant, and for that matter neither did anybody else.[88]

Draper applied for a renewal of his grant, expecting it to go through automatically, and when he was turned down, 'all hell broke loose'. He wrote to the newspapers and appealed to Eleanor Roosevelt, who was careful to get the facts of the case from Basil O'Connor before replying. His application never resurfaced.

Rivers took a delight in recollecting how Draper applied his theory of susceptibility to other diseases as well: 'He once told me that I was the type that would never get a duodenal ulcer – I want to tell you that I later had the granddaddy of all duodenal ulcers.'[89]

Declining donations to the President's Birthday Balls reflected both the lack of instant success in the scientific initiatives aimed at the prevention of polio and a waning of FDR's personal popularity – at least among the better-off – as he entered his second presidential term (and engaged in a damaging power struggle with the Supreme Court). Opposition to the New Deal had been widespread from quite early on, as the following – albeit jocular – correspondence between doctor and patient reveals.

'Dear Franklin,' George Draper wrote on 29 August 1934:

> There is a giant lying supine across the continent. He is paralysed in arms & legs – a victim of infantile paralysis. He is in the hands of a most earnest & conscientious doctor. But the doctor is over-treating him.
>
> My experience with the disease has convinced me that more cases with potential recovery possibilities are ruined by too vigorous treatment of the weakened muscles than are helped by it.
>
> I recall one case, known also to you, who was protected by pain from too early & too vigorous treatment.
>
> Would not the giant, stretched supine across the country, react more quickly & completely if the current prescription of massage & specialized exercises were interrupted for a period?
>
> Nature works in mysterious ways to heal wounds. The ideas of man often interfere with natural processes . . .[90]

'Dear George,' FDR replied from Hyde Park a few days later:

> I like your supine giant and I should like to stop the treatment were it not for the fact that sixteen million little cells out of the one hundred and twenty million cells in his body would lose their circulation, starve and die, if we were to stop the very gentle massage which keeps some blood running into these sixteen million little cells. That is the only treatment being given at the present time. Also, in cases like this, medical history proves that if treatment is suddenly stopped the giant is very apt to leap from his bed and either commit suicide or die of an epileptic fit.
>
> Enough said![91]

Not enough for Dr Draper, however, who was far from satisfied with this riposte. Six months later, he sent the President a telegram:

> Pediatricians have long been perplexed by difficulty of weaning infant from breast or bottle to teaspoon or cup. The shift often establishes permanent

> neurosis in subsequent adult. According to report in evening paper twenty two million citizen infants now hang on federal breasts. Can you wean them doctor and prevent national neurosis[?][92]

FDR once again responded good-humouredly:

> As a young intern you doubtless realize that the interesting transitional process, which you describe in your telegram, presupposes that the bottle, teaspoon or cup is not empty. Such vehicles of feeding, if empty, produce flatulence and the patient dies from a lack of nutrition.
>
> The next question on your examination paper is, therefore, the following:
>
> 'Assuming that the transitional period has arrived, where is the Doctor to get the food from to put in the new container?'
>
> If you will answer the above question intelligently, I will give you a passing mark and make it possible for you to obtain your Degree![93]

There is no record of a reply to FDR's challenge, so it must be assumed that George W. Draper MD flunked this particular exam. But his views, however light-heartedly expressed, were representative of his class, to whose interests FDR seemed more and more of a traitor. Roosevelt was 'ruining' the country.

One effect of FDR's reduced popularity was that his identification with Warm Springs came to be seen as detrimental rather than beneficial. His advisors, notably Basil O'Connor, stressed the need to formalise the divorce of the polio crusade from Warm Springs by creating a truly national organisation to incorporate the PBBC and carry the work a stage further; and on 23 September 1937 Roosevelt announced the formation of a National Foundation for Infantile Paralysis, the purpose of which was to 'lead, direct, and unify the fight on every phase of this sickness'.[94] The man appointed President of the fledgling foundation was none other than the Warm Springs supremo, O'Connor himself, 'the sly, black-haired lawyer in neat, rimless glasses . . . sucking on a cigaret in a holder, who as usual had brought things to a head'.[95]

The energetic Keith Morgan was promoted to be Chairman of the Committee for the Celebration of the President's Birthday in place of an aggrieved Colonel Doherty, about whom O'Connor wrote mysteriously to FDR's press secretary, Steve Early:

> It has always been my policy, so far as possible, never to burden the President or those around him with problems that could be solved without doing so. Let me say to you frankly that Keith Morgan didn't tell you anything! If you would like to take a couple of days off sometime, I think you could spend them listening to a very interesting story. All I need say to you now is that a very unfortunate and unsatisfactory situation has been well disposed of and that it should rest just where it is.

Early scribbled a note on the bottom of this letter: 'Have told Byoir no letter [of commendation?] possible.'[96]

Keith Morgan expressed his gratitude to FDR:

> It was ten years ago that I first came up to see you to discuss Warm Springs, infantile paralysis and Arthur Carpenter's case.
>
> Since then many things have happened. Our plans and dreams seem to be coming true at last.
>
> It has been grand to be associated with you in this work. This association plus your friendship and affection are the most prized possessions in my life . . .[97]

To Marvin McIntyre, FDR's other press secretary, Morgan wrote:

> The money that we receive will be used in a four-way attack on infantile paralysis:
>
> 1. Science Research
> 2. Epidemic Action
> 3. Telling to Doctors and Parents the NEW CARE now known to PREVENT crooked backs, curvatures of the spine, twisted bodies, and contracted limbs [see chapter 5].
> 4. Dollars to Orthopedic Centers (Hospitals, Clinics, etc.) for the IMMEDIATE Restoring of Human Wreckage.[98]

The phrase that was to become a byword for the polio crusade, 'the March of Dimes', was the inspiration of the Hollywood entertainer, Eddie Cantor. He was responsible for the idea of getting everyone to send their dimes directly to the President at the White House. At first, after Cantor had made his appeal on the radio, it looked as though the scheme had misfired. Tom Wrigley, public relations man for the new National Foundation, was summoned to Washington: 'Marv McIntyre said, "You fellows have ruined the President. All we've got is $17.50. The reporters are asking us how much we've got. We're telling them we haven't had time to count it."'[99] The next morning a truck brought twenty-three bags of mail to the White House and from there it snowballed; on the third day 150,000 letters came in, thirty times the normal daily delivery. Nobody could find the official mail in amongst all the dimes 'fixed with gummy tape, baked into cakes, jammed into cans, imbedded in wax and glued into profiles of the President'. According to the White House mail chief, 'The government of the United States darned near stopped functioning.' But the polio crusade was underway once again, the 1938 birthday celebrations eventually netting a record $1,823,045.[100]

On the research front, too, things were moving. Under the PBBC – according to Rivers, who is hardly an impartial witness – de Kruif had done pretty much as he liked, operating in a rather 'hit-or-miss' manner; sometimes he did not even tell the medical advisory committee who was applying for what. 'It was, to say the least,' Rivers remembered, 'a slipshod way of doing business, and it was done away with.'[101] O'Connor was happy to take advice from de Kruif and from the scientists de Kruif recruited on to the new research committee, but he was going to run the show himself, and in an orderly fashion.

Under O'Connor, de Kruif's star was on the wane and Rivers's was in the ascendant – the men from the Rockefeller Institute would finally be revenged on their tormentor when the upstart de Kruif resigned from the National Foundation in the early Forties after a row over the funding of a nutritionist whose work was only marginally, if at all, relevant to polio research.

'I'm only a layman, of course, and I'm not trying to tell you what to do,' O'Connor said to Rivers,

> but I think maybe we haven't been building our case from the ground up. Perhaps we have been trying to get a conviction with insufficient evidence. How about drawing up a list of research priorities, so that we can emphasise first things first and try and get somewhere for a change?[102]

Rivers set to work and came up with an eleven-point programme, which was then discussed in committee and put into an agreed order of priority. 'Production of a good vaccine', far from being top of the list, was relegated to number eleven, after 'Relation of constitution to susceptibility'.[103] The next decade (allowing for interruptions caused by the Second World War) would be taken up with the kind of homework which was the essential prerequisite to the development of an effective vaccine – such as ascertaining the nature of the virus. The pioneering work of Wendell Stanley, 'the father of virus chemistry' and joint-winner of a Nobel Prize in chemistry in 1946, in the crystallisation of the tobacco-mosaic virus (TMV) led to his announcement in 1935 that 'his crystals were protein and not living organisms but non-living molecules'.[104] This was 'a microbiological earthshaker', and one of Stanley's severest critics was his Rockefeller Institute colleague, Thomas Rivers, who could not accept that a virus was not a living cell like a bacterium. Further work by two English scientists led Stanley to revise his definition of TMV:

> TMV is a nucleoprotein, a combination of protein and nucleic acid such as is found in the nuclei of living cells. The proportions in TMV are 94 per cent protein and 6 per cent nucleic acid. This seemed to move the virus a little closer to the land of the living than Stanley supposed.[105]

It was not until 1955 that 'the bridge from plant to animal viruses was crossed'. That was when two scientists working in Stanley's laboratory succeeded in crystallising poliovirus. But it was already established that the 'midget microbe' was 'more akin to a cell nucleus than to the living cell as a whole'; that it was 'a borderland form of life, both organism and chemicle particle, depending' – as we now know – 'on whether it is in its host cell or in transit'.[106]

At Warm Springs, meanwhile, there had been further changes. When FDR had been in charge, the emphasis had been on fun; William Hassett, his Secretary in the White House from 1935, said of his boss at Warm Springs: 'One thing that he always did was to strive to keep a home atmosphere there. He fought any aspect of institutionalism. Wanted it kept one happy family with everything on a thoroughly informal basis.'[107]

But with O'Connor the emphasis was on efficiency; after all the publicity generated by the Birthday Balls, Warm Springs inevitably became a showcase for the after-treatment of polio, with all that that implied in terms of modern buildings and equipment. Despite FDR's wishes, it was to some extent 'hospitalised'.

Roosevelt was not alone in regretting the passing of the earlier, more free and easy, hand-to-mouth existence, when survival itself was at stake. In 1936, two key figures in the running of the place resigned – Dr Michael Hoke and Arthur Carpenter – and a third, Fred Botts, the 'Dean' himself, was only with difficulty persuaded to remain by FDR, who made him Registrar so that he would once again be dealing with people rather than things and would continue to write the warm letters, full of Shakespearean echoes, that Roosevelt cherished.

> My dear Mr President:
>
> What a glorious morning! A Sabbath calm seems to pervade the Foundation, broken only when the tantrums of a blue jay and gray squirrel converge on a 'free-for-all' far up in a pine tree. It's a feud that's come down through the ages, I believe, and only the Creator can stop it. The 'Colony' is waking up now. Footsteps sound on the porches – the attendants on the move, going from cottage to cottage, assisting in the chores-à-la-prophylactic. Soon the polios will be on parade with blowing plume and clanking sabre . . . Little José Alvarez emerges into the sunlight. He's the young sprout from Puerto Rico who arrived here one year ago, his polio misshapen body rivaling that of Oliver Twist. He couldn't speak a word of English. We bought a dozen Spanish grammars and dictionaries and set about becoming linguists with a vengeance; for he was placed in a cast up to his ears and we just *had* to understand some of his requests. But he's way ahead of us now, so we all 'speak English as she is spoke'. The rhythmic crunch of brace and crutch is stepped up a bit as the aroma of bacon and eggs filters on the morning air. And so the day's begun . . .[108]

Hoke's style was very different, but his complaint was much the same as Fred's: too much administration and paperwork brought about by Warm

Springs' excessive notoriety. 'The immediate thing', he scribbled in a note to Keith Morgan, 'is the scheme of admission of patients':

> The very, very difficult thing here – superlatively difficult – has been to keep the outside contacts straight with the polio people, senators, congressmen, public health workers, and God knows what other types who write in here thinking the great humanitarian President has founded a thousand bed hospital free for all applicants.[109]

Carpenter, who found himself at odds with O'Connor's style of management, moved on to study the feasibility of making local Birthday Ball committees the basis of a national organisation – what became the National Foundation for Infantile Paralysis – before retiring from the fray altogether. (Fragments of surviving correspondence suggest that Carp's subsequent life was far from trouble-free. Early in 1942, Basil O'Connor informed the President that 'Arthur Carpenter has had a complete mental breakdown due to several factors one of which is money,' and went on to suggest that a note from FDR addressed to the Veterans Hospital for Nervous and Mental Cases at Murfreesboro, Tennessee, where Carp had been admitted as a voluntary patient, 'would act almost as a cure'. FDR obliged and, two years later, wrote again to Carp: 'I am delighted to hear that you are coming along well and that you have taken up farming in Georgia . . .'[110])

The spotlight on Warm Springs during the first years of Roosevelt's presidency had revealed an anomaly that was to become something of an embarrassment to the administration. Just before the first Birthday Ball a congressman from Illinois wrote to Louis Howe in the White House to inquire if 'Negro children are to be benefitted by the Warm Springs Foundation'.[111] Despite its proclaimed openness to all races, colours and creeds, Howe learned that 'no negro has ever yet been greeted at the Foundation at Warm Springs', and his secretary, known as 'The Rabbit', wrote him a memo on 3 February 1934: 'In view of this fact and the additional factor that the Birthday Ball is over and that is probably what he wanted the information for, don't you think the best answer is that "we never received the letter"?'[112] (A week later, when a correspondent in Chicago asked if Jews were admitted to Warm Springs, Missy LeHand wrote a note to Steve Early saying, 'Of course they are,' and Early replied accordingly.)[113]

The Rabbit's evasiveness seems to have become the pattern for dealing – or not dealing – with such inquiries. In 1936, Walter White, secretary of the National Association for the Advancement of Colored People (NAACP), wrote to Eleanor Roosevelt:

> When we were asked, in 1934, to join in the President's Birthday Balls we regretfully had to decline when we could get no satisfactory answer to our request for information as to what would be the status of Negroes who might apply for treatment.[114]

During the 1936 election campaign, a self-styled 'white Democrat of the South' urged FDR to 'request that provision be made at Warm Springs, Georgia, for Negro children as well as white children suffering from infantile paralysis'. His letter was acknowledged 'in the absence of the President' by Steve Early, who said it would be 'brought to his attention upon his return to the city'.[115] But no further action was taken.

A white attorney in Albany, Georgia, who took up the case of a 'young colored boy . . . suffering from the effects of infantile paralysis', received no encouragement from Warm Springs; he was told: 'At the present time we have no facilities for caring for colored victims of infantile paralysis.' So he wrote to the President, on whose behalf Missy LeHand replied that she was sorry but 'I fear that there is nothing further that can be done at this end'.[116]

What finally incited the administration to act was the rumour that 'Republicans are circulating report throughout Ohio [and elsewhere], among members of the negro race, that negro children are not admitted to Warm Springs.'[117] Steve Early, one of the many southerners with whom FDR surrounded himself, wrote to Keith Morgan to ask if coloured patients in the various states were taken care of out of funds allotted to them by the Birthday Balls. 'I suppose', he went on grudgingly, 'there must be some provision made to prevent discrimination.'[118] Less than a month later, when the 1936 election campaign was reaching its climax, the *New York Sun* carried a damaging article headed, 'Warm Springs Bars Negro Boy', which drew attention to the fact that though individual Negroes had contributed to the foundation through the Birthday Balls (even if the NAACP abstained), they were not granted the use of its facilities.[119]

The boy in question was the nine-year-old son a black preacher, the Rev. J.S. Brookens. Brookens had been unable to get his son into Warm Springs and had written to the Secretary of the Interior, Harold Ickes, who passed the letter on to his adviser on Negro affairs, Robert C. Weaver, himself a black man. Weaver admitted that, 'after careful inquiry', he had been unable to find any institution which provided hospital service to Negro children suffering from infantile paralysis. But he promised to take up the matter 'with persons who are interested in Negro health', and to be in touch 'in the event that any provisions can be secured'.[120] When nothing further happened, Brookens wrote again, but by this time Weaver had become suspicious that something more than Brookens's son's health was at stake, as he wrote to Walter White:

> Events that have happened since the writing of the first letter by Dr Brookens and my answer lead me to believe that Dr Brookens is deliberately working with Republican forces to make political capital of a situation which has no place in politics . . .
>
> . . . It might interest you to know that a story similar to the one in the *New York Sun* appeared several weeks ago in certain Negro newspapers of the country, and I have the information that at least one of these papers, the *Chicago Defender*, received its 'facts' through the Republican Campaign Headquarters.[121]

As Walter White hinted in his letter to Eleanor Roosevelt, but was too polite to spell out, if Republican forces were exploiting this situation, it was because there was a situation for them to exploit; and Eleanor, who needed no convincing of that, passed the correspondence on to FDR with a note saying, 'They [Negroes] should have a cottage. What is the answer?'[122]

FDR replied to his wife via Missy: 'The President thinks this should not be answered by a letter. He will explain it personally. The Foundation has contributed to a negro place at Atlantic City where they do very good work.'[123] This last statement turned out to be untrue. Basil O'Connor wrote to Missy on 23 December 1936, after FDR had been safely returned to the White House:

> In reply to your note of the 18th as to whether or not the Foundation has given anything to the Negro Foundation at Atlantic City, the answer is 'No'. At the moment we are considering the whole negro situation very carefully . . .[124]

Three months later, O'Connor sent the President a digest of the deliberations of the trustees of the foundation along with a revealing covering letter:

> In connection with the attached letter about the negro people at Warm Springs, I can't help but think that its solution is the same that Dean Thayer made of the problem of admitting women to Harvard Law, which arose many years ago. After the matter had been discussed at great length by the people and by the dignified Board of Trustees of the Alma Mater, Dean Thayer said the problem solved itself because in the Law School there were no lavatories for women!
>
> Needless to say, I am not sending a copy of this letter to the other Trustees.[125]

O'Connor's official letter to the President, a copy of which he *was* prepared to send to the other trustees, was hardly on a higher level. 'We do get several inquiries on this colored question,' he wrote, 'but I think they can be fairly classified as coming from (a) professional colored promoters, and (b) sob-sisters connected with institutions such as Teachers College at Columbia University.'[126]

Eleanor's suggestion of having a cottage for Negroes at Warm Springs met with unanimous disapproval among those of the trustees who were prepared to commit their views to paper. Their comments were almost entirely negative, ranging from, 'the Foundation is not in a position to undertake the extra expenditure', to the blunt, 'My impression is that such a thing would not be desirable in Georgia.' A Warm Springs administrator wrote, 'Without facilities designed for the proper housing, we do not feel that we could make such patients comfortable both physically and psychologically,' and the surgeon who succeeded Hoke, Dr C.E. Irwin, felt that it was 'impractical and unwise to admit colored patients for treatment in residence'; a few were already treated as outpatients, though – Irwin added – 'I believe I am correct in saying that the Foundation has never extended any financial aid to any of these cases.' Cason Calloway, a trustee and Warm Springs's richest neighbour – and a friend of FDR despite his Republican sympathies – opted for the 'separate but equal' solution beloved of southern segregationists: 'it would be a fine thing to put in an entirely different location some medical aid for negroes who have been stricken with polio'.[127]

In passing on the comments of the other trustees and expressing his own views so freely, O'Connor must have known that they would more or less coincide with FDR's, if not his wife's – but then, from the first days of FDR's presidency, O'Connor had been one of the close advisers committed to the notion that they had to 'get the pants off Eleanor and onto Frank'.[128] Yet if FDR was guilty of the unthinking racism of his class and time, and fell easily into the more or less benign paternalism of the South, he broke the mould to the extent of encouraging – or not actively discouraging – his wife's increasing commitment to civil rights. When she asked if he minded her saying what she thought in public, he replied, 'No, certainly not. You can say anything you want. I can always say, "Well, that is my wife, I can't do anything about her." '[129] She was, in this matter as in others, his radical conscience prompting him to do what he could, whenever vital political interests did not stand in the way. But while he was happy to dedicate the building of the new Warm Springs Negro School in 1937, after it had been constructed with the support of the Julius Rosenwald Fund, blacks were never admitted to Warm Springs for treatment, except as outpatients, during his lifetime.

'Doc' O'Connor's lawyer-like cunning suggested a way out of this embarrassing political dilemma. If he could establish that 'the colored race was not very susceptible to this disease', the lack of facilities would be easier to explain away. He asked the retired Dr LeRoy Hubbard to investigate.[130]

New York 1916: mother and child in the heartland of the epidemic where houses were placarded to warn against this latter-day plague.

An Egyptian stele, dating from the eighteenth dynasty (1580–1350 BC) – evidence of the antiquity of polio as an *endemic* disease.

Loading an ambulance in New York, 1916: 'many mothers felt that sending their children to hospital was tantamount to condemning them to death or crippledom'.

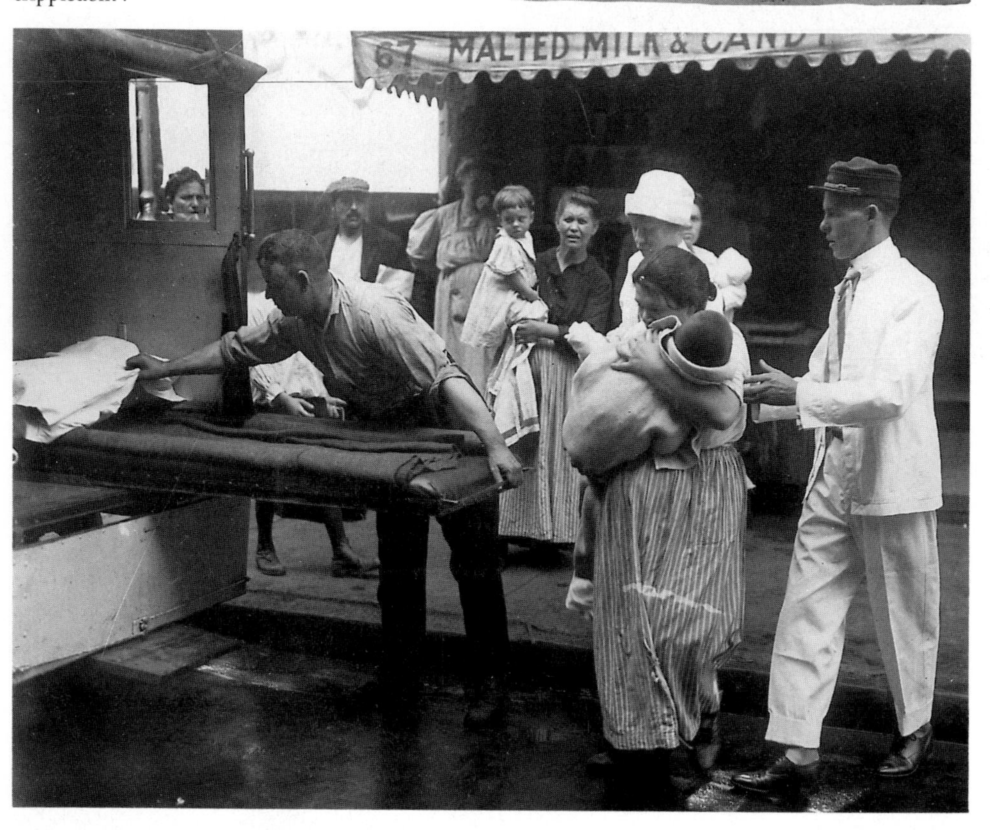

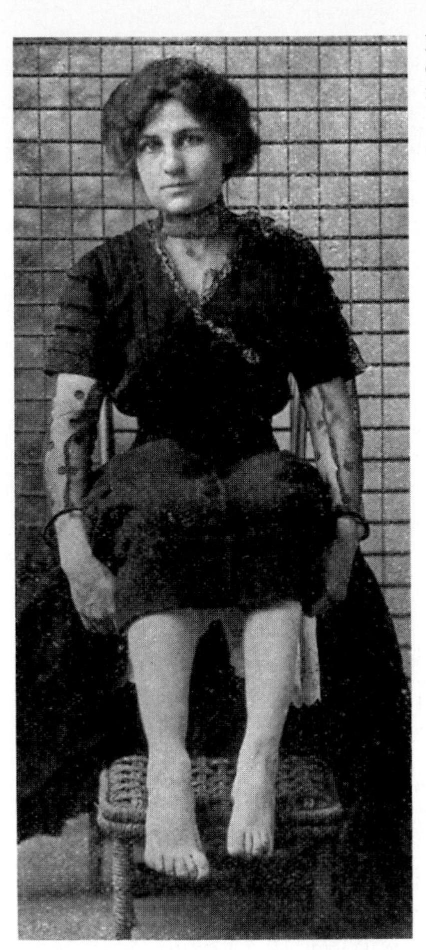

Modern methods of treatment: massage and the application of electricity to affected limbs were recommended in early manuals of Infantile Paralysis.

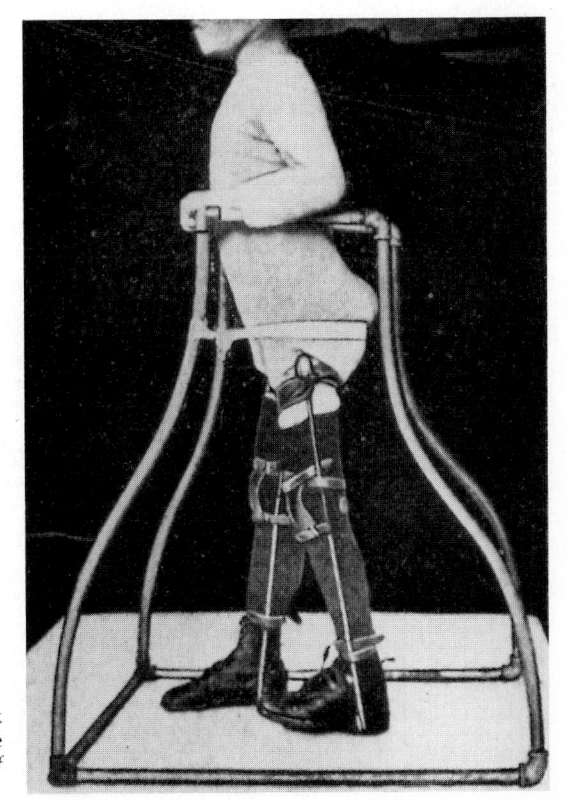

First steps: a paralysed child learns to walk again with leg braces and walking frame (from Robert W. Lovett, *The Treatment of Infantile Paralysis*, 1916).

'Swimming back to health': FDR with fellow polios and physiotherapists in the baths at Warm Springs in 1930.

Warm Springs 1939: FDR presides over Thanksgiving dinner which became a semi-official Founder's Day (note the rare appearance by Eleanor Roosevelt at Warm Springs).

Cosmetic walking: FDR leans on the arm of his friend and legal partner, Basil O'Connor, whose (first) wife looks on – note that FDR's right hand, presumably holding a cane, is out of view.

One of only two known photographs of FDR in a wheelchair.

Hollywood Bowl, 24 September 1932: FDR waits to emerge from his car while a secret-service man positions his legs and checks that his braces are locked at the knee. James Roosevelt rests a hand on the car seat as he prepares to provide a strong arm for his father.

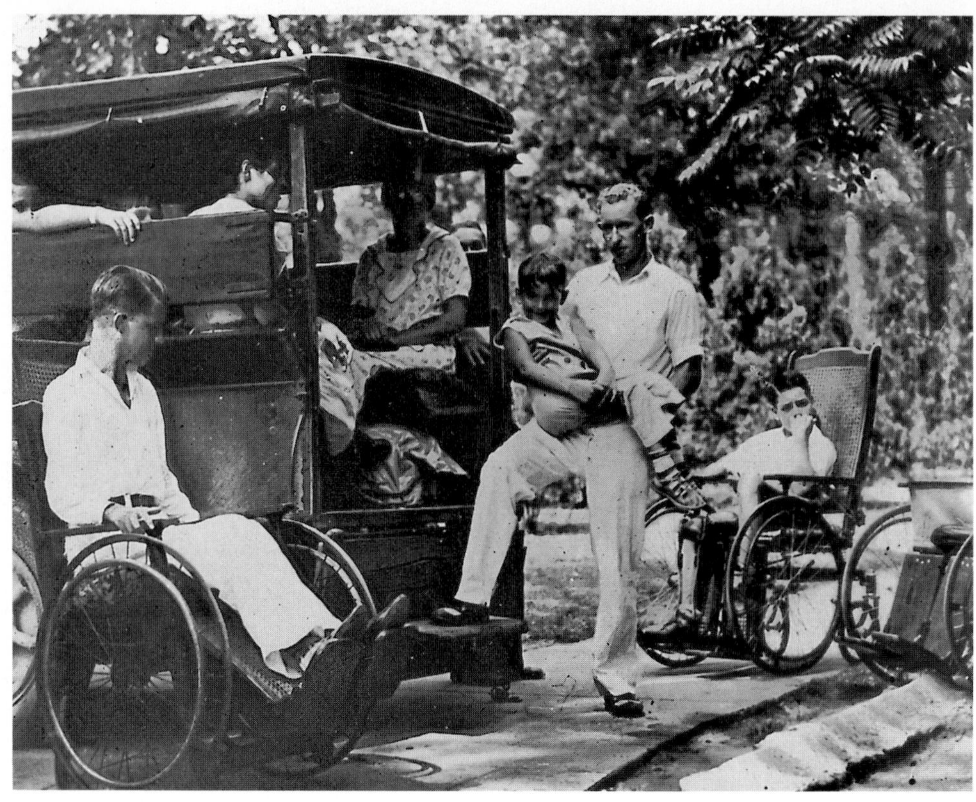

The first pool bus: a 'push boy' helps unload polios at Warm Springs' Roosevelt pool.

Occupational therapy: young patients working in the craft shop at Warm Springs.

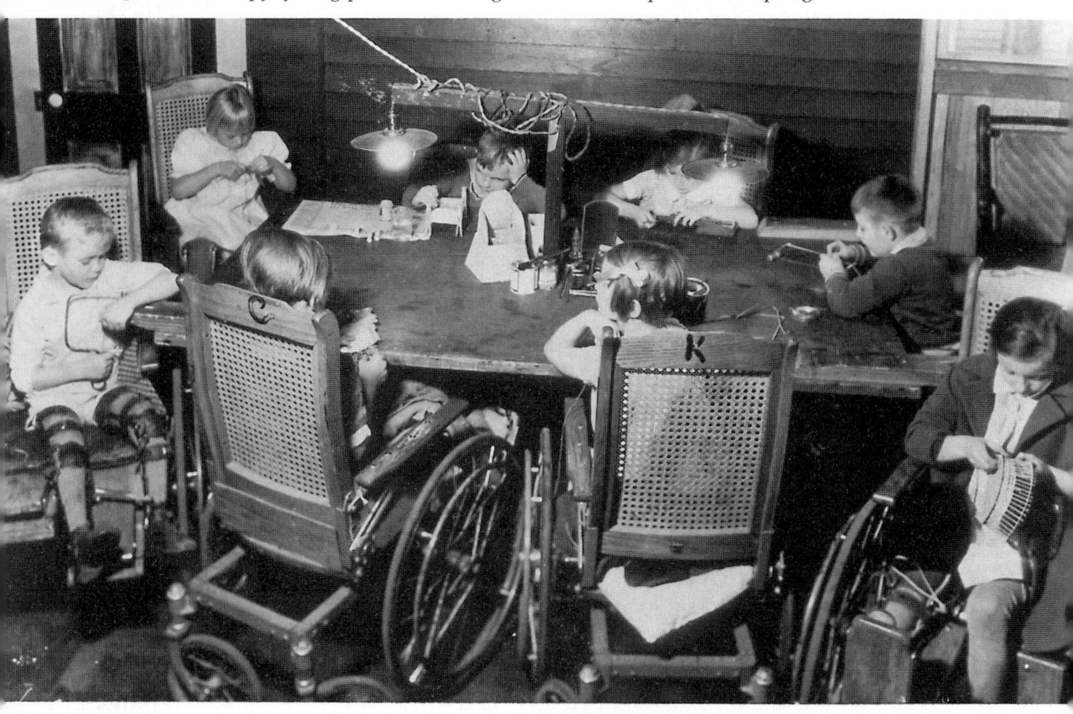

The Warm Springs 'Poliopolitan Opera Company' in action: Thanksgiving Day show, 1936.

Sun-lamp baths at Warm Springs: FDR was a great believer in the healing powers of the sun's rays, both natural and artificial.

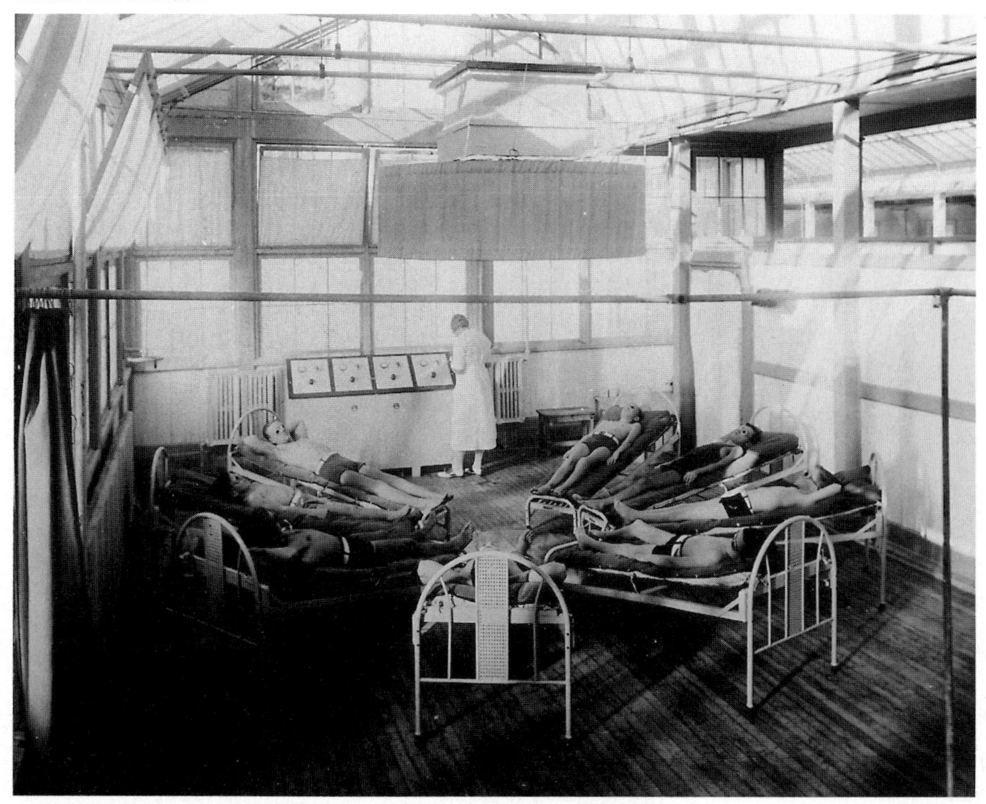

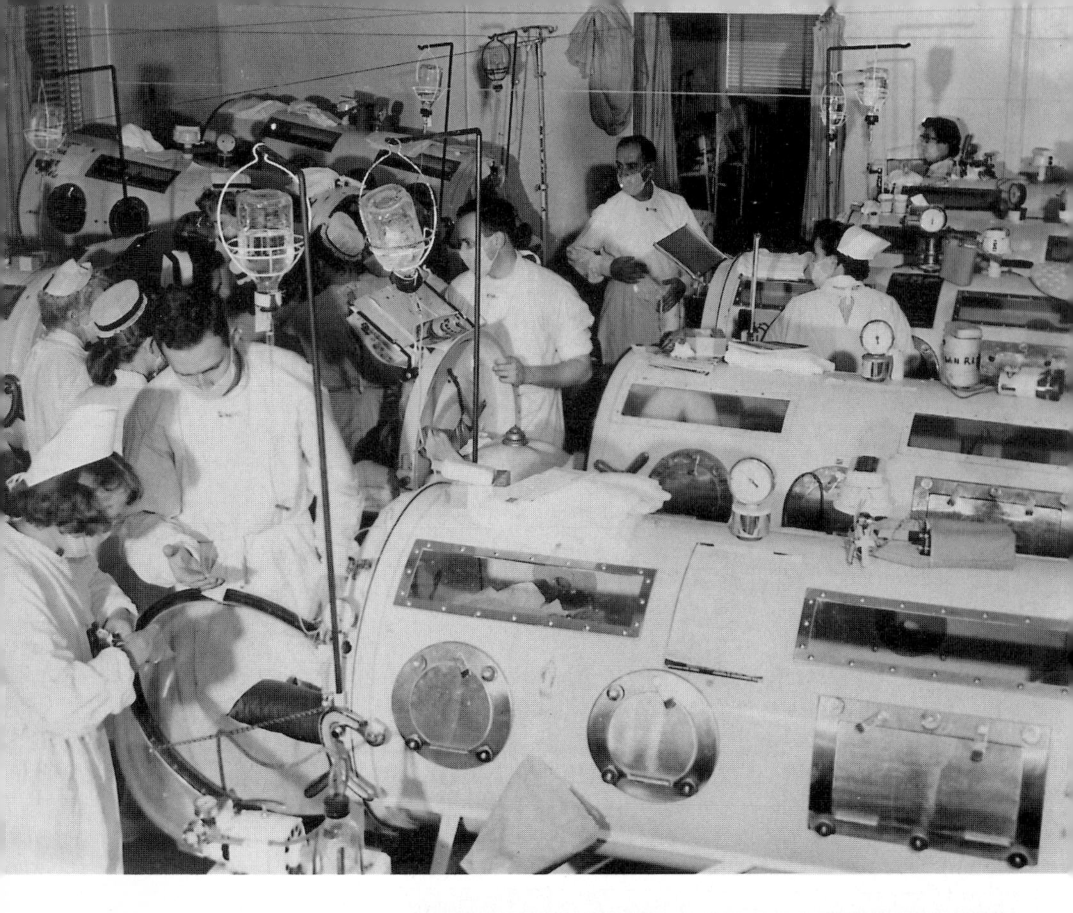

Iron lungs: Sister Kenny might call them 'torture chambers' but for many they were – and for some they remain – life-support systems.

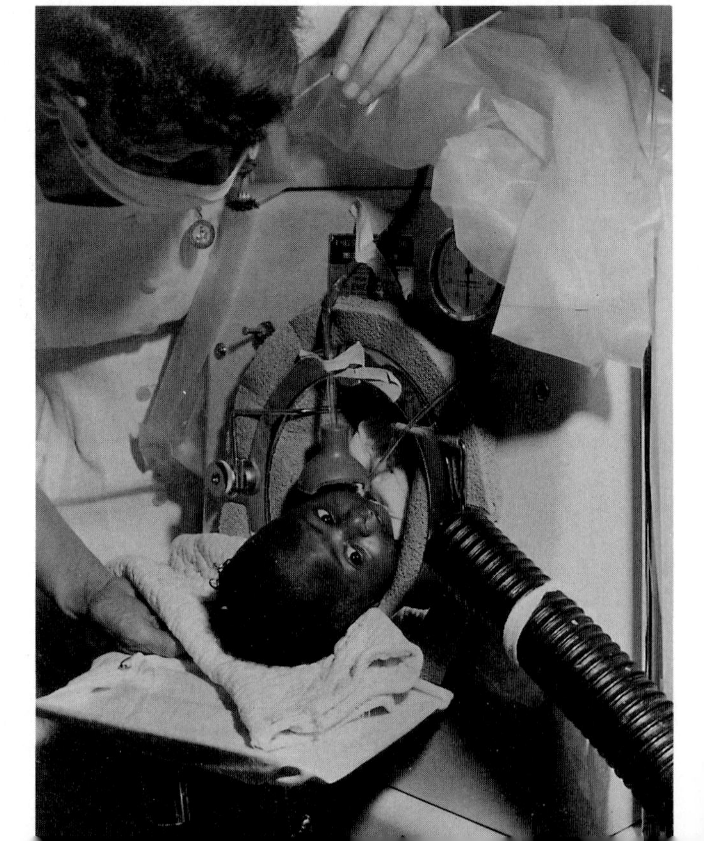

Where am I? A six-week-old polio patient in an iron lung in Houston, Texas – living refutation, incidentally, of the persistent myth that blacks did not get polio.

Hubbard examined reported cases of polio in the United States over the previous ten years in relation to the overall numbers of whites and blacks in the population and his findings were inconclusive. In 1936, the polio strike rate per 100,000 population was 7.2 in the case of whites and only 3.3 among blacks; but the year before it had been 1.7 for whites and 2.1 for blacks. Though the figures suggested a higher incidence among whites, it was not significantly higher and Hubbard concluded lamely: 'it may be stated from the figures analyzed that the attack rate in the colored race is *somewhat* lower than in the white, indicating that there may be *slightly* less susceptibility' (my italics).[131] With the benefit of hindsight, one would attribute this lesser degree of susceptibility to social and economic conditions rather than ethnicity: inferior sanitation meant that more blacks were exposed to the polio virus in infancy and therefore acquired an immunity denied to better-off and better protected whites. In that sense, polio was a great leveller.

When FDR received Hubbard's report, he noted on O'Connor's covering letter, 'I want to talk to Doc about this when he comes to Warm Springs.'[132] Obviously, something had to be done. The question of a cottage for coloured victims of polio at or *near* the foundation came up again and the new administrator at Warm Springs, Henry Hooper, was asked by Keith Morgan to study the feasibility of having one.

Hooper recommended, at the end of a long letter to Basil O'Connor, that 'in place of establishing accommodation for resident colored patients, the Foundation conduct investigations into the possibilities of establishing relations with an institution already equipped for the care of colored people'. O'Connor concurred with this segregationist solution and authorised Hooper to explore it further.[133]

The upshot was that, in 1939, Basil O'Connor announced, on behalf of the National Foundation for Infantile Paralysis, that it had decided 'to grant the sum of $161,350 to Tuskegee Institute [in Alabama, half a day's drive from Warm Springs] to provide the first polio center for negro people'.[134]

The Tuskegee Institute was founded by Booker T. Washington in 1881 and by the end of the century had become *the* educational centre for blacks in the South. The remarkable scientist and 'Peanut Man', George Washington Carver, who, like Booker T. himself, came up from slavery, was chosen as head of the Department of Agriculture at Tuskegee in 1896 and ended up staying there forty-seven years. Among the myriad uses of the peanut conceived in Carver's fertile brain was 'peanut oil therapy' for polio. His discovery of the beneficial properties of peanut oil in relation to Infantile Paralysis came about accidentally. He had developed a beauty treatment for women suffering from acne, only it had the unfortunate side-effect of fattening their faces:

> Interested, Dr Carver turned to a boy who had come to him to ask whether anything could be done about one of his legs. It was thinner than the other and it wobbled. Dr Carver located its wasted muscles; massaged them regularly with peanut oil. New life came to them; new flesh to the leg. The wobble disappeared. Later that boy played football at Georgia Tech.
>
> Later, too, Dr Carver learned that he had cured a leg deformed by infantile paralysis.[135]

He – and others on his behalf – made strenuous efforts to promote this 'miracle cure', while at the same time denying 'extravagant claims' for it. Even FDR, who had perhaps been put up to it by Basil O'Connor because of his plans to develop Tuskegee as a polio centre, endorsed the product to the extent of writing in a letter to Carver, 'I do use peanut oil from time to time and I am sure that it helps.'[136]

The Warm Springs 'Poliopolitan Opera Company', however, in one of its Thanksgiving Day shows, seized on this 'beauty treatment' as a suitable subject for satire. An ex-toreador known as G. Maximus (a reference no polio could possibly miss to the muscle that pads the buttocks, the *gluteus maximus*, a common casualty of the disease) becomes a Peanut Oil Vendor as a result of the Depression and touts his wares thus:

> In Cuba every merry maid
> Wakes up with this serenade
> Peanut Oil . . . Peanut Oil!
> If you haven't got banana oil
> Don't be blue
> Peanut oil will make you
> Beautiful too
>
> This is the right direction
> For that 'school-girl complexion'
> You'll 'get a lift' from
> Peanut oil . . .
> If you're looking for a moral
> To my song
> Fifty million pretty women can't be wrong.[137]

Carver knew that his treatment had helped individual polios but he was uncertain about which aspect was more effective, the oil or the massage. He expressed his doubts in a letter to a physician early in 1936:

> My method, which I feel is much more important than the oil itself, or certainly on a parity with it, is the thing I am attempting to work out. I make a study of the muscles, locating those that are functioning and those that are weak or that are not functioning at all as far as I can see. I let those muscles be the superficial or the deeper seated ones. Then I find the

> parts of greatest activity. And by graduated exercises, massages, and instructions to the patient, I have been quite successful in bringing about activity.[138]

If it was the massage and exercises, rather than the oil itself, that was doing good, then there was nothing particularly revolutionary, or miraculous, about Carver's treatment of polio. It was, as his biographer admits, 'not one of his sounder scientific endeavors'.[139]

What attracted Basil O'Connor to Tuskegee was not the Carver connection so much as the fact that there was already a crippled children's unit at the John A. Andrew Memorial Hospital on the institute campus, run by one of only two Negro orthopaedic surgeons in the entire country, Dr John M. Chenault.

Chenault accepted that there was a 'somewhat lower' incidence of polio among blacks than among whites, but pointed out that 'case fatalities have been relatively higher among Negroes': 'I firmly believe that this has been due to the notoriously poor treatment facilities available for Negroes and, as much as I hate to admit it, the failure of so many of our men to recognize the disease.'[140] Segregation laws meant that patients who had often travelled miles to seek medical attention had to wait – sometimes for hours – in a 'coloured' waiting room until the last white patient had been seen; and a diagnosis of polio could mean further discrimination, with the patient being shut away in the basement of a hospital and the parents being obliged to provide nursing care for want of any other. No matter how impure the motives of the whites involved in the establishment of the Tuskegee Infantile Paralysis Center were, its existence brought much needed relief and opportunity to a hard pressed people.

In the poverty-stricken economy of the South during the Depression the long-term hospitalisation of a child was still tough on families: they might have one less mouth to feed but they lost a helper in the fields and could rarely afford to visit the Infantile Paralysis unit. For the child, however, there was nothing but gain. A former patient called Louis Washington returned to Tuskegee to open a business, 'Lou's Stereo and Electronics', because 'he never forgot the encouragement and happiness he felt here while he was a polio patient'.

> He was loved and cared for; he received good food and enough of it, something he remembers he did not always have at home; he 'went' to school; he had books to read and to color; he was provided with things to 'take apart'; he was surrounded by people whose mission was to provide for the needs of the mind as well as the needs of afflicted limbs.[141]

There was a janitor who checked the building at night; he was known as the 'train man' and his visits were eagerly awaited by the children. As soon as they heard his train whistle, they called out, 'Come on, do the "train man".' He would come up to the third floor, where they slept, his train whistle getting louder all the time:

> Then peeping around the corner, to the delight of the children, he entered – shuffling and sliding his feet on the floor, moving his arms in a forward circular motion and, of course, making the wonderful train whistle. He got faster and faster as he came into the room – going to each child's bed – until he made sure each child got a special look at the train man. Then slower and slower, the 'train' moved until it pulled to a stop and gave one last blow of the whistle. Then the 'train man' turned out the lights promising to make another 'run' the following night. The children loved it![142]

The slogan 'separate but equal' usually meant 'separate and *anything but* equal', as the story of the Tuskegee syphilis study, initiated by the US Public Health Service in 1932, illustrates. This experiment 'to determine the natural course of untreated, latent syphilis in black males' in Macon County, Alabama, went on for forty years before adverse publicity put a stop to it. The study depended on the withholding of treatment, but in order to obtain the co-operation of its subjects in submitting to such invasive procedures as lumbar punctures the men were told they *were* being treated and were given placebos to strengthen the illusion. The US Public Health Service, using the facilities of the Tuskegee Institute and co-opting its director and hospital director, went to great lengths to prevent the subjects of its study obtaining treatment elsewhere, even persuading the army to exclude them from the anti-syphilitic treatment it automatically gave to the men it drafted during the war; and when penicillin became widely available in the early Fifties, it was not offered to those who would have benefited from it. By the time the experiment was halted in 1972, 'at least twenty-eight, but perhaps more than 100 [out of 400 test subjects], had died directly from advanced syphilitic lesions'. Not surprisingly, an investigatory panel appointed by the Department of Health, Education and Welfare condemned the study as 'ethically unjustified'.[143]

The National Foundation, once it had acknowledged its responsibility, behaved very differently. The Infantile Paralysis unit provided training as well as treatment, so that doctors, nurses, physiotherapists and brace-makers all learned their craft in a sympathetic environment. Much of the credit must go to Basil O'Connor, sceptical as he may have been initially. He was determined to make a success of Tuskegee, and he worked towards that goal with the same single-mindedness he brought to his work for Warm Springs and the National Foundation. And Tuskegee has not forgotten 'Doc' O'Connor: the nurses' hall of residence there still bears his name.

5
An Angel Abroad

We Irish have a saying about our females that they are frequently 'devils at home and angels abroad' . . . Sister Kenny fits this description perfectly.

Basil O'Connor: confidential memo to FDR, 10 May 1943

Sister Kenny knew little of the pathology of poliomyelitis . . . An anterior horn cell might have been on the head of a buffalo for all she knew.

Dr Stanley Williams, cited in Desmond Zwar: *The Dame: The Life and Times of Dame Jean Macnamara, medical pioneer* (1984)

She treated the doctors in her classes like some medical men treat nurses. 'Can you teach me to wiggle my ears?' one doctor once mockingly asked her. 'It isn't necessary,' she answered. 'Any ass can do that.'

Victor Cohn: *Sister Kenny: The Woman Who Challenged the Doctors* (1975)

In the Polio Hall of Fame at Warm Springs only one woman is commemorated, the scientist Isabel Morgan (daughter of the Nobel prizewinner and originator of the gene theory, T.H. Morgan), who was part of the impressive polio research team at Johns Hopkins University. The men, with the notable exceptions of Franklin D. Roosevelt and Basil O'Connor, are all doctors and scientists too. There is no place for the outspoken nurse from the Australian outback who revolutionised the treatment of polio in the teeth of powerful opposition – medical, political and personal. Not that one should feel too sorry for this strapping, upright Irish-Australian 'Sister' (in the nursing rather than the religious sense): she was no mean bruiser herself, often pre-emptively lashing out at a potential opponent. She was also a tireless propagandist – worthy rival, in this respect, even to O'Connor – the subject of a schmaltzy Hollywood movie and for many years second only to Eleanor Roosevelt (and once first) in the US women's popularity stakes, as conducted by Dr Gallup.

She created herself as she went along, mythologising her past and, in the process, lopping six years off her age. Her most serious biographer, Victor Cohn, former science editor and medical correspondent of the *Washington Post*, knew her from his days as a reporter on the *Minneapolis Tribune*, when 'covering polio was like baseball' – a matter of counting the number of cases. Intrigued by her personality – she could be 'a real pain in the ass', he found, but once, when he brought her a book, 'she melted' – and challenged by the controversy surrounding her work, Cohn spent years trying to fathom both.[1] He came up with a sympathetic, but far from hagiographic, portrait of 'The Woman Who Challenged the Doctors', and a thoroughly researched and informed assessment of her contribution not just to the treatment of polio but to modern medicine in general.

For many years, Australians suppressed all connections with their convict past. But times and attitudes change, and nowadays to have convict forbears is more a matter of pride than shame. 'Convicts', after all, came in all shapes and sizes and many of them had less to be ashamed of than their prosecutors, or their sometimes sadistic guards and tormentors. There were rebels as well as criminals, protesters as well as murderers and, of course, those who defied easy categorisation – poaching to feed a family not being the same thing as poaching for sport or gain. Elizabeth Kenny's maternal grandfather, 'Big Jim' Moore, described in *Sister Kenny* as 'a giant of a sea captain', was also, Cohn has since learned, a giant of a convict.[2] Much has been written about Australian 'mateship' and contempt for authority, of the outcast as hero (in the Ned Kelly mould), and Kenny may have inherited as well as imbibed from the atmosphere in which she grew up in Queensland the aggressive egalitarianism that so scandalised the rigidly hierarchical and male-dominated medical world of the Twenties, Thirties and Forties.

The often-told story of her first encounter with infantile paralysis has a legendary air about it. Summoned to a remote settlement, the bush nurse discovered a distraught mother with a two-year-old blonde child whimpering in pain, her body twisted and her eyes full of fear. The nurse, baffled, sent a telegram to her doctor friend and adviser, Aeneas John McDonnell, who replied succinctly: 'Infantile paralysis. No known treatment. Do the best you can with the symptoms presenting themselves.' After praying for guidance, the nurse returned to her patient, who was obviously still in agony, and tried various forms of heat treatment before hitting upon the one that brought relief and allowed the child to sleep – strips of heavy woollen blanket, placed in boiling water, then wrung out dry and wrapped around the affected limbs.[3] Elizabeth Kenny had discovered the 'hot packs' forever associated with her name.

In her ignorance, her next step was gently to move the affected limbs until the child could begin to do it for herself. She treated half a dozen stricken children in this fashion and they all recovered the use of their limbs. When she reported this to Dr McDonnell, he was amazed and got her to demonstrate what she had done on one of his own patients, a small boy whose legs were strapped on to splints. Kenny removed the splints, used heated strips of blanket to ease the muscle pain and then started to teach the boy the movements she claimed he had 'forgotten'. According to Kenny's account, McDonnell told her, 'Elizabeth, you have treated those youngsters for symptoms exactly the opposite of the symptoms recognized by orthodox medical men.' He showed her textbooks, in which she was surprised to find no mention of the condition she had seen and treated: *spasm*. He went on, she would have us believe, to say:

> Your heart will be broken and your spirit will be crushed, for medicine is not kind to its reformers. But the day will come, if you have the courage . . . when your work will be recognized, and the great cities of the earth shall gladly welcome you.[4]

The idea of the deaf and rather brusque McDonnell saying anything of the kind is implausible. But part of the potency of the Kenny myth derives from the fact that she convinced herself of its validity before she set about persuading others.

These events took place in 1911, but the follow-up was delayed for two decades, partly because of the intervention of the First World War, in which she came by her title of Sister, though she had had no formal training as a nurse (a secret so well kept that, with the exception of one intemperate Australian doctor, none of her detractors ever used it against her in later years), but more because she was as yet unaware of her vocation, whatever she later claimed. Between 1915 and 1918 Kenny served mainly on ships, ferrying the wounded back home from the European and Near Eastern theatres of war, before herself being invalided out of the service with heart trouble which was to dog her for the rest of her life – though one doctor who treated her maintained that her 'attacks' were hysterical, the product of stress and overwork, and there was nothing actually wrong with her heart. In the Twenties she took the unusual step of adopting a young girl from a broken home, not so much as a daughter for herself, more as a *replacement* daughter for her widowed and lonely mother. She also designed and marketed a stretcher for which she collected an annual payment for years.[5]

She began to treat polio patients again at the beginning of the Thirties. They were all chronic cases that had long since been written off by orthodox medical practitioners, and she acquired a reputation for achieving unlikely recoveries. At Townsville, in Queensland, she held an open-air clinic under an awning for up to thirty patients. But she was ambitious and wanted government recognition for her rehabilitation work. Her attempt to persuade the hospital doctors of Brisbane to abandon their splints and plaster casts set the pattern for many future encounters with officialdom: her words were greeted with at best incomprehension and, more commonly, derision; her demonstration of her method – until she became adept at it – with embarrassment.[6] Her conviction that she had a kind of gift laid her open to the accusation of quackery, and long after her method of treatment had been universally accepted, the aura of the charlatan, on the one hand, or of the 'miracle-worker', on the other, clung to her; and her ambivalence about her powers did little to dispel it.

As a result of political pressure, and a measure of medical endorsement, Sister Kenny eventually set up clinics in Brisbane and in Sydney, as well as in smaller places like Townsville and Toowoomba, but still she had to fight every inch of the way. Dr (later Dame) Jean Macnamara, in Melbourne, Victoria (the state worst hit by polio epidemics), was the leading Australian authority on infantile paralysis, and she was wholly committed to the orthodox orthopaedic practice of immobilising polio patients for weeks and months in order to avoid subsequent deformities. She was also a very different kind of woman, emollient where Sister Kenny was abrasive, diplomatic instead of confrontational, but equally convinced of her own rectitude. 'I could take away the splints, too', she told her family, 'but I don't know what the patient would be like in ten years.' She was a keen gardener and used a horticultural analogy to justity her approach: 'A splint or a caliper is like staking a rose bush so it will grow straight.'[7]

In October 1935, when the Queensland government announced that a royal commission would investigate the Kenny method of treatment of infantile paralysis, it was inevitable that one of the commissioners would be Jean Macnamara. Though Aeneas John McDonnell was another, the dice were still loaded against Kenny: to expect a group of doctors to approve of abandoning their own tried and tested methods in favour of the half-baked theories of an untutored nurse, who poured scorn on their treatments and publicised preposterous claims for her own, was tantamount to asking them to commit professional suicide.

Their report was not published for another two years, during which Kenny spent some time in England, where the London County Council proved willing to try out her methods and Queen Mary's Hospital for Children at Carshalton, on the Surrey Downs, set aside some wards for her treatment. British orthopaedic surgeons and Sister Kenny both demanded – though for opposite reasons – earlier access to patients than the system of initial isolation in fever hospitals permitted.[8] At Carshalton Kenny got to demonstrate her skills only on early convalescents.

In England at that time outbreaks of polio, though they gave cause for concern, were of less than epidemic proportions: between 1926, when 1,159 cases were reported, and 1938, when notified cases again topped the thousand mark, the annual average was 640.[9] The experimental clinic at

Queen Mary's, Carshalton, was set up in July 1937, and received a cautious welcome in a *British Medical Journal* editorial six months later:

> An inspection of Sister Kenny's treatment at work shows that it embodies many re-educative manoeuvres which could be used with advantage in centres where poliomyelitis is treated on orthodox lines. As a distinct system of treatment the method is obviously still *sub judice*, and a final opinion must be withheld until a larger series of fully documented cases have been examined and reported on by an impartial committee of experts.[10]

The committee of experts, impartial or otherwise, that sat in judgment on the Kenny clinic at Carshalton pronounced that her system was 'not the last word in the treatment of poliomyelitis'; but it had 'no difficulty in recommending – certainly for an extended period – the continuation of Miss Kenny's system of hydrotherapy in the poliomyelitis unit of this hospital'. It considered that 'very early attempts to initiate voluntary movements and also early and frequent passive movements are harmless but of unproved value', and that 'under certain conditions . . . splints can be dispensed with in the treatment of the early stages of the disease'.[11] Faint praise perhaps, but better by far than the outright condemnation of the Queensland Royal Commission, which claimed that her abandonment of immobilisation was fraught with danger and would produce 'a harvest of spinal deformities'.[12]

Despite the commissioners' rejection of her treatment (only McDonnell did not sign the report, pleading ill health; he died of cancer in 1939), Sister Kenny still had the support of the three key state governments – in Queensland, New South Wales and Victoria. Her work expanded, and an outbreak of polio in and around Melbourne took her into Jean Macnamara's home territory. But Macnamara was a shrewd operator and encouraged the government of Victoria to try Kenny's treatment alongside her own in the Children's Hospital in Melbourne; that way they both received funds.[13]

Kenny's work in Melbourne attracted favourable notice from two orthopaedic surgeons there and, back in Brisbane, she finally got to treat some acute cases of polio with excellent results. But the commission's report still worked against her, the outbreak of the Second World War took away a number of doctors and nurses, and her Brisbane clinic had to be closed for a while. A prophet without honour in her own country, she might have more influence abroad – or so a group of her medical supporters believed. They petitioned Queensland's Premier to send her on a mission to the United States, which he and his health minister, who had funded her clinics to the tune of several thousand pounds, were only too pleased to do. When the latter told a friend they were sending Kenny to America, he added, 'Bloody good riddance'.[14]

A piece of equipment to which Sister Kenny took exception in the same way as she objected to plaster casts, splints and calipers, was the respirator, or 'iron lung' as it soon came to be known in several languages – in German and French, as well as in English. Prototypes of 'a body-enclosing apparatus to produce ventilation by pressure changes within the chamber' were devised by – among others – Dr Woillez of Paris, whose hand-operated bellows-driven 'Spirophore' won him a silver medal at the Le Havre exhibition of life saving equipment in 1876 and then seems to have disappeared without trace, and by Dr Steuart in South Africa in 1918. Steuart constructed an air-tight wooden box, sealed at the shoulders and waist with clay and powered by a motor-driven bellows, specifically for the treatment of polio, in which he seems to have had some success. But the first mechanical respirator to be at all widely used was devised by Philip Drinker, an engineer working at the Harvard School of Public Health, in 1928.[15]

Drinker was watching a colleague measuring the breathing of an anaesthetised cat lying in a metal box with a rubber collar around its neck and its head exposed, when he had an idea for an experiment. He paralysed a cat's breathing muscles and then pumped air in and out of the box in which it was incarcerated up to its neck, and discovered that the animal could be kept alive like that for hours. The implications of this for humans were obvious and Drinker approached the Consolidated Gas Company of New York for funds. (Two years before, Consolidated Gas had financed a commission – of which Drinker had been a member – to develop 'improved methods of resuscitation' in the hope of reducing the number of suits brought against it by people accidentally gassed due to a leaking pipe, or by the families of attempted suicides who stuck their heads in gas ovens.) The company handed over $500 and Drinker and his colleague Louis Shaw constructed a man-sized respirator which opened and shut like a drawer but otherwise worked on the same principle as the cat box, with a rubber collar preventing any escape of air around the neck. That was the tricky part; Drinker's brother Cecil, a Harvard professor of physiology, was convinced that it would never work: any collar tight enough to act as an airseal would impede the flow of blood. But Philip persisted and proved his brother's fears to be groundless when he successfully submitted himself as the first human guinea pig.[16]

His sister takes up the story:

> On his way through the wards [of the Children's Hospital in Boston], Phil saw children dying of suffocation, induced by polio; he could not forget the small blue faces, the terrible gasping for air. The respirator had not been designed specifically for infantile paralysis. Yet when the machine was perfected, the first patient happened to be a little girl from Children's

> Hospital, suffering from severe polio and expected to be in respiratory difficulty very shortly. Phil had the machine moved into the ward near the child's bed so she could see it and get used to the loud whine of the motor. Early next Sunday morning the hospital called Phil. By the time he reached the child she was in the machine, unconscious, but the staff had been afraid to turn on the power. Phil started the pump, and in less than a minute saw the child regain consciousness. She asked for ice cream. Phil said he stood there and cried.[17]

Unfortunately, the eight-year-old girl did not survive for long; at the end of a week she died of pneumonia. But the life-saving potential of the iron lung had been established.[18]

The question that initially bothered the doctors at the Children's Hospital was a moral one: once they had succeeded in saving a polio life by means of the machine, would the patient be obliged to remain in it for ever?[19] The experience of a second respirator patient, a man who was treated at Boston's Peter Bent Brigham Hospital, set their minds at rest on that score: the patient recovered his ability to breathe and at the end of a year he was walking again with calipers and a stick.[20]

Not all patients made such quick recoveries. Frederick B. Snite, a young man from a wealthy American family, contracted the disease in China in 1936; the paralysis spread and he was soon having difficulty in breathing. By serendipity it turned out that he was just minutes away from what was almost certainly the only iron lung in the entire orient; it was one of sixteen original Drinker respirators and had been purchased by the China medical board, an organisation set up by the ubiquitous Rockefeller Foundation, and shipped to the Peiping Union Medical College in China. Snite, the first polio patient to use it, described it as sounding like 'a threshing machine with a cold'. He would be moved into quieter and more sophisticated machines (Jack Emerson, the son of Haven Emerson, New York's Health Commissioner in 1916, designed an improved model), but he would never again breathe unaided. He lived in an iron lung for the rest of his life; but he travelled, married and fathered three daughters, and became in the process something of a celebrity – second only to FDR among polio survivors in the United States.[21]

Despite such success stories (and wealth, of course, played no small part in this one), doctors continued to feel guilty at saving the lives of so severely handicapped people – but they really had no choice, since the extent of enduring disability could not be accurately predicted at the beginning of the illness. In epidemics during the Thirties, John Paul recalls, other 'agonizing decisions' had to be made:

> For instance, although three or four patients with respiratory difficulties might be on hand and waiting, there often was only *one* respirator available

> – what to do? whom to choose? the patient with the severest disability, who possibly would die anyway; or the patient with lesser disability and a better prognosis?[22]

Logistical problems of this sort were meat and drink to Basil O'Connor. As soon as the National Foundation came into being, respirator centres were established and national meetings on 'respiratory physiology' organised. Dr Paul pays tribute to the National Foundation for bringing order out of 'what promised to be a most difficult, chaotic, and distressing situation'.

> The foundation was not only a pioneer in the establishment of respirator centers for the care of patients and for the special training of physicians, technicians, and nurses, but it also organized mobile expert teams which could be rushed to the scene when epidemic aid was needed.[23]

In view of their life-saving capability, it may be wondered why Sister Kenny called iron lungs 'torture chambers', and her medical collaborators declared baldly: 'The Kenny technique does not permit the use of respirators.' Though over-dogmatic, her approach was logical; she objected to respirators for the same reason as she objected to plaster casts and splints: they impeded natural recovery and created a dependency that was hard to break. In Brisbane, she put her principles into action:

> A boy in a respirator was rolled into her ward. She stood over him all evening, watching nervously. At last she turned to a young resident doctor. 'This is the first time I've ever done anything without a doctor's permission. But I want you to feel utterly relieved of all responsibility.' Removing the child from the tanklike machine – 'It was a heart-stopping moment,' an observer says – she sat up all night, treating his affected muscles with hot packs. Muscle tightness relieved, the child regained normal control.[24]

But even Sister Kenny had to resort to respirators in cases of bulbar poliomyelitis, where the virus-damaged cranial nerves in the upper spinal cord severely affected breathing and swallowing and no amount of hot packs could restore these functions. As with other aspects of her treatment, she erred on the side of optimism, stressing the advantages to the patient of active participation over passive dependency. But sometimes that did not work, and then she was enough of a pragmatist to fall back on conventional techniques.

While Kenny was battling against the orthodox medical practice of splinting in Australia, the people at Warm Springs were regarding it as a revolutionary new development.

On 29 May 1936, Arthur Carpenter wrote to Basil O'Connor: 'Columbus was looking for the Indies when he discovered America. I think the Foundation's voyage of exploration and pioneering in the after-treatment of infantile paralysis may turn out much the same way.' During the past five years, he went on, there had evolved at the Children's Hospital School at Baltimore a programme of after-treatment so different from that practised at Warm Springs that 'if they are right, we are wrong'. It was passive rather than active treatment, but there was a lot more to it than just staying in bed. The patient was put in splints and braces to protect the weakened muscle groups and things were done to stimulate the circulation and avoid atrophy. *Preventive* orthopaedic surgery – the avoidance of deformities necessitating subsequent operations – was what it amounted to, and Carp suggested that a team of three independent doctors should be invited to spend some time at Baltimore and then at Warm Springs to assess the relative merits of the two methods. 'Coming back to my parallel about Columbus', he wrote, 'we may find that the right answer is not in the direction in which we are looking.'[25]

In response to this letter, Dr LeRoy Hubbard prepared a résumé of the evolution of the after-care treatment of polio, which was circulated to all the trustees of the Warm Springs Foundation. Hubbard paid tribute to the pioneering work of Dr Lovett of Boston, who 'started a new era in the after treatment of Infantile Paralysis by physiotherapy'. The exercises devised by Dr Lovett and his associate, Wilhelmine G. Wright, were now done underwater both at Warm Springs and in Dr Charles Lowman's 'Underwater Gymnasium' in the Children's Hospital School at Los Angeles. In the latter, however, rigid splinting was practised out of the water and swimming was discouraged, except under the direction of a physiotherapist. In Baltimore, the head physiotherapist at the Children's Hospital, Henry Kendall, who worked closely with his wife, Florence, used the pool only for walking exercises; in the early stages of treatment he employed 'only light massage and contraction of the muscle fiber without motion' and used 'various kinds of apparatus to keep the muscles in what he calls "the natural position", to prevent any stretching':

> This method is practically the same as that which has been used for some time by Dr Jean Macnamara of Australia and Mr Kendall is using some of the splints which were devised by Dr Macnamara.[26]

In July, Dr C.E. Irwin, the chief surgeon at Warm Springs, went to Baltimore and, though he was impressed with the treatment there, he recommended 'caution in advocating a line of care so different in principle from that on which this institution is founded'. He was well aware of the limitations of hydrotherapy on its own, but he still regarded exercises in water as 'a valuable adjunct . . . a complement and supplement to other proven methods' and 'a pleasant release from . . . forced inactivity', and he emphasised the psychological advantages of the freedom of movement underwater exercises provided.[27]

Dr Kristian G. Hansson of New York makes this last point even more forcefully in a paper published in the *Journal of the American Medical Association*:

> Nothing can replace the therapeutic pool for the security that the patient feels when supported by the water. One is dealing with physically handicapped children who are very apprehensive of falling and doing harm to themselves. This fear is entirely eliminated in the water, and their morale, which is often low after a long illness, is boosted enormously. The pleasures of crippled children are few, and they look forward to their pool treatment with a great deal of interest and joy . . . Nothing explains better the psychologic effect of the underwater exercises than the plaque which was placed on the wall of the pool at the Hospital for the Ruptured and Crippled. It says 'This pool is given to the physically handicapped children that in pleasure they may regain their health'.[28]

Yet Hansson, too, applauded the Kendalls of Baltimore for bringing back 'the logical and rational view of the muscle reeducation of patients with infantile paralysis' expressed in 'two fundamental principles: the importance of rest and the protection of the weakened muscles'. His main objection to their treatment was the psychological damage done to a child by a lengthy period of inactivity. He gives an example from his own experience of a girl of seventeen living on a small farm in Connecticut whose paralysis was slight 'but her mental attitude was that of a village cripple'. Hannson sent her to Warm Springs, and 'in six months she was walking without assistance, had regained her mental balance and had taken up art, for which she was well suited'.[29]

The Baltimore method got the US government seal of approval when the Kendalls were invited to prepare Public Health Bulletin No. 242, on *Care during the Recovery Period in Paralytic Poliomyelitis*. This document became, for Sister Kenny, the epitome of everything she deplored and she attacked it at every possible opportunity.

The extent to which the Kendalls' view became the orthodoxy is manifest in nursing textbooks of the period such as Jessie L. Stevenson's *Care of Poliomyelitis*, published in 1940, the year in which Sister Kenny first set foot in the United States. 'The function of orthopedic nursing in the care of poliomyelitis', Stevenson writes, 'is *essentially that of prevention* [of deformity]' (my italics).

> Well-meant sympathy of relatives and friends who think 'it is cruel to tie the poor child down on a frame' sometimes creates a difficult home situation which puts additional strain on the mother who is trying her best to

> follow the physician's advice. Sometimes the nurse can help them to understand that it is even more cruel to allow him to develop lameness and deformity . . .[30]

Dissenting voices were heard in the United States, even before the advent of Sister Kenny. As early as in 1934, an Illinois doctor, Robert D. Brown, read a paper at a convention of the American Congress of Physical Therapy in which he ascribed his own speedy recovery from polio at the advanced age of fifty-four to hot packs – 'From my waist to my toes I was wrapped in a blanket wrung out of hot water' – and the application of electric currents. Five and a half years later his legs were still 'queerly shaped, but apparently as strong as before'.

> The point I wish to make is my belief that treatment of the paralysis of poliomyelitis should be begun early, contrary to the usual advice to give complete rest, and even plaster casting the affected parts for several weeks or months. It seems logical to me that the earlier treatment is begun the better.[31]

Another dissenter, Milton Berry of the Milton H. Berry Institute in Van Nuys, California, sounds like a forerunner of Sister Kenny in more ways than one. He writes, 'I am not a psychiatrist, mental or divine healer – just a muscular engineer or a physical re-educator – but no one of average intelligence could share my experiences of the past forty years without being impressed with the importance of a proper mental attitude, along with hard work, will power, and proper handling of the marvelous physical mechanism of the body, if one is to remain well or to regain health after its loss'. He rails against 'orthopedic doctors', whose 'failures' he contrasts with his own record of success, and he uses language as arcane as Kenny's when he talks of 'the Biophysics of Locomotion from the Stand Point of the Cripple'.[32]

Like Sister Kenny in another respect, too, Berry was no mean self-publicist, putting out a monthly journal ostensibly stuffed with 'Facts concerning Infantile Paralysis, Its Crippling Effects and Methods of Treatment' but actually consisting largely of testimonials from satisfied customers.[33] Satisfied customers there were, however, so Berry, whose two sons followed in his footsteps, may not have been such a quack as his publicity leads one to suppose.

On 16 April 1940, Sister Kenny and her adopted daughter Mary – who, since the death of Kenny's mother, had been initiated into the mysteries of the Kenny treatment and put to work – disembarked at San Francisco and began their transcontinental journey to New York by train. The sister was nearly sixty (though she told journalists she was in her early fifties), and far from confident of a welcome in the United States.

Her first meeting with Basil O'Connor hardly reassured her. He could promise her nothing in the way of support from the National Foundation and his loyal lieutenant, Dr Thomas Rivers, refused even to meet her. Like many bigoted people, Rivers was proud of his prejudices; he had never met Sister Kenny, but he knew he 'couldn't stand her'. Why? Because 'the virologists that I knew in England told me all about her, and so I was well acquainted with her work before she ever came to America'.[34]

From New York, Kenny went with her adopted daughter (whom she introduced as her ward or niece to avoid awkward explanations) to Chicago, the headquarters of the American Medical Association. There she met the influential editor of the *Journal of the American Medical Association*, Dr Morris Fishbein, another sceptic. 'She came in wearing that hat that made her look like Admiral Nelson', he recalled. 'She *looked* like a screwball.'[35]

Another cool reception like that and she would have been ready to head for home. But Dr Melvin Henderson of the Mayo Clinic in Rochester, Minnesota, to whom she had a letter of introduction, proved more open-minded than either Rivers or Fishbein. He consulted with a colleague and together they recommended that she go to Minneapolis, where there were many more polio cases than in Rochester. There Kenny began to convert a few physicians to her way of thinking, less by what she said – her use of such terms as 'spasm' and 'mental alienation' was unfamiliar – than by what she did. Her self-promotion and her peculiar theories of the disease were forgotten as soon as she got to work on a patient.

One of the first Minnesotans to receive her attentions was the eighteen-year-old Henry Haverstock Jr, who had contracted polio just at the outbreak of the Second World War. For nearly six months he lay at home with a frame over his bed equipped with pulleys and ropes with handles to enable him to heave himself up. There was little else that he could do. Then, in February 1940, his father, who was a lawyer in a big firm and 'had clout', got him into Warm Springs. Henry liked it there; it was a beautiful place with big green lawns, and they had regular film shows. Once he saw Roosevelt drive through in his car with special hand-controls, smiling and waving to people. Every day he was taken down to the pool by push boys and lifted on to a treatment table just below the surface of the water, and a physiotherapist would get to work, moving his limbs and urging him to do what he could for himself, which was not much. On his arrival at Warm Springs he had been fitted with braces:

> Two full-length leg braces, and some kind of metal corset. And I had a splint on this thumb – the left thumb . . . When I left Warm Springs four

> months later, they told me I'd never walk, and they were right: I never would have, with those braces on. They weighed about fifteen or twenty pounds, you know. Nobody could walk with that stuff.[36]

His parents drove him home to Minneapolis. 'But heck', he now says, 'it looked like a blank wall ahead.' His father carried him upstairs to bed, and it seemed that he was doomed to remain there for the rest of his life.

Dr John F. Pohl, whose patient he became, told Henry's parents, 'There's some woman here from Australia. I don't know if she has anything but the boy won't walk again and it's worth a try.' The Haverstocks agreed and a sceptical Dr Pohl brought Kenny round to see his new patient. Henry remembers them all trooping upstairs to his bedroom:

> She had that big black hat on, that you've seen in pictures. And she pulled out a bunch of letters, right there in my bedroom. She was going to read 'em – one from this Dr McDonnell – but we waved the letters aside . . . and asked her to show us what she could do. Somebody mentioned that I should be carried down and put on the dining-room table. So Dad carried me down – he was kind of a husky guy – and put me on the table. And Sister Kenny could see the braces there; she looked at them and she says, 'You don't need these any more.' So – that was good news. This is where I get emotional . . . Anyway, she started lifting my limbs, one after another. She'd lift my leg and she'd say, 'Now, Henry, think of this spot here, the quadriceps right above the knee there, think of that. Think of pulling . . .' And she'd do the same thing with the feet, the arms and the hands [*breaks off in some distress*]. I do this every time, when I get to this point . . .

This is a seventy-year-old man, recalling events of more than fifty years ago. But every time the same thing happens: the emotion, even when recollected in tranquillity, is too much for him. He has a friend, he says, 'who was on the Death March over in the Philippines . . . He gives talks on it, and he has written a book. But he breaks down every time he talks – those guys getting murdered . . .' Henry is not unhappy to talk about his experience – rather the reverse – but he cannot regard Sister Kenny dispassionately. He says simply, 'To me she was an angel.'

He was taken into the Abbott Hospital and later transferred to the Minneapolis General Hospital, where Kenny opened a clinic. She put him in a pool – a Hubbard tank – once in a while, but the emphasis, in and out of water, was on the training of individual muscles: 'Her theory was that the muscle, no matter how weak it was, should be trained to its potential . . . "Muscle alienation" was one of her big words. And "muscle re-education": re-educating the pathways . . . With her you became a real muscle expert.' But her greatest strength, according to Henry, was that 'she was very good psychologically'. She told him he was brilliant, that he would be walking 'better than normal' and she kept on saying it until he came to believe it himself. In fact, he never walked without crutches; but he soon discarded the braces, just as she had said he would. His recovery enabled him to go to university, study law and follow in his father's footsteps as a successful attorney, as well as lead a normal family life, marrying and fathering three daughters.

His father and Sister Kenny were the two most important influences in Henry's life. His father was 'a very independent cuss', he recalls, who 'wouldn't tolerate any messing around'. A fiercely upright man, Henry Haverstock Sr sounds like a typical Victorian paterfamilias, undemonstrative but demanding, a difficult role model for a severely crippled son. Sister Kenny, on the other hand, though she too could be fearsome, was a wholly benign influence on Henry: she gave him confidence; she said to him, 'You've got brains, you can do this.' He believed her and, as a result, tried harder. Of one thing he is certain: without her intervention he would never have walked again – 'I really wouldn't have.'[37]

Henry Haverstock Sr was so grateful to Kenny for what she did for his son that he became an active and influential supporter of her cause. At a talk he gave at the end of 1940, he listed four reasons 'Why Sister Kenny is worthy of our confidence':

1. She talks, thinks and breathes polio all the time.
2. She has studied it and worked with it for years.
3. She positively,
   unmistakeably
   and without fear of any contradiction
   brings strength and power back where others say there is little or no hope
4. She asks for and seeks no monetary reward.[38]

Dr Pohl was impressed, too: 'Henry's case made the reserved Pohl an evangelistic Kenny convert.'[39] He was the third of a trio of Minneapolis and St Paul doctors who became her mediators, translating her often opaque terminology into acceptable medical language and promoting her method of treatment in a stream of papers for various medical journals; the other two were Drs Wallace Cole and Miland Knapp. They learned from her, but she also learned from them. One of the things that impressed observers was that, however dogmatic Kenny might appear to be, she was quick to pick up useful information and adapt and refine her technique, even if she did not always acknowledge the provenance of her ideas.

In mid-January 1941, Dr Pohl arranged a demonstration of the Kenny method, which leading physiotherapists such as Alice Lou Plastridge from Warm Springs and the Kendalls of Baltimore came to Minneapolis to attend. Henry and Florence Kendall sent a lengthy report to the National

Foundation for Infantile Paralysis. 'Unfortunately', they wrote, 'we did not have the opportunity to observe actual treatment by Sister Kenny or her assistant.'[40]

Sensing, perhaps, that the Kendalls came less to observe than to patronise her, Kenny behaved badly. She opened proceedings in her usual manner by reading out testimonials from a number of Australian and English doctors, which did not impress her visitors. They noted sourly that 'her lecture consisted mainly of her life history and the story of her success against great odds'. A discussion of terminology led to 'more confusion than clarification', and the Kendalls regarded her knowledge of muscle function as 'very incomplete' and even 'inaccurate'. Worst of all was 'her failure to recognize the value of grading the power in muscles (as determined by muscle testing)', which suggested to them that 'she should not be entrusted with teaching the principles of muscle actions'.

By the third morning, when Kenny was due to demonstrate her treatment, the tension was palpable. In an effort 'to be agreeable', Mrs Kendall told Sister Kenny that 'we thought she had a contribution to make in her treatment to relieve spasm in the acute stage, and that in regard to other principles, we could get together on many things by being open-minded and discussing the problems'. That way Mrs Kendall was sure they would find plenty of common ground.

'Mrs Kendall', Sister Kenny replied testily, 'you say there is no difference between my treatment and yours, and I maintain they are entirely opposite.'

From that point the meeting degenerated into a one-sided slanging match, in which Kenny maintained that she had nothing to prove and that the Kendalls were talking about a disease that did not exist, and the Kendalls stood on their dignity and said that they 'preferred to discontinue any discussion until the doctors arrived'. When they reminded Kenny that Dr Pohl had invited them to observe treatment, 'she said she didn't care what Dr Pohl had invited us to do, she did not intend to show us any treatment'.[41] Nor did she. Dr Knapp, who witnessed the discussion, described it as 'like trying to get an agreement between Winston Churchill and Hitler'.[42]

Sister Kenny's behaviour was boorish, but she was surely right to insist on the incompatibility of the two methods of treatment. No matter how sophisticated their approach, the Kendalls were apostles of immobilisation, which was anathema to Kenny. They might have agreed on secondary issues, such as the limited value of underwater treatment in early convalescence, if Kenny had taken a more conciliatory line, but that would only have masked their disagreement over fundamentals. Kenny no doubt felt that the Kendalls were intent on assimilating her treatment and subsuming it into an all-embracing orthodoxy, and she was not going to be assimilated or subsumed by anyone.

Alice Lou Plastridge deferred writing *her* report until she had come back, along with the new Director of Physical Medicine at Warm Springs, Dr Robert Bennett, to take a proper look at the treatment while Kenny herself was absent on a visit to Australia. In some respects, Plastridge was as critical of Kenny as the Kendalls were: she deplored the element of showmanship in her presentation of case histories, and thought her a rotten lecturer, 'neither scientific nor logical'. But unlike the Kendalls, she had no doubt that Sister Kenny knew her anatomy. Above all, Kenny was getting such good results 'that it does not seem possible that they could all be a matter of chance'.[43] Bennett also approved, though he too tried to assimilate Kenny. He wrote to Dr Miland Knapp:

> I was quite surprised at the entirely new conception of the Kenny system I got by spending the afternoon with Mary Kenny. As you realized long before this, under the ceremony and ritual and fanaticism which surrounds the so-called Kenny system – (it amounts to a cult rather than a system) – there is very excellent treatment which, oddly enough, conflicts very little with the orthodox methods. I agree thoroughly with her early attempts to relieve muscle pain, and release muscle spasm by carefully applied hot packs, and while I cannot fully concur with her method of positioning the patients, I believe we have much to learn from it. The routine of exercise for muscle re-strengthening and re-coordination is commendably thorough, but cannot be said to be original with Sister Kenny.
>
> I hope that someone, like yourself, will eventually present this method of treatment in a logical, scientific manner, so that even Americans can understand it.[44]

In 'a preliminary report', published in the journal of the AMA in June 1941, Knapp and his colleague, Wallace Cole, did precisely that, recording their observations of twenty-six cases treated by Sister Kenny.[45]

Sympathetic laymen, such as the Mayor of Minneapolis, Marvin L. Kline, felt no obligation to temper their enthusiasm with scientific caution. 'The polio ward', Kline recalled, 'once a place of tomb-like silence, was now filled with laughter! One look, and we knew this granite-faced woman had something good.' A little girl who had lost the use of her thigh muscles solemnly informed a now smiling Sister that 'she had spoken crossly to her *quadriceps femoris* for being so lazy, and she was sure it would do better today'. Kenny encouraged her gently before moving on. 'A nine-year-old boy shouted that he wanted his *tibialis anticus* – a lower leg muscle which lifts the foot – put in shape with all haste so he could kick a certain ward companion in the *gluteus maximus*.'[46]

By 1942, Sister Kenny seemed to be carrying all before her. In December 1941, the medical advisory committee of the National Foundation officially endorsed her method of treatment; and the foundation published a booklet providing essential details of this method, prepared by the trio of doctors from the Twin Cities and with a preface by Basil O'Connor himself. In October 1942, the *Reader's Digest* printed a glowing account of her triumph. Kenny treatment had become 'the treatment of choice'. Dr Philip Lewin, an orthopaedic surgeon at Northwestern University Medical School, was quoted as saying, 'Continuous rigid splinting is not only on its way out, it is out,' and Dr Philip Stimson of Cornell considered that respirators no longer had any use except for 'keeping a patient alive until he can get the Kenny treatment'.[47]

With hindsight, it is easy to see that 'the fashion for treating weak limbs by early and prolonged immobilization had gone too far', as Dr John Paul puts it. 'Incredibly, *immobility* had become such a fetish that it was proposed even as a preventive measure for "keeping the paralysis away".'[48] Kenny put a stop to all that and by 1947 the thousands of splints that the National Foundation held in reserve for use in epidemics had been sold for scrap. One foundation official remembered going for a walk near Warm Springs and coming upon a black man's plot of land: 'Here he was using our old polio splints . . . for bean stalks. He had them pyramided, and beans growing on them.'[49] Sister Kenny would have approved.

Dr Bennett of Warm Springs read a paper at the thirty-sixth annual meeting of the Southern Medical Association at Richmond, Virginia, in November 1942, in which he conceded that 'deformities that we formerly considered due to the pull of a normal muscle against a paralysed one are now considered to be caused by the unrelenting pull of the involved hypertonic muscle against one primarily uninvolved', and went on to say, in the discussion following his talk, that Sister Kenny's most original contribution was that 'she has advanced and proved to us the far-reaching early and late effects of muscle spasm'.[50] Another doctor, Plato Schwartz, at the University of Rochester, set out to find scientific evidence of muscle spasm, using an electromyograph to measure it. This was the first scientific investigation of Kenny's concept of the disease and, insofar as it confirmed the existence of spasm, it 'well bore out the preliminary opinion of physicians who "did not always agree with her explanations, but did applaud her results"'.[51]

Margaret Ernest had had polio as a child. The residual weakness in her left leg was not serious enough to prevent her getting around 'almost as well as the next person', but her mother met Sister Kenny and arranged for Margaret to go and see her in her apartment nearby; that was in May 1941. Kenny always insisted on having a doctor present at a consultation, and Dr Pohl was there when Margaret saw her, though she 'hardly noticed him'. Kenny was not as large as she appeared, but she still 'seemed to fill the room with her presence'. Margaret scarcely had time to wonder whether or not she was the 'quack' that people were saying she was before the examination was over and Kenny was asking her to come back in two days' time. When Margaret returned, she was still uncertain. But as Kenny explained why the muscles in her left leg would never regain their strength, all Margaret's doubts vanished and, though she already had a job, she found herself agreeing to become Kenny's secretary.[52]

Margaret stayed with Sister Kenny for four years, first as her secretary, then as her 'confidential secretary' with a secretary of her own, before leaving to join the Red Cross towards the end of the war. It was a demanding job; Kenny did not seem to need much sleep, usually going to bed at one in the morning and getting up at five, when she did her writing and made early-morning phone calls to people – for whom they often provided a rude awakening. In the evenings she caught up with her correspondence, with Margaret's help.

> She considered most social activities a waste of time. They took her away from her work. She always believed that her campaign against polio was more important than social amenities. It was ever work, work, work; not to forget fight, fight, fight if she felt the circumstances merited it.

Kenny was a fighter, but she did not want for provocation. Early in their relationship, Margaret accompanied her to see a prominent doctor who, throughout their visit, kept his feet on his desk and a lighted cigar in his mouth, and barely bothered to conceal his belief that she was a quack. When they left his office, he called out to her to shut the door. Margaret was more shocked by his behaviour than by Kenny's fiery response. Yet his incivility did not deter Kenny from approaching that doctor again when the work demanded it.

Sister Kenny liked to be early for engagements, and if she was travelling by rail she would get to the station half an hour, or even three quarters of an hour, before the train was due and have her favourite refreshment, a root beer float known as a 'Black Cow'. Then she would relax every muscle of her body and remain virtually motionless till the train came in. After she became famous, such solitary relaxation was no longer possible; wherever she went she was mobbed by reporters and cameramen and other inquisitive people. Occasionally she would surprise Margaret by suggesting that they go out to dinner or a film:

> For a time, I couldn't figure what movie could take her away from her work. She never gave a reason, but it always seemed to follow a day when she had been fractious. That type of action was characteristic of her. Words came hard but deeds were easy. Only in matters concerning her treatment did she become voluble to any extent.

Money interested her only as a means to promote her work; she was indifferent to her own needs. She gave what she got from Hollywood to charity and during the Second World War handed over the royalties she still received from the stretcher she had invented to the Australian government; and she expected others to give – of their money and of themselves – as selflessly as she did. She often quoted the Bible from memory, though her religion was very much a private affair. In the end, 'her work was her life, the acceptance of her treatment the only "thanks" she ever wanted'.

She was frequently invited to speak at functions, and she generally refused, since she resented time spent away from her work; but one organisation was particularly insistent and appealed to Margaret Ernest to intercede on its behalf. As a favour to Margaret, Kenny agreed to talk at one of its lunches. Shortly afterwards, 'the Minneapolis Exchange Club voted to pay Sister Elizabeth Kenny $500 a month to stay in Minneapolis and carry on her fight'. Margaret still takes a quiet pride in her contribution to the cause of the woman she came to idolise once she had overcome her initial doubts.[53]

Kenny was guest of honour at the 1943 President's Birthday Ball; she was awarded honorary doctorates at prestigious universities; and in June she finally went to the White House for lunch. There is a photograph of her shaking hands with FDR; he is seated, though whether in a wheelchair or an ordinary chair it is impossible to say since the photographer has followed the rule of not revealing the extent of his disability. But Kenny is smiling down at him from under the feathered hat that Dr Fishbein said made her look like Admiral Nelson, while FDR grips her hand and smiles up at her warmly. In the background, standing between them, is Basil O'Connor; he is also smiling, but seemingly to himself as he watches his old friend and partner exercise his famously boyish charm on a not unsusceptible lady. Behind the formal pose the photograph perfectly captures the disposition of each of the three leading polio crusaders.

Up to this point, O'Connor had been supporting Kenny through the University of Minnesota, but the new Kenny Institute, which had opened in Minneapolis the previous December, was proving expensive and when the city appealed to the National Foundation for more money, O'Connor stalled. At the time, he was involved in a dispute with Kenny over the position of two of her therapists – one of whom was Mary Kenny – sent to Buenos Aires in response to an appeal from the Argentinian government for help in a polio epidemic there. O'Connor wanted the physiotherapists brought home and Kenny did not like 'the thought of leaving Argentinian or any other polio patients stranded'.[54] The physiotherapists stayed on, financed by the Argentine government when the National Foundation stopped paying them.

O'Connor also refused to fund the Kenny Institute, saying there was no need for it now that the Kenny method was on everyone's agenda. There was now an open rift between the tough Irish-American lawyer and the equally tough Irish-Australian nurse. Kenny publicly dissociated herself from the NFIP and set up her own rival Sister Elizabeth Kenny Foundation. A National Foundation historian described its 'marriage of convenience' to Sister Kenny as 'an uneasy coupling from the outset, the relationship . . . always stormy and the protracted separation proceedings . . . sensational'.[55]

O'Connor's policy, pursued with predictable subtlety and skill, was to separate the method from its originator, to isolate Sister Kenny while he praised and promoted her treatment. As he wrote in a 'special article' published in *Archives of Physical Therapy* in April 1944:

> The establishment of a Kenny institute in Minneapolis as the only place where the Kenny method would be taught has been suggested. But, of course, it's impossible to train all the Kenny technicians we require at any one place – in Minneapolis or elsewhere. And it would be equally impossible for any one person to supervise the various centers of teaching now supported by the National Foundation for Infantile Paralysis.
>
> The ultimate aim is to make *whatever is sound* in the Kenny method a part of the curriculum of every medical, nursing and physical therapy school in the country – and that aim will be accomplished. No one institution can have a monopoly on the teaching of the Kenny method. While it is Sister Kenny's contribution to humanity, for humanity's sake it must be available to all [my italics].[56]

So much guile behind the innocent-sounding words! Sister Kenny might protest that what passed for Kenny training in most parts of the country was but a pale shadow of the real thing, a brief course which failed to engage with her concept of the disease, and even contradicted it. But wasn't the lady protesting too much when *your* National Foundation had spent half a million dollars of *your* money to promote *her* method and do everything within its power 'to loosen the grip that infantile paralysis has on our children'?[57] Despite her enormous popularity, which showed no sign of waning, Kenny – though no mean antagonist herself – was practically powerless against so shrewd an adversary. No wonder she kept threatening to leave the country and return to Australia.

Albert Deutsch, 'one of the finest sociological and medical journalists of the time',[58] recognised that Kenny's 'extreme sensitivity' was 'largely defensive, resulting from decades of unfair attack'. But at the beginning of 1944 even he felt obliged to point out that the fault was not entirely one-sided; she was a hard person to work with, impatient of criticism and authoritarian in her attitude towards her medical co-workers: 'It isn't easy for physicians, who traditionally regard nurses as subordinates, to knuckle under and take orders and instruction from a nurse'.[59]

The fact that she appealed 'directly to the public for support, over the heads of medical men' did not go down well either. But retribution was at hand, in the form of yet another negative evaluation of her treatment, this time from a committee of medical men chaired by the orthopaedic head of the Mayo Clinic, Dr Ralph K. Ghormley, who had once been an assistant to the great Dr Robert Lovett. It was published in the *Journal of the American Medical Association* by a gleeful Dr Morris Fishbein, who had told Deutsch that far from inviting Sister Kenny to leave the United States, as she had accused him of doing, he had 'merely suggested that she might visit other countries, where her work is not so well known as it is here. Which' – Deutsch commented – 'sounds like some more of the Fishbein sledgehammer subtlety.'[60]

Ghormley *et al* condemned Kenny for her opposition to muscle testing; they accused her of misrepresenting the results of her own and other forms of treatment; they were sceptical of her claims to prevent paralysis and deformities by early treatment; they questioned the value of hot packs in cases where spasm was minimal or non-existent; they denied that her discovery of a means to diagnose the disease early would prevent paralysis as she claimed; they suspended judgment on the question of whether her treatment prevented scoliosis; and they deplored the widespread and misleading publicity associated with her name. The only good thing they could bring themselves to say about her approach was that 'this has stimulated the medical profession to reevaluate known methods of treatment of the disease and to treat it more effectively'.[61] Summing it up at an AMA meeting in June 1944, one committee member said: 'This report proves that what is good in the Kenny treatment is not new, and what is new is not good.'[62]

In view of the fact that after Kenny's death Ghormley gave a far more favourable account of her treatment, it is difficult not to agree with Dr John Paul's verdict that 'some prominent American orthopedic surgeons . . . attempted to discredit Miss Kenny's reputation – almost, it would seem, out of professional jealousy'.[63]

In the war of words over Sister Kenny and her treatment, disinterested comment was at a premium. Dr Robert Lawson, a paediatrician in North Carolina on a year's leave of absence to study neurotropic viruses, particularly poliovirus, went away from a visit to the Kenny Institute with mixed feelings, as he admitted in a letter to Dr John Pohl. He agreed that the Kenny treatment was 'a definite improvement over the treatment of a few years ago', though, in his view, there was 'not that much difference'; but he thought 'the type of claims made for this treatment by the lay people connected with the Institute' damaging – and Kenny herself far from guiltless in this respect: 'Miss Kenny is an amazing woman with a rather unfortunate Messiah complex which has caused her to make claims that cannot be substantiated. Her conversation with me showed this curious mixture of common sense and utter rot'.[64] Dr Lawson calls it a Messiah complex; Basil O'Connor told people Kenny had 'a Jehovah complex'; and the sociologist J.E. Hulett Jr sees her in Weberian terms as a 'charismatic' leader of a healing cult:

> All through the story of her life as presented to the public there runs an implication of her being 'set apart' for this work, that she has experienced a special calling for her task . . . There is the conventional situation where she is forced to 'make her choice' between romance and marriage and service to humanity; she had a small vision; and at one time she was near death from heart trouble but recovered after having been given up by her doctors in Australia . . .
>
> One of the ways in which a cult leader responds to the leadership situation is to undergo an evolution of ideas regarding the theoretical basis for the movement. This leader's conception of her role has evolved from that of a simple pragmatist to that of the astute hypothesist. Whereas at first she merely applied a pragmatic discovery to the treatment of the disease, she now has developed a whole 'new concept' of the disease which she describes as 'the direct opposite' of the concept previously held by the medical world.[65]

It was puzzling to many doctors and commentators alike why Sister Kenny was not content simply to be the originator of an effective treatment for a baffling disease. According to Hulett, she 'makes frequent appeals to "science" for support and justification', while at the same time revealing 'a certain impatience, to say the least, with the deliberate and methodical practices of the scientist'. In a radio broadcast in 1943 she contrasted her 'gift' – 'not a new treatment of a well known disease, but a new concept of the disease' – with the 'over one hundred million dollars' spent on research. 'The result of this research', she informed her listeners, 'made no progress in the pathology of the disease and the further knowledge of the symptoms of the disease or its treatment'.[66] This was the kind of 'utter rot' that Dr Lawson was referring to in his letter to Dr Pohl.

Hulett stresses the 'difficulty of communication between cultist and scientist', but admits that Kenny herself 'has perceived this difficulty' and that each of her textbooks, written in collaboration with different doctors, 'represents a greater effort than the preceding one to put her views into what she describes as "scientific terms"'.[67] Like her resistance to being labelled a 'miracle-worker' (when a physician whose patient she had treated said, 'You have taken me back to the Middle Ages. The day of miracles has not passed,' Kenny replied, 'It's not a miracle, doctor. It's simply understanding the symptoms of the disease'[68]), Kenny's eagerness to work within the system and earn the approval of doctors and scientists suggests an unease in the role of cult leader.

The Nobel prizewinning Australian virologist, Sir Macfarlane Burnet, who worked closely with Dame Jean Macnamara (as in 1931, when together they established that there was more than one strain of poliovirus) and might therefore be considered a natural enemy of Sister Kenny, describes a two-hour interview with her shortly before her death as 'the most interesting and concentrated opportunity I have ever had to sense the quality of another human being'. He made notes of it immediately afterwards, so his account is worth quoting at length:

> I met her in the lobby, a large, white-haired, slow-moving woman dressed like old-fashioned royalty in purple with a wide-brimmed hat. She had a strong-featured, heavy-fleshed Irish face that confirmed the legend of her determination and persistence against opposition but she looked tired and there was a tremor of her right arm. She spoke softly and at length with a curious overlay of accents, an occasional American intonation and sometimes an Australian lapse in grammar.
>
> She was very much aware of herself as a public figure and at once made it clear that she knew all there was to be known about polio. She had treated many more cases than anyone else in the world – she gave the precise number, 7,828 – and no one else was in a position to speak with her authority.
>
> Her ideas on the pathological nature and development of paralysis in polio were highly individual, heretical and, unless all reputable research men were grossly deceiving themselves, completely untenable. On the other hand I felt certain that in the handling of a neglected polio patient she could be supremely successful in making the best of the victim's disabilities and potentialities. Her confidence on this aspect was overwhelming and convincing but I could feel a basic uncertainty in her dissertations on what the virus does in the disease. She was pathetically eager to show me the letters that a few not-very-well-known pathologists had written to her in support of her views . . .
>
> In many ways she was a supreme egotist but how could she help it? Perhaps two-thirds of her anecdotes took the form: Such and such a doctor showed her a paralysed child, saying that its only hope was to have some drastic operation on the muscles. She said: 'With your permission, doctor, I shall cure that patient in five minutes.' And with two minutes' manipulation of the contracted muscle the child's deformity disappeared . . .

Burnet was impressed with her despite her 'untenable' scientific theories and her boasting. Writing many years after her death, when she was 'almost forgotten by the world', he concludes: 'there was an air of greatness about her and I shall never forget that meeting'.[69]

Dick Owen was twelve years old when he went down with polio in 1940, the year that Sister Kenny came to Minneapolis. Unfortunately, he was several hundred miles away in Indianapolis.

> So I was treated by the old traditional method, which was immobilisation in an outfit called a Bradford Frame, which was a piece of canvas with ropes, that slung me up with a bare bottom, and then with some things called Toronto Splints – I think if I were Toronto I'd try and get my name removed from those. But Toronto Splints theoretically kept you in a neutral position, which was a bend of the knees, flattening of the foot, and then with the legs spread. So it was a little bit like that picture by Leonardo da Vinci that has the man spread out in a neutral position . . .[70]

Dick was immobilised for nine or ten months, from October to the following June or July, during which time he grew six inches, to five foot eight or nine inches, which he has remained ever since, and got so fat – 'food is love, so they fed the child that was just lying there' – that he weighed 178 lb, which he has never again approached. Then he was fitted with 'one long leg brace, one short leg brace, a pelvic band to hold the brace to the hip' and crutches; when he was helped to his feet, he had to adjust to 'the weird perspective of looking down at the chairs and furniture that were so far below me it was terrifying'.

Fortunately, there were two independent-minded physiotherapists at Indiana University, who were curious about Kenny. They went to see her in Minneapolis and came back converted. As a result, Dick 'was started on the Kenny method two years late', and managed to reduce 'about 16 or 17 lbs of bracing down to a very lightweight brace on one leg'. Then Kenny visited Indianapolis in person:

> She was a fascinating woman – I mean, to me she looked about seven foot tall and was huge. I remember she had on black stockings and a huge black hat, and she had this very humble – humbled – group of physicians walking behind her: the dean of the medical school, the professor of orthopedics, the professor of surgery, the professor of pediatrics, all non-humble men. But she was blasting away . . . She was imposing as a person, and she had an idea that was working, which was very bothersome to people.

Dick learned to walk with Canadian crutches – a sort of elbow crutch with leather cuffs round the forearms. Eventually, he discarded those, too, and for the next thirty years walked with a single stick. He became a doctor not so much because of the polio, though that certainly helped to concentrate his mind, but because medicine was in the family tradition: five generations of doctors, there are, from his great-grandfather to his son.

He set out to be a surgeon, which he now regards as over-compensation, or 'part of a big denial mechanism'. He took a surgical internship and felt very out of place: 'I didn't really like to see blood flow; and I didn't like to worry about patients dying in the middle of the night.' When he discovered the spinal cord injury rehabilitation programme at George Washington University, in Washington DC, where he was training, he switched to rehabilitation medicine. That was still a new branch of medicine and Dick's board certificate number is 343, whereas now his youngest associate's is 'two or three thousand and something'.

In the late Forties, when Dick applied to study medicine, 'it was very unusual for a disabled person to be entered in a medical school; so much so that most of my life I've been an unusual phenomenon in being a crippled doctor, so to speak'. At first, he did not altogether accept this role:

> My denial mechanism told me that there was nothing wrong with me. People's vision told them that there was something wrong with me. And so I had an entrée to disabled people that I took advantage of, you know – a bad trait in a person to take advantage of other people – but I made use of something, more like an actor than a human being. Until I was about forty years old I didn't think of myself as having a disability.

In his forties, Dick suddenly took up wheelchair athletics – basketball in particular – and found himself relishing the speed and violence of it; he could bump into people without having to say he was sorry: 'It freed me to be a different person.' It also changed his perception of his function as a doctor, and others' perception of him – including his patients':

> When we were together in athletics, they weren't my patients. They often called me 'Doc', but they didn't look upon me as a doctor. And often times when we would be playing out-of-town wheelchair basketball teams and somebody would fall out of a wheelchair injured, and I'd take a look at them and dispense some medical advice, they'd say, 'Gosh, I didn't know you were a *doctor*!' – meaning that they didn't think a doctor ought to play mean like that.

Over the years walking became more difficult for Dick; he found that he 'could barely walk six blocks without wanting to rest'. So he began using a wheelchair in places like museums and art galleries, where there were long distances to be traversed or an exhausting amount of standing required; this gave him an insight into what disabled people experienced, and made him realise that he was, indeed, one of them.

> I would see a person that was doing something dumb, as I would have done ten years earlier, like struggling against getting a brace, or using two canes instead of one, or using a wheelchair. It was so much easier to say, 'Wait until you have the experience, you'll find it so delightful to really *move* fast.'

'I wouldn't have chosen polio,' Dr Richard Owen now says. 'But I think it was probably good for me. And it's a touch of history that most people don't get to have that I've actually ended up here' – 'here' being the Sister

Kenny Institute (now a part of the Abbott Northwestern Hospital complex in Minneapolis), of which he is Emeritus Director.[71]

Asked to sum up Sister Kenny's contribution to rehabilitation medicine, Dick Owen hands over his notes for a talk he has just given on the subject to the Rotary Club:

> SK made use of early mobilizing. Stretching before any other exercise. Recognizing that the coordinated use of muscle is not possible without pain-free range of motion. Retraining motor control is far more important than increasing strength. Coordinated skill must proceed strengthening. One can function in the presence of significant weakness if maximal efficiency in use of the available strength has been learned.[72]

Dr Owen was amazed to discover, when he went to Warm Springs in 1955, twelve years, that is, after the American Orthopedic Association – 'a group of trusty men who wouldn't endorse anything, not even a toothpaste' – had endorsed Sister Kenny's method, that people were still being immobilised if their monthly muscle testing, which was anyway approximate, registered the tiniest diminution in the strength of, say, their quadriceps. 'But every day you didn't walk', he says, 'your brain forgot how to walk.' Kenny's practice of early reactivation, though it was '*murder* in the eyes of pediatricians' of her era, is 'one of our principles in rehabilitation now'.[73]

The history of medicine, as one commentator remarks, 'is replete with instances where the most prominent authorities of the time, through their influence, retarded innovations'.[74]

6

A Planned Miracle

The Salk polio vaccine, described by Basil O'Connor as a 'planned miracle', is typically a product of contemporary living. Here The Man in the Long White Coat, who pores over his ideas and bits of recondite data, hoping to add something of interest to the sum of knowledge, has found himself working shoulder to shoulder with the promoters and organizers – The Man in the Grey Flannel Suit and The Organization Man – who want to put useful knowledge across with the utmost impact. Best man in this marriage of science and modern communications is Dr Jonas Edward Salk, a quiet, serious-minded, warm-hearted young man, the most publicised virus hunter of all.

Greer Williams, *Virus Hunters* (1960)

You men of science will forgive me if I entertain the suspicion that the expert committees of the [National] Foundation have laboured under the rude sort of stimulus that only a hard-bitten lawyer could provide.

From a speech by Chief Justice Earl Warren at the
New York Academy of Science dinner honouring
Basil O'Connor on his 65th birthday in 1957

The press proclaimed Salk a national demigod, while some colleagues, resentful of all the attention he received, suspected him [of being] a demagogue. The vaccine became a perfect cause for an age in which ideology was suspect. The scientific atmosphere of the 1950s was fraught with Cold War overtones. The vaccine, an affirmation of American scientific and technological progress, was viewed as a triumph of the American system. American science, pragmatic and purposeful, demonstrated the continued viability of the promise of American life.

Allan M. Brandt, 'Polio, Politics, Publicity, and Duplicity:
ethical aspects in the development of the Salk vaccine',
International Journal of Health Services, Vol. 8 No. 2, 1978

On 4 December 1941, Keith Morgan wrote FDR a typically upbeat letter, thanking him for receiving the many state and committee chairmen of the National Foundation at the White House and rehearsing the fund-raising achievements of that year's birthday celebrations, which had raised over two million dollars. But since 'the past three years have been the three worst years in the epidemic history of Infantile Paralysis', it was still 'ONLY a beginning' and Morgan outlined his plans to make FDR's sixtieth birthday an even more spectacular event.[1]

Just two days after receiving this letter, on Sunday, 7 December – 'a day which will live in infamy' – FDR had other, more important things than the size of the latest polio epidemic to consider: the Japanese had bombed Pearl Harbour and, in addition to his other responsibilities, the President was now Commander-in-Chief of a nation at war.

At the National Foundation headquarters in New York officials were wondering, 'Does this mean we should call off the January drive?' Basil O'Connor had no such thoughts. 'We're never going to call off the war against polio,' he told his staff. 'Go on. Put on the campaign.' And the foundation officials were agreeably surprised to find that raising money in wartime was even easier than in peacetime: in January 1942 they collected nearly five million dollars; and the takings mounted annually until, in 1945, they reached nearly nineteen million.[2] But the number of cases was also inexorably rising.

There was an outbreak of polio among Japanese Americans herded into the euphemistically named Colorado River War Relocation Project at Poston, Arizona, in 1943.[3] In the summer of 1944 the worst epidemic on the east coast since 1916 struck the little town of Hickory in the foothills of the Blue Ridge Mountains of North Carolina with such force that existing hospital facilities were soon overrun; the National Foundation came to the rescue, like a National Guard for polio, and almost overnight a small summer camp for underprivileged youngsters was transformed into an emergency Infantile Paralysis hospital. The foundation – never knowingly undersold – published articles and pamphlets proclaiming 'The Miracle of Hickory':

> The heart of 'The Miracle of Hickory' is perhaps best expressed in the words of one of the patients. She was just a young girl whose college plans were interrupted when she was stricken. Swathed in steaming hot packs from shoulders to toes, she said – 'I didn't pay much attention to the annual "March of Dimes" Appeal before this; it wasn't very important to me. I gave something to it, but it didn't mean much. I'll pay a lot of attention to it now – it's one of the most important things in my life!'[4]

In retailing such stories the publicity machine of the National Foundation was merely playing its part in the war against polio, a war which FDR himself consciously coupled with the larger national effort. On 1 December 1944, he wrote in an open letter to Basil O'Connor:

> For two successive years the American people, at war abroad, have had to combat infantile paralysis at home . . .
>
> . . . Victory is achieved only at great cost – but victory is imperative on all fronts. Not until we have removed the shadow of the Crippler from the future of every child can we furl the flags of battle and still the trumpets of attack. The fight against infantile paralysis is a fight to the finish, and the terms are unconditional surrender.[5]

Despite this show of support from the President, Basil O'Connor – in his capacity as Chairman of the board of trustees and Treasurer of the Warm Springs Foundation (in July 1944 Roosevelt had appointed him head of the American Red Cross in addition to his other charitable commitments) – had to plead with the relevant authorities that Dr Robert Bennett should not be drafted into the army but be allowed to remain at Warm Springs, where he was doing important war work.

'At the moment', O'Connor wrote, 'we have approximately 15 or 20 Army and Navy patients at Warm Springs and the tendency is for that particular figure to increase.'[6]

Bentz Plagemann, who contracted polio in Naples while serving in the navy as a senior pharmacist's mate, arrived at Warm Springs on 1 November 1944 and could hardly believe his good fortune. To have stepped out of the war into this 'small principality' was like discovering 'that villa to which the characters of the *Decameron* had retreated to outlive the plague'.[7]

Plagemann's induction course began when he was visited by the Chief Physiotherapist – or physical therapist, as they had come to be known in America. Alice Lou Plastridge did not mock his ambition to be a writer, merely pointing out that Sir Walter Scott might be a better model than Byron since he was a polio too.

> I did not know why, but the way she said the word 'polio' fell strangely on my ears. I had thought I was familiar with the word, but she had robbed it of its harshness, its diagnostic sound. She made it sound like a secret order.

Plastridge's positive reinforcement went further: she told Plagemann that 'only a particular group of people contract polio'; these were people with 'a highly organized central nervous system, which usually means talent, or special ability of some kind'. No longer a serial number, or a casualty listed under the letter 'P', Plagemann basked in the feeling of being human again. It was only later that he began to question whether 'a psychiatrist, for example, might approve of that identity with a special group which Miss Plastridge was holding out to me'. At the time:

> I seized upon the idea, and when I looked at Miss Plastridge I seemed to see upon her face that light which we see upon the faces of all dedicated people.

> Perhaps she was a saint, I reflected, a saint in starched white linen, her symbols not the gridiron and the rack, but a tape measure and a muscle-test chart.

His sense of religious awe increased when he was taken to the indoor pool, which 'resembled a cathedral' and seemed to him to contain the frequently invoked '"spirit" of Warm Springs', a phrase which suggested 'the presence of a tangible being who resided there, like a sybil of ancient mythology. And for me, always, the pool was the temple where the spirit was, and Mrs Schlosser [his physical therapist] her high priestess'.

A girl called Janie, who frequently returned to Warm Springs for operations and treatment, explained its appeal to Plagemann thus: 'The people who walk are the abnormal ones because we outnumber them. It gives us the upper hand.' It was different in the outside world, where the able-bodied were dominant. Janie was at a disadvantage there; she had watched her high-school friends and class-mates grow up: 'They had boy friends. They became engaged. They married. They used to ask me to their houses for dinner, and when their babies were born I went to see them.' But she was not happy. She solved the problem – or so she told Plagemann – by changing her friends: 'Every year I go to the candy store and meet the new high-school kids.' Plagemann pretended surprise to find that she herself was no longer a high-school kid, and Janie 'threw back her head and laughed'.

Encouraging romantic attachments as a way of bolstering self-esteem was an integral, if informal, part of the process of rehabilitation at Warm Springs; so too was humour. Warm Springs anecdotes sprang up like urban myths. There was the tale of the fat man who, when asked if he had Infantile Paralysis, said no, he had been hired to test the wheelchairs; there was the story of the young patient who called out to a car-load of gawping tourists, 'If you come back at five o'clock you can watch us being fed'; and, favourite of all, there was the 'rehearsed "miracle"', in which a push boy, wearing borrowed leg-braces and posing as a patient, sat motionless in a wheelchair by the fountain while visitors gathered and another push boy fetched him a cup of water:

> After drinking greedily, the cap held for him, he began to shake, jerk, jump; a glad smile broke forth on his face, and with a shout of delight he leaped from the chair, shedding braces, and danced about the pool crying, 'I'm cured! I'm cured!' while the real patients sat all about him, helpless and weak with laughter.

These anecdotes have a common denominator of turning the tables on the able-bodied – laughter as the means by which the weak are revenged on the strong.

Into this highly charged atmosphere strode the boys of the President's Marine Corps bodyguard, harbingers of FDR's eagerly awaited

Thanksgiving Day visit. While the officers danced with the pretty physical therapists, the marines circled among the patients. 'As a sailor', Plagemann writes, 'I would have hesitated to admit the boast of the Marine Corps for its reputation with the ladies.'

> But I gladly gave credit where credit was due that night. May it be forever an honor to the Corps; an added luster to its bright shield, that, to a man, they saw in the young ladies of Georgia Hall, not polio victims, but members of the opposite sex. They brought with them into the Hall the warmth and memory of a fuller life; they brought stars to the eyes of the girls . . .

The President himself arrived by train. His already legendary status was enhanced by newsreel shots from the 1943 Tehran conference in which he was shown seated between Churchill and Stalin, the Big Three disposing the future of the world. Among the usual crowd gathered to greet him at the station was the mother of one of the Warm Springs patients, who later told Plagemann and others of the awe she had felt when she saw FDR being carried down the ramp from the train like one of 'the great rulers of the ancient past whose people had not allowed them to touch the earth with their feet'. Someone else in the crowd was struck by FDR's resemblance to the postwar Woodrow Wilson, so grey and prematurely aged was his face. Yet once he was seated in his own car, with his dog Fala beside him, in the familiar Georgia landscape, 'a weight seemed lifted from his shoulders'.

The polios awaited his arrival outside Georgia Hall, their wheelchairs drawn up in a semicircle in the porticoed entrance:

> When we saw the cavalcade round the corner of the drive, his car in front, the cars of the secret-service men behind, our throats tightened and we were quiet and did not look at one another.
>
> As he turned into the circular drive before us, he slowed his car, and when he reached us he stopped. Then, sitting in silence, he smiled. He raised his arms above his head, clasping his hands like a victor, and in our chairs we smiled back at him, and wept.

'Mr Roosevelt enjoyed his last Patient Show following the Founder's Day dinner in 1994,' writes Adam M. Lunoe of the National Foundation in his 'Notes on Georgia Warm Springs Foundation'.

> It was called the 'Spirit of Warm Springs' and was planned to be divided in three parts:
>
> 1. Past
> 2. Present
> 3. Future

> Last-minute wartime plans for Mr Roosevelt made it necessary to shorten the play and ironically 'The Future' was eliminated.[8]

Bentz Plagemann attended this dinner and describes it and the effect it had on him in such a way that it acquires overtones of the Last Supper, with FDR Christ-like in his simplicity among his disciples:

> You did not feel that it was the President coming into the room, nor even the Squire of Hyde Park, but simply a polio in a wheel chair. Here the man of many roles reserved for himself the privilege of being himself. He smiled, waved his cigarette holder, and was wheeled to his table, followed by the members of his party.[9]

Prominent in FDR's party was a beautiful blonde whom Plagemann identifies only as 'the Famous Movie Star'. In this account she serves as a foil to the President, affected and artificial where he is spontaneous and natural. The high point of the show was the sketch 'involving the Skeleton, the Fat Lady, and Franklin Roosevelt', the familiar story from the prehistory of the foundation of Fred Botts & Co. that FDR himself had so enjoyed telling. Plagemann noticed that the President made notes during the skit.

The mood changed when the evergreen Fred Botts himself appeared in his wheelchair, 'dressed in an absurd, ill-fitting woman's dress, a hat with feathers, lipstick on his face', and sang, to the President's and everyone else's delight, 'Walter, Walter, Meet Me at the Altar'. Then the reason for the President's note-taking became apparent; he spoke extemporaneously: 'Yes, he remembered those early days . . . He had always wanted to come back here. The war had kept him away; the war had kept so many men away from the places they loved most.'

> And then he spoke of polio. The silence in the room was harsh. On the faces of the able-bodied there was a look of discomfort, as if they had no right to be there. He wanted a great deal of money used for research, the President said. Research, when he spoke of it, became something real, something more than a sheet of notes and figures from a laboratory. So much had been done to improve the after-treatment of polio; he admitted that. But after-care was secondary. The disease must be met on its own ground. It must be conquered.

Plagemann was writing at a time when Roosevelt's disability was referred to only obliquely, if at all. But were not his friends doing him a disservice 'by not stressing the influence of his disability upon his character and decisions'? Perhaps it would only be in retrospect that one might see such a man whole.

> But we who sat with him in our wheel chairs, opposite his wheel chair, learned, possibly with surprise even for us, that the President was, like

> ourselves, a polio first, and man and President after that. Only we could sense the endless tedium of his days, the being lifted in and out of chairs, of bed, bathtub, pool. And know that war and politics and the glory of his fame and career were outside of all this, strung like beads on the thread of the waiting days. And our admiration was touched with compassion because we knew, by comparison with ourselves, how great the heart was which sought such noble compensation.

The party over, the President sat by the double doors and was introduced by Alice Lou Plastridge to all the patients who wheeled past. Plagemann heard someone ask FDR how long he planned to remain at Warm Springs.

> 'Only a few days,' he said, responding with his ready smile.
> 'But I am coming back in the spring', he added, 'and then I am going to try to stay for two weeks.'
> Now all the world knows that when Franklin Roosevelt came back in the spring, he came back forever.

In addition to the Famous Movie Star, who had impressed Plagemann so unfavourably, the President's party at the Founder's Day dinner had included the recognisable faces of 'Mr Hassett, his secretary . . . and Basil O'Connor, who, in his double capacity with the Red Cross and the National Foundation for Infantile Paralysis, might almost be included in Anatole France's "moment in the conscience of mankind" '.[10]

O'Connor and Hassett were both anxious about the President's health. In September 1944, FDR had finally been roused from the torpor which marked the early stages of his campaign for re-election for an unprecedented fourth term by Republican questioning of his fitness to continue in office; he delivered his famous 'Fala' speech, in which he said that while he and his wife and family were all used to having attacks made on them, his dog Fala was not – and did not take kindly to them. After that, he subjected himself to a punishing schedule, culminating in a fifty-mile drive in an open car through four New York boroughs in the cold and heavy rain, without noticeable ill-effects. In reality, however, as astute observers like O'Connor and Hassett could tell, FDR was exhausted; even the Founder's Day visit to Warm Springs, which in the past might have been enough to restore his energy, did little to revive him. Apart from anything else, it was too brief.

In January 1945, immediately after his inaugural, FDR flew to Yalta for another meeting of the Big Three; and when the new Vice-President, Harry S. Truman, stood in for him at the annual President's birthday lunch in the White House, Basil O'Connor alarmed him by asking, 'Are you getting ready to take over?' Truman replied, 'What do you mean? Are you serious?' And O'Connor said, 'I'm perfectly serious. It may be six months or a year, I don't know when, but it's coming.' 'My God,' was Truman's only response.[11]

On 2 March, shortly after his return from Yalta, FDR made a unique public reference to his condition when he reported to Congress:

> I hope you will pardon me for the unusual posture of sitting down during the presentation of what I want to say, but I know you will realise that it makes it a lot easier for me in not having to carry about ten pounds of steel around on the bottom of my legs.[12]

Basil O'Connor was with FDR at Warm Springs in early April, and left on the morning of Sunday the eighth to prepare for his forthcoming visit to the Pacific as head of the Red Cross. He recalled going into the Little White House and telling FDR: 'I want to say something to you, and I know it won't do any good. Your only chance is to go away for ninety days and do absolutely nothing. We don't need you to win the war – the war's in good shape. But we will need you after the war.' FDR replied, 'If only I could give it some thought.' O'Connor was reluctant to leave, but FDR assured him he would be all right. The two men shook hands and O'Connor departed.[13]

When a friend phoned him in New York at noon on 12 April and asked after the Boss, O'Connor replied that 'it can happen any minute now'.[14] He was right. An hour or so later FDR – surrounded by attentive women, including two cousins and his old love, the recently widowed Mrs Winthrop Rutherfurd, née Lucy Mercer, as well as her friend Madame Elizabeth Shoumatoff, who was at that moment painting his portrait – complained of a sudden headache, keeled over in his chair and never regained consciousness. At 3.35 p.m. – by which time Mrs Rutherfurd and the portrait painter were several miles away – he was pronounced dead of a cerebral haemorrhage.

Bill Hassett wrote, 'Presently I was called to the phone by Doc O'Connor in New York, who, with no hint of what had happened, phoned to inquire about the Boss.'[15] O'Connor remembered it differently: he had phoned *after* he had heard the news.

He went to Washington to pay his respects when FDR's flag-draped coffin arrived. 'While I was standing there [he recalled], Truman jumped out of his car. He said, "Do you remember what you told me on January 30?" I said, "Very well." And he just walked away.' O'Connor is alleged to have stayed two or three weekends at a hotel in Washington with no other purpose than to 'be near Roosevelt's atmosphere. . . . He had to get it out of his system.'[16]

If anyone imagined – and some people did – that the National Foundation for Infantile Paralysis would not outlive its founder, they were reckoning without Basil O'Connor. He had been present at the Founder's Day dinner in Warm Springs at which FDR had reaffirmed his commitment to the conquest of polio 'on its own ground' – i.e., through the research efforts of medical scientists. And for O'Connor the achievement of this goal became a sacred trust.

During the Second World War, polio research inevitably lost out to other more urgent, war-related scientific tasks. But a US Army Epidemiological Board Commission on Neurotropic Virus Diseases, consisting of three doctors – two of whom, Dr John Paul and Major (later Colonel) Albert Sabin, were already prominent researchers – was sent to Cairo in 1943 to investigate viral infections, including poliomyelitis. Through Tom Rivers's influence the virus commission (as it was more familiarly known) was partly funded by the National Foundation. The occurrence of paralytic polio among British and American servicemen in the Middle East, where poliomyelitis was not even supposed to exist, was at first sight puzzling. The official Egyptian view, shared by some British and American medical officers, was that the troops 'had brought their own disease with them'. But the virus commission found that polio was far from unknown in the Middle East.

'Poliomyelitis', the medical director of the Children's Hospital in Cairo told Dr Paul, 'is not rare in Egypt. It is never epidemic – always sporadic. It is a disease of poorer classes . . . The cases are as a rule mild, involving only one limb – rarely a fatality. The patients are seldom seen during the acute phase of the illness. Instead the child is brought to the Hospital weeks after the onset with the complaint that its leg is weak and doesn't seem to be getting any better . . .'[17]

This explained the absence of official polio statistics: 'The patients who survived and came to the hospital late in convalescence for treatment of their paralysis were *not* reported as poliomyelitis.' There was polio in Egypt – and in Malta, India and the Philippines – but it was, as it had once been in the West, largely confined to children and endemic rather than epidemic. This provided John Paul and his colleagues with a crucial insight into the aetiology of the disease. In the era following Pasteur, when sanitary methods were introduced in the more advanced countries of the world, notably Scandinavia and the United States, and improved standards of personal hygiene – and purer food and water – became the order of the day, the effect on polio was far from favourable. While other diseases might be substantially reduced, outbreaks of poliomyelitis (as we have seen) increased dramatically. Bearing in mind the contrast with Egypt, medical scientists could now provide a plausible explanation:

> What happened was that the circulation of polioviruses became more spotty and intermittent as the twentieth century progressed; children arrived at school age and even adolescence without having been exposed

> or infected, i.e. they remained as susceptibles. Accordingly, they became increasingly vulnerable, and when, inevitably, the virus was introduced after an interval of some years, it spread rapidly through an awaiting susceptible population much as happens with measles. The result was periodic *epidemics* of the disease.[18]

As late as the Thirties it has been admitted by experts that 'no circumstance in the history of poliomyelitis is so baffling as its change during the last two decades of the nineteenth century from a sporadic to an epidemic disease'.[19] Now, as a byproduct of the displacement of people due to the exigencies of war, the mystery had been solved. Polio was not, of course, *caused* by the advances in hygiene and sanitation that had taken place in the West at the turn of the century; but these measures interfered with the natural acquisition of immunity in infancy that had previously protected the bulk of the population from paralytic polio.

It was the puzzle of why 'polio epidemics hit good sanitary environments and clean healthy kids with more force than it did slums, where the kids might have been scrawny and dirty'[20] that so long delayed the recognition in the United States that polio was, as John Paul asserts, an '*enteric* viral infection'. In 1937 Paul and his Yale colleague Trask had examined the faeces of polio patients during the acute stage of the illness and found such quantities of virus as to suggest that it multiplied in the alimentary tract. This provided 'strong evidence that intestinal wastes might be a potential source of periodic community virus distribution'.[21] But when – at a meeting of the American Epidemiological Society held in Baltimore in April 1938 – Paul had suggested that polio 'just might, after all, "deserve to be considered as an intestinal disease"', his words were greeted with 'a chilling silence'.

> In the audience were several men that had had considerably longer experience with poliomyelitis than the speaker. To them this was an old and outmoded heresy. Such men included Haven Emerson, Lloyd Aycock, and James Leake, among others. One member of the Public Health Service at length arose and said, 'For some years now we have been following the good work that the Yale unit has been doing, but now, although I am loath to say it, they have apparently gone off the deep end.'[22]

Prior to the Yale unit's discovery – which was soon confirmed by other investigators – there had been just one American voice crying out in the wilderness that the 'portal of entry' of the virus was not the nasal passage but the alimentary tract, and that belonged to Dr John A. Toomey, a paediatrician in the Municipal Hospital in Cleveland, Ohio (who was to be commemorated, as we shall see, in a way he can never have imagined). Toomey had been funded by the President's Birthday Ball Commission because, as de Kruif put it, 'it was deemed unwise to exclude any worker from a chance to prove his point on the grounds that his idea was unorthodox'.[23] Unfortunately, his insight was not matched by his research technique. But by 1940 'examination of stool specimens had become the method of choice in tracing clinical poliovirus infections'.[24]

Another epidemiologically puzzling feature of the disease explained by the disclosure of the adverse effect of improved hygiene and sanitation on the spread of polio was the steady rise in the age of susceptibility over half a century – and since it was well known that 'the older a susceptible individual is when infection is acquired, the more likely the illness is to be serious',[25] the outlook for the future was grim.

Karl Marx grumbled about philosophers' dedication to interpreting the world when the point – as he saw it – was to change it. Basil O'Connor, though in no sense a Marxist, felt the same way about scientists; even before FDR's death he had become impatient with the apparent lack of progress in research funded by the National Foundation and was pushing for action. In 1944, in an internal memo to his medical director, Dr Don W. Gudakunst, he asked, 'What, if anything, have the grantees . . . discovered since January 3, 1938? If the answer is "nothing", let's say so; if it is "something", then let's set it forth clearly and intelligently.' Gudakunst's lengthy and unflattering reply – he found polio research workers 'self-satisfied' and 'jealous of prerogative', their vision often 'limited', their backgrounds 'narrow' – only confirmed O'Connor's fears.[26]

Gudakunst died in 1946, but it was on his recommendation that Dr Harry Weaver, a professor of anatomy at Wayne State University in Detroit, was that year appointed Director of Research – a job title which set alarm bells ringing in laboratories all over the United States and provoked a confrontation between O'Connor and his Chief Scientific Adviser.

Tom Rivers took the line that 'you can't "direct" research without offending scientific tradition, estranging the researcher, and defeating your own purposes.'[27] (O'Connor's admiration for Rivers was not unqualified; he once said of him that 'he was extremely learned but not what you'd call exceedingly bright'.[28]) Weaver set out to prove Rivers wrong, but Rivers was a tough nut to crack. 'Why', Weaver recalled,

> old Tom Rivers acted as if I were a foreign agent during my first two years around the Foundation. He challenged everything I said. If I had said he was good-looking, he would have argued with me. Later, of course, the evidence began piling up in favor of experimental vaccination and he came around.[29]

Rivers acted as intermediary between the scientists and the National Foundation, and it was not always a comfortable position to be in. But in conversation with Saul Benison, Rivers paid this tribute to Weaver:

> . . . to my mind he was one of those who later made [an] extraordinary contribution to the successful development of polio vaccines . . . he had a wonderful quality of being bold. In research you often need a person like Harry Weaver around, you know, someone to kick over the traces, or someone to encourage people to see what the grass is like on the other side. In other words, a catalyst. Harry Weaver performed that function beautifully.[30]

Weaver began by organising a series of conferences, which served the dual purpose of acquainting him with developments in polio research and bringing together key scientists in an atmosphere conducive to the exchange of ideas. It soon became evident that, as he put it privately to O'Connor, 'if real progress were to be made, more exact methods of research would have to be instituted, objectives would have to be clearly defined, procedures and techniques would have to be developed to permit attaining these objectives, and individual groups of workers would have to sacrifice to some extent their inherent right to "roam the field", so to speak, and concentrate their energies on one or, at most, a few of their objectives'.[31] This kind of talk was anathema to scientists, and Weaver knew better than to alienate the workers on whose efforts both he and O'Connor depended if they were to achieve their goal. He could not openly direct research but had to be 'an infiltrator, an *agent provocateur*'.[32]

In reviewing the state of the art, he came to the – at that time still quite radical – conclusion that 'prevention, as contrasted with cure, was the only promising avenue of approach'.[33] He questioned the conventional wisdom that poliovirus would grow only in nervous tissue: too many virologists, from the Swedes in 1911 to the Yale doctors Paul and Trask, had found it in human faeces. The notion that the virus would grow only in the spinal cords of monkeys had long been a stumbling block to the production of a vaccine (as well as casting into doubt the likely effectiveness of one).

Weaver's next step was a practical one, and it demonstrated his ability to get what he wanted done by persuading scientists that it was in their interest to do it – backed, of course, by the huge resources at his disposal. Macfarlane Burnet's discovery in 1931 that there was more than one type of poliovirus had been borne out in later work by Paul and Trask; and, in 1948 and 1949, David Bodian and Isabel Morgan at Johns Hopkins and John Kessel and Charles Pait at the University of Southern California independently tested a number of strains of poliovirus and found that they all fell into one of three distinct types. If there was to be an effective vaccine against polio, it would have to provide immunity against all three types. But what if there were more than three? To answer this, Weaver set up the poliovirus typing programme, of which he later said, 'I know of no single problem in all of the medical sciences that was more uninteresting to solve. The solution necessitated the monotonous repetition of exactly the same technical procedures on virus after virus, seven days a week, fifty-two weeks a year, for three solid years.'[34]

Such a chore was not likely to attract the leading names in virology – though they would happily sit on a supervisory committee – so Weaver had to select up-and-coming microbiologists who would be flattered to be asked and see it as a means to other ends. One such was Dr Jonas Salk at Pittsburgh, whose previous work had been on influenza. Salk's was one of four laboratories involved in the virus typing programme, but he quickly established himself as its leader and spokesman. He was ambitious and highly intelligent. Sir Macfarlane Burnet had met him at Ann Arbor, in Michigan, in 1943 and recalled 'that no one had ever picked my brains about influenza so expertly as he did'.[35]

In 1948, John F. Enders and his Harvard colleagues, Thomas Weller and Frederick Robbins, succeeded in growing poliovirus in non-nervous tissue cultures – a breakthrough for which they would receive a Nobel Prize in 1954.

Enders, a man of independent means, was in the great tradition of the 'amateur' – in the double sense of doing something for love more than material gain and of being, or appearing to be, rather casual about it. Enders played up to this image in the self-effacing way he talked and in the manner in which he dressed – he generally wore a bow-tie and a waistcoat with a gold watch chain across his midriff – but it was a role, not the reality. Despite the fact that he came late into the field, having switched from English to medicine and then abandoned his MD course in order to study microbiology under the great Hans Zinsser, there was no shrewder virologist than he. If he was also lucky, it was only in the sense that 'you make your own luck' and that all scientific discoveries require an element of good fortune. He was thoughtful and, above all, patient. 'We waited', was how he explained this vital breakthrough.[36]

Using techniques pioneered by the English husband-and-wife team, Hugh and Mary Maitland, in Manchester, who had succeeded in cultivating the cowpox virus in a mixture of minced hen's kidney tissue and blood serum in 1928, Enders and Weller found they could grow the mumps virus *in vitro*, i.e. in test tubes. Neither they nor Robbins (Weller and Robbins had been Harvard Medical School classmates in 1940) wanted to work on polio, and it was largely by chance that they found themselves doing so. They had not sought money from the National Foundation, but because they were working in the Infectious Disease

Research Laboratory of the Children's Hospital in Boston they had received part of the $200,000 grant the Harvard Department of Bacteriology had been given for a five-year 'study of viruses'.[37] They also happened to have some polio virus stored in the deep freeze, which Enders had been sent some years before. So one day when Weller, who was trying to grow chickenpox virus, had prepared more flasks of human tissue culture than he had virus to put in them, Enders suggested seeding them with the polio virus. 'It was in the back of my mind', he said, 'that, if so much polio virus could be found in the gastrointestinal tract, then it must grow someplace besides nervous tissue.'[38]

Sabin and Olitsky had already succeeded in growing polio virus in nervous tissue culture in 1936, but Enders, Weller and Robbins were the first to cultivate it in non-nervous tissue. They did a series of tests designed to prove the virus was not merely surviving but also multiplying in the tissue culture; they carried out neutralisation tests with immune serum; and they went on to cultivate strains from both other types of polio virus – and were successful on all counts.

When Enders casually confided his initial breakthrough to John Paul two months before the results were published in the magazine *Science*, the latter failed to appreciate its implications, as he admits: 'At least it did not appear to me as an electrifying piece of news.'[39] Paul would make up for this lapse – as when he writes:

> Most significant of all – and this was a feature which crowned the whole tissue-culture story – was the highly practical discovery that multiplication of polioviruses was accompanied by a characteristic kind of change within the infected cells. It represented a specific injury that could be readily recognized under the high power of the light microscope, and thus only simple observation of the cultures was required to detect whether viable and growing poliovirus was there or not! This was the characteristic which made all the difference. It opened the door to quantitative determinations as to how much poliovirus existed in a given culture; to a way in which neutralization tests could be performed in vitro, and to countless other advances. In brief, it led to an enormous increase in the facility with which polioviruses and other viruses could be handled for a variety of purposes.[40]

Jonas Salk was quicker off the mark: he was the first to learn the new technique of virus cultivation and adapt his laboratory to take full advantage of it. He could see that the way was now clear for the development of a vaccine. (So too could Enders, but when his colleagues broached the subject he argued that the laboratory facilities at the Children's Hospital in Boston were inadequate for such an undertaking, thereby signalling his unwillingness to get caught up in something that was beneath the dignity of an independent researcher such as himself.)

Isabel Morgan had already demonstrated that it was possible to immunise monkeys against polio by means of a vaccine in which the virus was inactivated by formalin, and it was this path, pioneered by the unfortunate Brodie, that Salk chose to pursue in his – as yet undisclosed – determination to produce a human vaccine. Despite the tragic loss of lives resulting from Kolmer's mid-Thirties experiment, conventional wisdom still favoured live-virus vaccine, so Salk's was the unorthodox – if scarcely revolutionary – option. His experience with influenza no doubt influenced his choice; the flu vaccine on which he had worked with Dr Thomas Francis Jr, under whom Salk had served his virological apprenticeship in New York and Michigan, was made with inactivated virus.

Weaver's call for 'group planning and a pooling of ideas and resources' under the banner of the National Foundation encountered resistance, not just from 'old Tom Rivers' but from other scientists as well. As one complained, 'We are beginning to look like a troupe of trained seals.'[41]

The National Foundation might be promoting a good cause, but did this license it to ride roughshod over its scientists? Another grantee grumbled in a letter to John Paul: 'Are we now employees who are ordered about?' It was more than their independence that was at stake; it was their legitimate ambition. Altruism was all very well and co-operation might be a fine thing – as de Kruif had proclaimed more than a decade earlier – but you did not work your butt off just to have someone else steal the glory.

The foundation felt that it lacked time for scientific niceties, especially after the meeting of its immunisation committee at the chocolate town of Hershey, Pennsylvania, in March 1951. There, by prearrangement, Dr Paul asked a colleague an innocent-sounding question: 'Dr Koprowski, did you have some data to present at this time?' And Hilary Koprowski, who was not a foundation grantee but worked for the pharmaceutical company Lederle Laboratories, dropped his bombshell: he had fed live-virus vaccine to twenty volunteers without mishap. As he later recalled:

> I reported just after lunch, and everyone was somnolent. Tommy Francis listened to me droning away and said to Jonas Salk, 'What's this – monkeys?' and Salk answered, 'Children!' Francis sat up and gasped, 'What?' Albert Sabin got all perturbed and said to me later, 'Why have you done it? Why? Why?'[42]

However premature it was to immunise human beings with Koprowski's vaccine (and later tests conducted in Belfast would suggest that it had indeed been premature) news of it alerted the National Foundation to the seriousness of the commercial competition. The foundation might encourage, if not insist on, co-operation among its grantees, but the last thing it wanted was for a firm like Lederle to steal a march on it in the still unacknowledged race to produce a viable vaccine.

An aspect of Koprowski's trial that bothered a number of people, including Tom Rivers – if his later testimony, disputed by Koprowski, is to be believed – was his use of institutionalised children as 'volunteers'. 'An adult can do what he wants', Rivers said, 'but the same does not hold true for a mentally defective child. Many of these children did not have any mommas or papas, or if they did their mommas and papas didn't give a damn about them.'[43] At the same time, Rivers recognised that what Koprowski was doing was standard practice; indeed, Salk would later test his vaccine on the boys of the Polk State School for retarded males (as well as on polios at the D.T. Watson Home for Crippled Children); and Sabin would try his out on the inmates of a penitentiary, who made up the other category of 'volunteer'. Rivers said, 'I don't even know that you can actually call a prisoner a volunteer. I believe that, although prisoners are usually told that they will get nothing out of volunteering as guinea pigs, deep down they believe that they may get a commutation or reduction of their sentence.'[44]

In Britain such procedures were prohibited (as they would later be in the United States), so the editorial in the *Lancet* which reported Koprowski's experiment could afford to be smug:

> Koprowski *et al* tell us in a footnote that for obvious reasons the age, sex and physical status of each volunteer are not mentioned. The reasons must be more obvious to the authors than to the reader, who can only guess, from the methods used for feeding the virus, that the volunteers were very young and that the volunteering was done by their parents. One of the reasons for the richness of the English language is that the meaning of some words is constantly changing. Such a word is 'volunteer'. We might yet read in a scientific journal that an experiment was carried out with 20 volunteer mice and that 20 mice volunteered as controls.[45]

The reference to mice is facetious, of course, though the use of animals in medical experimentation provokes the same strong feelings today as the use of captive humans, particularly children, aroused after a war in which the Nazis had flouted the canons of human decency by subjecting people whom they conveniently categorised as subhuman to atrocious tortures in the name of medical research.

At an international conference in September 1952 Pope Pius XII made a speech on 'The Moral Limits of Medical Research and Treatment', in which he condemned experiments which 'transgress the right of the individual to dispose of himself'.[46] Such pronouncements reflected a genuine and heartfelt concern. But attempts in America to draw up a code of ethics for the use of human subjects for medical research in line with the Articles of the Nuremberg Tribunal created as many problems as they solved. The Nuremberg code (which stated: 'The voluntary consent of the human subject is absolutely essential. This means that the person involved should have legal capacity to give consent') was drawn up to deal with war crimes involving human experimentation, not clinical research designed to benefit humanity.

'The cold truth', John Rowan Wilson writes with reference to Koprowski's volunteers, 'was that if the grounds of selection were too rigid, there was simply no way of trying out the vaccine at all.'[47] And John Paul argues that if just two of the articles of a code of ethics based on Nuremberg principles 'had been followed to the letter during the early trials of either the inactivated or the attenuated poliovirus vaccines, not a single dose of these vaccines could have been administered . . .'[48]

The Second International Poliomyelitis Conference organised by the National Foundation for Infantile Paralysis was held at the University of Copenhagen in September 1951. 'Some of those present at the first Congress, held in New York three years ago', the *British Medical Journal* reported, 'felt that despite a vast expenditure of moneys and monkeys research was bogged down and that the future was not too hopeful. This year at Copenhagen, on the contrary, there was an atmosphere of optimism: useful leads were opening up in several directions.'[49]

There were scientific and political reasons for this improvement. The *BMJ* recognised the 'most exciting' work of Enders, Robbins and Weller, which Enders presented to the conference in a characteristically lucid paper, but failed to detect the hidden hand of Harry Weaver, opening up those leads 'in several directions'. For example, Dr J. Salk, 'on behalf of a large group of US workers', as the *BMJ* put it, reported the findings of the poliovirus typing programme: that out of a hundred strains tested every one conformed to one or other of the three known types of virus, the overwhelming majority – eighty-five per cent – to Type I.

On the sea voyage to Copenhagen, Salk had found himself on the same ship as John Enders, Albert Sabin and some of the other delegates to the conference; and once, he remembered,

> Albert came and put his arm around me and talked about establishing a polio research institute. How great it would be to have Joe Melnick and me with him, Melnick being the other 'bright young man' in polio work. We'd have meetings every morning and we'd decide what to do and what not to do. One big happy family. I was rather laconic about it, I suppose. I was busy laying plans for my own work, and the prospect of submitting to Albert in a kind of institutionalized typing-committee arrangement was not what I had in mind for myself. He really is a remarkable fellow. During the

> voyage on the *Stockholm* it became obvious to anyone who had not heard of it before that I was a nice young whippersnapper from Pittsburgh, going to Denmark to report on some drudgery I had performed. I might have failed abysmally, it seemed clear, if Albert had not been up in the flies, pulling the strings and setting the standards.[50]

Salk and Sabin, their names linked in undying enmity, had more in common than either was prepared to admit. 'Both were viewed by some as intruders in the medical establishment,' Jane Smith writes in her book on polio and the Salk vaccine. 'At a time when anti-Semitism was an open feature of academic life and "ambitious" was a popular synonym for Jewish, both Sabin and Salk were widely regarded as very ambitious people.'

The reason Sabin became such a master of condescension (as illustrated in Salk's anecdote) is probably that he himself had been patronised by the likes of Tom Rivers, who had greeted the elegantly attired young man on his return to the Rockefeller Institute after a year in England with the words: 'You can't turn a cheap East Side Jew into an Englishman in one year!' The fact that Sabin was an immigrant from the Russian-Polish border and had grown up in New Jersey rather than Manhattan was beside the point as far as Rivers was concerned; the 'Jewboy' was putting on airs and had to be put in his place. Rivers's cheerful condescension did not affect his estimation of Sabin's ability, or even his likeability – or Salk's, for that matter. One was 'the smart Jew', the other 'the young Jew'. But as Smith comments, 'No one has ever recorded any sympathy between Salk and Sabin based on their shared distinction, in the eyes of some others, as "the Jews"'.[51]

Sabin and Salk might both be outsiders from an ethnic point of view, but Sabin was fully integrated into the scientific community by virtue of a long and distinguished career, whereas Salk was seen as something of an upstart – not least by Sabin, who never missed an opportunity to pull rank on him. At this stage Sabin was not trying to exclude Salk from the élite; he merely wanted to dictate the terms of admission – the main one being a proper deference to him, Albert Sabin. In this, he was behaving according to the stereotype of an assimilated immigrant (Salk was not an immigrant, though his parents, like Sabin, were Russian Jewish immigrants – his father was in the New York rag trade). Salk's error was to imagine that he could short-circuit the system, strike out on his own and still gain acceptance among his peers.

If the journey out, and the conference itself, brought home to Salk that he was still seen by Sabin and his confreres as being on probation, the return journey on the *Queen Mary* saw the beginning of a friendship which would lift him out of the ruck and facilitate his ultimate triumph without in any way reducing – in fact, it increased – his pariah status among fellow scientists.

Basil O'Connor was informed by Harry Weaver that two National Foundation grantees, Enders and Salk, were on board; so as a matter of courtesy he invited them to join his party and have their meals at his table. Salk was delighted; Enders, whose independent means and superior New England status made him resistant to the razzmatazz surrounding the foundation and the glitzy lifestyle of its President, less so.

O'Connor's younger daughter, Mrs Bettyann Culver, a thirty-one-year-old mother of five, was accompanying her father. The previous summer she had phoned him in New York to say that she thought she had got 'some of your disease'. When he had asked what she meant, she had told him that her doctor had diagnosed polio. O'Connor was appalled; Bettyann was his favourite daughter and, like any other parent, he admitted: 'I'd never dreamed of polio hitting us.' After she had got over the acute stage, Bettyann had gone to Warm Springs, where someone who knew who she was had seen her in a wheelchair and asked in amazement, 'Are you a polio?' to which she had replied, 'No, I'm my father's spy here.' She had made good progress in a little over a year, but she was depressed by the slowness of her recovery and her father hoped that the cruise on the *Queen Mary* would cheer her up.[52]

O'Connor and Salk had met at scientific gatherings without making much of an impression on one another. But the leisurely transatlantic voyage drew them together. Salk's instinctive kindness and consideration towards Bettyann encouraged her and earned her father's gratitude; but it was his breadth of vision and his ability to give as good as he got across the dinner table that piqued O'Connor's interest. Weaver had praised Salk to the skies, and O'Connor could now see why. 'Before that ship landed', he said later, 'I knew that he was a young man to keep an eye on.' O'Connor prided himself on his knowledge of scientists, and this one struck him as a bit different.

> He is aware of the world and concerned about it. He sees beyond the microscope. He takes an overview. He tries to see how things fit together, not just in the laboratory but in the whole shooting match. He's a generalizer and a synthesizer. For someone with a legal mind like mine this is impressive. I am less impressed by the sort of mind that gets bogged down in fringe details.[53]

When Sabin had offered to take Salk under his wing, the latter had reckoned it was more as a means of stifling him than of promoting him; Sabin was too much of a rival to be considered disinterested. But O'Connor was an acceptable father-figure, in that he could provide the means whereby Salk could realise the dream they held in common; it would be a symbiotic relationship, a genuine partnership in which each provided what the other lacked. Other scientists might – and would – accuse Salk of selling out, renouncing respectability, by associating so closely with a layman like O'Connor. But Salk would have none of it:

> Not respectable? By what social or scientific standards? It is precisely O'Connor's assistance that makes it possible for me to do more than could possibly be done without his assistance. From the standpoints of science and society, then, O'Connor is an invaluable facility, a rare item of human equipment. Name another like him . . . Obviously I honor his great qualities but it is irrelevant to speak of 'indebtedness'. We do not do favors for each other like a pair of ward-heelers. We try to get work done in biology.[54]

Was Salk protesting too much? Perhaps, but this remarkable friendship produced remarkable results, and it arose as much out of spontaneous sympathy and likemindedness as out of a concert of interests. If O'Connor detected 'an echo of his own ambitions' in this 'young man making the same journey from obscurity to achievement, overcoming the same obstacles that faced anybody who was not Anglo-Saxon, not Protestant, not the child of wealthy or educated parents', then Salk was surely also thinking of himself when he described O'Connor as combining 'self-interest and social interest in ideal proportions'.[55] Whatever the truth, Salk could now set about preparing a vaccine in earnest, secure in the knowledge that he had a very powerful backer indeed.

Three of the four cornerstones on which a successful vaccine would rest – the ability to grow poliovirus in non-nervous tissue culture; confirmation that there were no more than three basic types of virus; and the discovery that a formalin-inactivated vaccine worked on monkeys, producing antibodies without ill-effects – were now in place. The fourth was the realisation that the route by which the virus travelled into the central nervous system (CNS) was not along the nerves, as had previously been thought, but through the bloodstream.

Early attempts to detect viraemia (the presence of virus in the blood) had failed because blood samples were taken too late, after the virus had penetrated the CNS and done its worst, and on the few occasions when viraemia was found its significance was not understood since it did not fit in with the prevailing view of the pathology of polio. But in September 1951 a preliminary field trial of gamma globulin – that fraction of the blood (or serum) containing antibodies, isolated by the Harvard scientist Dr Edwin Cohn during the Second World War and found to be highly effective against infections – was conducted by the Pittsburgh epidemiologist Dr William McD. Hammon during an outbreak of polio at Provo, Utah.

The results of this and two subsequent trials during epidemics the following year suggested that gamma globulin did provide a limited and temporary protection against the disease. The significance of this was not lost on virologists: both David Bodian at Johns Hopkins and Dorothy

Horstmann at Yale did tests on monkeys that revealed the presence of poliovirus in the bloodstream during the early, pre-paralytic phase of polio. This was 'the virus' point of vulnerability – its Achilles heel, as it were'[56]: vaccine-induced antibody in the bloodstream would prevent paralysis. With the fourth cornerstone now slotted into place it was only a matter of time before someone came up with a vaccine.

The immunisation committee of the National Foundation met in New York on 18 April 1952, when Bodian and Horstmann's findings were – to their astonishment – front-page news. Bodian found Tom Rivers, from whom he was prepared to put up with a great deal, in a cranky mood. As he made his report on viraemia Rivers challenged him so often that Albert Sabin felt obliged to intercede. But when Bodian pointed out that 'antibody could break the chain of infection from the alimentary tract to the nervous system', Rivers burst in once again.

> 'I would like to make a statement,' he said. 'In the first place, I don't want to hurt Dave's feelings, but I wouldn't like for him to call monkeys "he".'
>
> *Bodian*: 'What?'
>
> *Rivers*: 'You used the term "he" for a monkey.'
>
> *Bodian*: 'Chimpanzee.'
>
> *Rivers*: 'I think it's proper to say "it" when you are speaking about an animal. I yet do not accept chimpanzees and cynomolgus as equal to man, and it kind of grates on my feelings of propriety to have someone keep on calling the monkey a "he".'
>
> *Bodian*: 'Well, he wasn't a she.'
>
> *Rivers*: 'Well, it is an it.'[57]*

What Rivers was actually insisting upon, in his roundabout way, was the distinction between man and animal in terms of disease; he was arguing 'that Bodian should not talk about viremia in poliomyelitis but about viremia in *experimental* poliomyelitis'. It worried him that the newspapers were saying that a vaccine was just round the corner; he was appalled at the prospect of people clamouring for non-existent vaccine and inadequately tested gamma globulin; and he was fearful of the effects of that kind of pressure on scientific rigour.

> I firmly believe [he said] that we have in the near future, a matter of a year or two, a solution to many things that we want a solution to, but if we are pushed ahead of time and before we are ready, we will make a terrible mistake and we will get into trouble that we will regret as long as we live. We have got something big here, and unless we handle it right it will be terrible, and I for one don't like to be pushed into that position.[58]

* See Appendix: Monkey Business

When he stopped nitpicking and came to the point, Rivers showed why he was a natural leader and why O'Connor was right to have such faith in him that he was wont to say, 'If I had to defend what we did on the Salk vaccine before an intelligent jury, all I'd say is, "Gentlemen, we had Tom Rivers on our side. I rest my case."'[59]

Despite its rather grim name, the D.T. Watson Home for Crippled Children was by no means a grim place. Sited on a pleasant hill in the wealthy suburb of Sewickley Heights some miles to the north-east of Pittsburgh, it overlooked the Ohio River. Its Medical Director, Dr Jessie Wright, was an authority on polio rehabilitation and a formidable woman, 'well over two hundred pounds of iron determination, demanding of her staff and sometimes hard on patients who weren't putting forth enough effort to get their muscles moving'. It may not have been Warm Springs but, in the words of Jane Smith, 'If you had to learn to bear the burden of paralysis, this was a good place to do it.'[60]

The D.T. Watson Home's close links with the National Foundation meant that when in the spring of 1952 Jonas Salk approached Jessie Wright for permission to test his vaccine on the polios in her care, she was predisposed to co-operate. Salk explained that he wanted to

> take blood from polio patients, find out the type of antibodies they had, and then inoculate them with vaccines of the same types to see if the vaccines would raise their antibody levels. Since the subjects already had antibody and were immune to another paralytic attack from the one type of virus, the experiment would be as safe as it could possibly be.[61]

Lucile Cochran, the Nursing Superintendent and Administrator of the Watson Home, recalled that both staff and parents were very much in favour of trying it:

> It may seem peculiar, but we had no sense of making history. We just enjoyed being part of the project. And the parents and children and staff were so wonderful. Nobody said a single word to an outsider. The experiments remained an absolute secret.[62]

They were also an absolute success: there were no ill-effects, and blood samples taken after a few months and tested in the laboratory demonstrated that the booster effect on the antibody levels was even greater than Salk had dared hope; the shots produced more antibody than the disease itself had done. But he later admitted to a reporter. 'When you inoculate children with a polio vaccine, you don't sleep well for two or three weeks.'[63] Nevertheless, he was confident enough to try out his vaccine on inmates of the Polk State School who had not had polio – with equal success. A bonus was that by the end of the year the antibody levels were holding up in both sets of subjects.

So far, among the interested parties, only Weaver, Rivers and O'Connor (who had suffered a heart attack in June, but by September had resumed his usual punishing schedule) were in on the secret. 'But when the December bleedings turned out as they did', Salk recalled, 'we knew it was time to move toward more extensive tests. To do this we needed the support of the polio-research community.'[64]

The immunisation committee of the National Foundation met once again at Hershey, Pennsylvania – which had the merit of offering a hotel comfortable enough for Basil O'Connor, as well as being far removed from the beat of prying pressmen – towards the end of January 1953. Salk and Sabin travelled there together by train from Pittsburgh, not as a sign of a new-found amity but because Sabin had been visiting Salk's laboratory and picking his brains about tissue-culture methods. During this visit Salk had invited Sabin home to dinner but had not taken him into his confidence over the tests he was about to report on, though Sabin had probed him and let him know of his own determination to develop a live vaccine as soon as he could.

Sabin learned the results of Salk's inoculation of 161 human beings with 'various Formalin-inactivated poliovirus preparations' – Salk was reluctant as yet to talk of a vaccine, let alone the 'Salk vaccine' – along with his colleagues on the immunisation committee, sitting around a conference table at the Hershey Hotel. Enders was there but both Rivers and John Paul were absent this time, though they read the minutes of the conference and undoubtedly discussed what happened with others. Two virologists who were present, Dr Joseph E. Smadel and Dr Thomas B. Turner, pressed for an early field trial along the lines of the gamma globulin tests held at Provo, Utah. But Albert Sabin and John Enders opposed such precipitate action. As Rivers put it, 'Albert not only looked directly at a question, he looked around it and examined every possible facet, including a few theoretical aspects that didn't occur to others.'[65] But whereas Sabin's opposition might be considered suspect, due to his own interest in producing a vaccine, Enders's could not be regarded as anything other than principled, and he recommended that Salk proceed cautiously so as not to 'jeopardize the entire program'.[66] When John Paul heard the news he wrote to congratulate Salk and offered him further 'unsolicited advice':

> I want to add my voice to those of others, not only in praise of the well planned and carried out experiments, but particularly for the forthright stand taken as to the future [Salk himself was resistant to a field trial at this stage]. You must not and no doubt will not be railroaded into doing anything that you yourself have not planned or desired.[67]

Already, Paul comments, 'there was a certain uneasiness in the wind that the vaccine trials were about to be "taken over"'.[68]

Salk responded courteously to John Paul's letter, saying he would heed his 'wise words of caution';[69] it was Sabin's reaction to his report that got under his skin.

> It was almost as if he were trying to minimize what I had said. His interpretations made my work seem incredible, of no meaning or significance. We hadn't done this and we hadn't done that and this was premature and that was unsubstantiated. I remember asking him later, 'Why do you constantly emphasize the negative?' He answered that this was 'the scientific way of doing things'. John Enders was gentlemanly, but not more favorably disposed to what we'd been doing. And I remember Howard Howe coming to me and saying, 'I feel as I would if I had just lost a child,' referring, no doubt, to the [inactivated] vaccine he had been experimenting with. It was a tense meeting and I was by no means the tensest person there.[70]

Professional rivalry and clashes of personality, though they were real enough, should not be allowed to obscure the ideological divide between the proponents of live/attenuated vaccine and those of killed/inactivated vaccine. The majority of virologists favoured the live-vaccine option and regarded killed vaccine as, at best, a stopgap. The most obvious advantage of live over killed vaccine was that it was administered orally, whereas killed vaccine had to be injected, not just once but three times, at intervals, to protect against all three types of poliovirus; mass immunisation would be much simpler if the necessity – and fear – of the needle could be removed. Then there was the question of the length of immunity conferred by the killed vaccine, which had yet to be established; if it lasted only one to three years before a booster injection was required, that would create problems for public health authorities. Live vaccine, by contrast, should provide long-term, if not lifelong, immunity, just as the disease did – since the whole point of it was to infect with the disease but in such a mild form as to render it harmless.

With both vaccines there were problems of safety: in the case of the killed vaccine it was the near-impossibility of ensuring the inactivation of every minute particle of virus, however confident Salk was that his method provided a 'margin of safety'; with the live vaccine there was the danger of reversion to virulence when it passed through a body. Within each virus type some strains were more virulent than others, and this raised another question: was it preferable to utilise the virus strain that held more dangers but provided better protection, or one which might be safer but less effective?

The great fear of the live-vaccine camp was that Salk's killed vaccine might be sufficiently successful in the short run – whatever its long-term prospects – to put a stop to further experimentation. The National

Foundation funded research on both sides without discrimination, but if, in its eagerness to find a solution, it threw its entire weight behind the front–runner, which – as live-vaccine supporters believed it would – turned out be not much better than gamma globulin as a prophylactic, then the opportunity to produce something of lasting value would go begging. Seen in this light, the spirited resistance of Sabin, Enders, Paul and company cannot be dismissed simply as the refusal of a rump of orthodox, jealous and spiteful scientists to recognise a younger genius and admit him into their midst.

The National Foundation's thinking is clealy revealed in a memo Harry Weaver sent to O'Connor following the Hershey meeting:

> There is no question of the facts that with additional research: (1) A still more effective poliomyelitis vaccine could be produced; (2) We would be better informed as to the kind and frequency of untoward effects that might result from the use of the vaccine; and (3) We would be better informed with respect to the best route of inoculation [intradermal or intramuscular], and the best time for administration, of the vaccine to obtain maximum protection against paralytic disease . . .
>
> If such research is carried out, a very considerable amount of time will elapse before a poliomyelitis vaccine is made available for widespread use; with the result that, in the interim, large numbers of human beings will develop poliomyelitis who might have been prevented from doing so had the vaccine been made available at an earlier date.[71]

This was the crux of the problem: the number of cases of polio in the United States was mounting yearly; it reached an unprecedented 57,628 in 1952 and stretched both the resources of the National Foundation and the patience of March of Dimes contributors who were constantly being promised jam tomorrow. Those scientists who argued for 'more haste, less speed' were safely ensconced in their laboratories, far from the front line of the fight against polio, masterminded by General O'Connor in his office in New York. O'Connor was not reckless in his demand for action; he listened carefully to his scientific advisers. But if the time was right, and the action responsible, he was in favour of it. So, too, was Harry Weaver. Rivers called it 'sharing responsibility with Dr Salk', and backed Joe Smadel's call to arms at the special meeting convened by the foundation in February:

> I was sure that Jonas had an inactivated vaccine that was safe for children. If I didn't think that I would never have allowed the Foundation to use my name to call the February meeting. I can tell you that if I had a kid I wouldn't have hesitated for one minute to inoculate him at that time with

> Salk's vaccine. Damn it, do you know that at this meeting, Salk wouldn't even call his vaccine a vaccine; he kept calling it an inactivated preparation. Now, what I have just said doesn't mean that I was ready to run off half-cocked for a field trial in the spring. It simply means that, like Samuel Johnson, I realized that nothing would ever be accomplished if we waited to overcome all possible objections.[72]

Cautioned by one set of virologists not to take precipitate action, and accused by others of delaying things ('What are you waiting for?' Joe Smadel asked. 'Why don't you get busy and put on a proper field trial?'[73]), Salk could be forgiven for wanting to bury his head in the sand. 'Each school of thought was unhelpful, really,' he said, mildly enough. 'The sense of being caught between two lines of fire made me ache to retreat to Pittsburgh and continue my work in my own way. I was naïve to think that this was possible.'[74]

The question of safety was paramount and it was decided at the February meeting not to organise an immediate and large-scale field trial but to allow Salk to go ahead at his own pace and vaccinate a further 400 or 500 children in Alleghany County, Pennsylvania, his home base. The results of his earlier tests were due to be published in the *Journal of the American Medical Association* in March, and Rivers volunteered to write a letter on behalf of the assembled medical eminences, to be published in the journal, pointing out 'why a polio vaccine could not be ready for mass distribution until much later than the polio season of 1953'.[75]

But in the words of Robbie Burns, 'The best laid schemes o' mice an' men/Gang aft a'gley.' Planned publication of Salk's vaccine test results and Rivers's cautionary letter in the sobersided *Journal of the American Medical Association* was overtaken by an embarrassing leak in, of all places, the 'Broadway' syndicated gossip column of Earl Wilson. NEW POLIO VACCINE – BIG HOPES SEEN, the headline proclaimed. Salk was alarmed; he had already had a taste of what press publicity, even of a relatively innocuous kind, could do to his reputation among his fellow scientists. On 9 February, *Time* magazine had carried an article on the imminent prospect of a polio vaccine under a picture of Salk, which had prompted Sabin to write him a letter – which managed to be simultaneously insinuating and vaguely threatening – spelling out the choice of loyalties facing him:

> Although it was nice to see your happy face in *Time*, the stuff that went with it was awful – I knew you couldn't possibly have had anything to do with it, for if you did they would have gotten the story straight. Please don't let them push you to do anything prematurely . . . John Paul talked to me over the phone today – and his reaction is pretty much like mine. Tommy Francis, John and I expect to discuss it this Saturday when we're in Washington for a meeting . . . The foundation has made unwarranted and

premature promises before and there is nothing much we have been able to do about it. However, this is the first time they have made a public statement based on work which the investigator has not yet completed or had an opportunity to present it at a scientific meeting or in a scientific journal. It is good to know that you will do the best possible job, regardless of what is said by others.[76]

If this was a warning shot fired across his bows, how much worse would follow Earl Wilson's unexpected revelations? Some of Salk's colleagues already thought of him as a publicity-seeker and 'glory hound'. In his anguish he went to New York to see Basil O'Connor and made what was, in the circumstances, an extraordinary offer. He suggested that the National Foundation sponsor a broadcast in which he would tell people the facts and explain why they should not expect a vaccine immediately. O'Connor was delighted with the idea: what better example could there be of that productive partnership between scientist and layman which was one of his favourite themes; and from a fund-raising point of view, what could be more appealing than a few honest and modest words from the horse's mouth? No sooner agreed than arranged. O'Connor would introduce 'The Scientist Speaks for Himself' on the CBS national radio network at 10.45 p.m. on Thursday 26 March, two days before the publication of Salk's article in the journal of the AMA.

In a bizarre attempt to pour oil on troubled waters – can he really have been so naïve? – Salk sent Sabin a draft of his broadcast 'on the assumption that he, of all people, would be most alert to the problems against which it was directed'. Sabin was not amused by this flagrant breach of scientific etiquette. 'He phoned me, incensed,' Salk recalled. 'Told me I was misleading the public. Urged me not to do it. I was flabbergasted.'[77] Salk went ahead and made the broadcast which, if it was not a bid for fame, nevertheless made him famous.

'From the moment of his first announcement,' Charles L. Mee Jr, a polio casualty of 1953 (just too early for the vaccine), wrote thirty years later, 'such an outpouring of hope and gratitude attached to him that he came to stand, at once, as the doctor-benefactor of our times.' Only in retrospect could he be described as 'an insignificant-seeming fellow with big ears, a receding hairline, and a pale complexion';[78] at the time he was 'a real-life Dr Kildare',[79] and 'handsome' and 'modest' were the words most frequently used to characterise him.

Back in Pittsburgh, Salk tried to pretend that nothing had happened and got on with vaccinating people (including his wife and three small sons, the foundation's Dr Hart Van Riper and *his* three children – his wife had been paralyzed with polio when Van Riper was a paediatrician in Florida, an event which had dictated his move to New York and to the National Foundation – and Harry Weaver and his son). But O'Connor and Weaver now had the bit between their teeth and there was no stopping them. Salk struggled to retain control, but the reins were slipping from his grasp and such authority as he had was courtesy of the National Foundation, which needed him as much as he needed it. The attitude was, 'let's get this show on the road', and the precipitating factor was the formation of a new committee.

Years later, Saul Benison put the key question to Tom Rivers with the studied objectivity of a legal counsel:

> Dr Rivers, in May of 1953 the National Foundation organized a special Vaccine Advisory Committee to advise the Foundation with respect to the field trials of the Salk vaccine. Why was this necessary when the Foundation already had in being a special Immunization Committee composed of the leading virologists in the country?[80]

Rivers answered in a roundabout way. He began by pointing out that the immunisation committee was mainly composed of foundation grantees who had 'a special stake' in polio research: 'As is natural, each man's opinion was a bit biased in favor of what he was doing, especially when it came to a question between his work and that of another.' The immunisation committee, he maintained, was formed for discussion purposes, so that its members and the foundation could be kept informed of what was going on; it was 'never an executive committee', whatever some of its members imagined. He admitted that 'when the Vaccine Advisory Committee was formed they felt that they were being bilked out of making decisions on the Salk vaccine', and that they resented this – 'as a matter of fact, they were pretty vocal about it'. Finally he came to the point:

> Mr O'Connor designed the Vaccine Advisory Committee as a small executive committee whose duty it was to inform the Foundation what was going on, scientifically speaking, and *to devise a program for action in developing a vaccine.* We didn't think that it would be proper for anybody who had a personal stake in immunization research – whether it was Salk, Sabin, or anybody else for that matter – to be allowed to vote for anything [my italics].[81]

Rivers gives the game away when he goes on to speak of Smadel and himself being able to provide a 'disinterested opinion'. Smadel had made repeated demands for a field trial of the Salk vaccine; and old man Rivers, though he favoured a live-virus vaccine as the *ultimate* solution, was firm in his support of Salk's vaccine as an interim measure. John Enders, despite his enormous prestige and despite the fact that he did not have 'a personal stake in immunization research' – indeed, he shunned it – was *not* invited to join the vaccine advisory committee; he had shown his colours all too clearly. Harry Weaver put the matter more succinctly: 'We formed the Vaccine Advisory Committee to break a log-jam. The Immunization Committee was not able to function with the necessary dispatch. It could get entangled for months in technical debates.'[82]

The architect of the mass trials was Harry Weaver; but he was not around to see them through. When it came to a question of status, administrators could be as prickly as scientists; Dr Hart Van Riper, the Medical Director of the National Foundation, bitterly resented the fact that Weaver, nominally his subordinate, consistently ignored him in making his plans. It cannot have helped that Weaver had easy access to O'Connor, who saw him as a man after his own heart, a man of large vision and even larger self-confidence; or that the two of them sometimes dined together informally. But even O'Connor was beginning to feel excluded by Weaver.

For Van Riper it was a case of: 'Either Weaver goes or I go.' But O'Connor, who did not want to lose either man, washed his hands of the whole business, went off to Europe and left Van Riper to sort it out. Van Riper remembered it as the only occasion upon which O'Connor had ever let him down. Within hours of O'Connor's departure, Van Riper and Weaver were at each other's throats; Weaver said he would resign rather than do what Van Riper wanted, and Van Riper accepted his resignation on the spot.[83]

Weaver attributed the deterioration in relations to the design of the field trial: 'Jonas and O'Connor never really bought the idea of a double-blind study.'[84] And indeed the nature and organisation of the field trial became matters of considerable dispute – provoking another resignation, that of the epidemiologist Dr Joseph Bell, who had been brought in from the National Institutes of Health to direct it – before the foundation finally accepted that some outsider of sufficient stature and independence of mind had to be commissioned to do the job. Dr Thomas Francis, who was touring Europe with his wife when the call came, was initially reluctant to handle such a hot potato and would only do so on condition that he was given absolute authority to do it in his own way. As Jonas Salk's mentor and one of the élite of research scientists, he knew both Salk and the foundation – and the tricks it could play – well enough to insist on that.

The problem was that O'Connor was less interested in an epidemiologically sound field trial – mention the word epidemiology to O'Connor, Rivers recalled, 'and it's tally-ho'[85] – than in vaccinating as many children as possible against polio. His idea of a field trial was very simple: you compared the number you vaccinated with an equal number who were not vaccinated and by adding up the cases in each group you discovered how effective the vaccine was. Furthermore, he had promised a number of states that this would be the method used: children in the second grade would be vaccinated and children in the first and third grade used as controls.

This unsophisticated approach appalled epidemiologists and had led to Dr Bell's early departure. Francis, when he finally agreed to take control of the field trial, insisted on using the double-blind technique, compromising only to the extent of allowing O'Connor to keep his word in those states where he had already promised the cruder method would be used. The double-blind method involved the use of placebos which were indistinguishable from the vaccine itself, not only to parents and children but to the physicians giving the injections as well, thereby minimising the risk of any skulduggery; no one except Francis and his staff at what became the Poliomyelitis Vaccine Evaluation Center at the University of Michigan, Ann Arbor, had access to the code that revealed which child had received the vaccine and which the placebo. In effect, there would be two sorts of trial running simultaneously.

Another complicating factor was the vaccine itself. In a sense, Salk was right to fight shy of the word and prefer such euphemisms as 'inactivated preparation'. There was not a single finished product, an established vaccine; it was constantly evolving as Salk and his staff kept coming up with modifications and improvements. This was fine so long as the vaccine was being prepared in his own laboratory, but for a field trial as extensive as the one planned it had to be manufactured by commercial companies working from 'protocols' – the pharmaceutical equivalent of recipes, to use the homely language favoured by Salk himself in press interviews – and some of the companies involved were having difficulties inactivating the virus according to Salk's prescription. Salk admitted that he was better at stating the principles from which his vaccine was derived and giving concrete examples of how these principles were applied in his laboratory than at writing the specifications required by both the pharmaceutical companies and the Laboratory of Biologics Control (LBC), the section of the National Institutes of Health responsible for licensing vaccines for commercial distribution.

Things came to a head in March 1954 when five of the first six batches of commercial vaccine checked at the NIH were considered to have caused polio in monkeys and were therefore ruled unfit for human use. Dr William Workman, director of the LBC, called for a postponement of the field trial, precipitating a crisis at the National Foundation – since any postponement at this late stage would have meant putting it off for a year to avoid running into the polio season (and thus introducing a random element through cases occurring naturally). At a meeting in Washington, Rivers got 'madder than hell' and O'Connor had to kick him under the table, then take him aside and send him back to New York, while cooler counsels prevailed. David Bodian, whose unrivalled expertise in the pathology of polio was widely recognised, saved the day by declaring that, whatever lesions the monkeys had, they were not caused by polio. The field trial went ahead.[86]

Although O'Connor was reluctant to relinquish one iota of control in what promised to be his hour of glory, he was too shrewd not to realise that the Public Health Service had to be actively involved in an experiment on this scale; if things went wrong and it came to a question of blame, he did not want the finger pointed at him. For its part, the PHS had been happy to let the National Foundation foot the bill for polio research since that enabled the federal funding agency, the National Microbiological Institute of the NIH, to direct its resources into other important fields not so well covered. But as one commentator puts it, 'Once the National Foundation moved from giving out money to giving out medicine... everything changed.'[87] The free distribution of gamma globulin in the summer of 1953 led to interventions by the authorities, who were alarmed by the spectre of socialised medicine that it raised. By getting them on side in advance of the vaccine field trial, O'Connor hoped to avoid a repeat performance in 1954.

The sociologist Melvin Glasser, who had worked for O'Connor at the Red Cross (which O'Connor had left in 1947) and got on well with him, was brought into the foundation to plan the future of the organisation once polio was conquered – such was O'Connor's confidence in Salk. But before he could begin to think about long-term prospects Glasser was thrown into the maelstrom and made administrative director of the field trial,

> and a few hours later some guy came and asked me how many monkeys I wanted per month and whether I wanted rhesus or cynomolgus. I told him I didn't know I wanted any monkeys and didn't know the difference between a rhesus and a cynomolgus. But I learned. I had to, and not just about monkeys. The basic problem, of course, was to get three shots of either trial vaccine or placebo into upwards of a million children in communities representing a cross section of the country. Also, blood samples had to be taken from about one of every fifty. So we had to begin educating the half million or million children who were going to be injected, and we had to remember the additional children – perhaps three-quarters of a million – who would have to be observed as *uninjected* controls. Therefore we had to consider not only about 2 million children but more than 3 million parents and had to do whatever was necessary to help them know what the field trial was about, so that the parents would request the children's participation.
>
> We also had to take into account the recruitment, instruction, and education of the professional and nonprofessional volunteers required for the program. Eventually twenty thousand physicians and public-health

> officers, forty thousand registered nurses, fourteen thousand school principals, and 220 thousand nonprofessional volunteers were involved.[88]

When the question of hiring clerks to fill in the myriad forms was raised, O'Connor demurred. 'What makes you think our volunteers can't do it better? You can't hire ten thousand clerks with that kind of ability. But our people will do it for free and do it better.'[89] This was where the phenomenal grassroots organisation of the National Foundation in hundreds of local chapters, all voluntarily run, paid off handsomely. A sociological study published in 1957 invokes Weber's concept of the 'routinization of charisma' to explain how the foundation conformed to the pattern of 'social movements [which] originate as the following of a charismatic leader, and develop in time into large-scale organizations'.[90]

O'Connor's identification with the American people, his constant references to '*your* National Foundation', was by no means a cynical manipulation of a gullible public by a powerful leader; it should not be forgotten that he, too, was a volunteer. However large his expense account, he was not paid a salary – at least not until 1959. So he was not asking anything more of the humblest volunteer in the remotest local chapter than he himself was prepared to give. The organisation might be run on authoritarian lines – with O'Connor reserving the right to make all policy decisions – but the importance of the grassroots contribution was never underestimated, and local initiative was to a certain extent encouraged. For instance, the Mothers' March on Polio – a brilliantly effective technique for drumming up support – was dreamed up by volunteers in Wausau, Wisconsin, in 1951.

In the Cold-War, conformist America of the Fifties this quasi-military approach to fund-raising was no deterrent to participation. A majority of volunteers were established members of their communities in the sense that they had lived in them for many years; they were overwhelmingly middle-class, but not upper-middle- or upper-class.[91] These were the people O'Connor knew he could rely on to help out during the field trial, and towards whom he felt a responsibility commensurate with his power – as Melvin Glasser recognised:

> I'll never forget a meeting in the fall of '53, when we were having problems about vaccine procurement and the design of the field trial and whether the Foundation should have any voice in the evaluation of results and whether the vaccine would cause kidney damage. O'Connor sat there writing figures on a yellow pad and listening to the scientists go around in circles. Suddenly he said, 'I have just figured out that during this coming summer thirty or forty thousand children will get polio. About fifteen thousand of them will be paralyzed and more than a thousand will die. If we have the capacity to prevent this, we have a social responsibility here that none of you have been talking about. Let me remind you that we are supported by

> the people and it is our duty to save lives, no matter how many difficulties may be involved.' This stopped the debate in its tracks. It was one of the most impressive things I've seen O'Connor do. It dramatized the strength of a people's voluntary agency.[92]

Once the proponents of a live vaccine, who had been in the habit of speaking their minds at meetings of the National Foundation's immunisation committee, realised that they had been outmanoeuvred, if not outwitted, and that – in John Paul's words – 'the Vaccine Advisory Committee had as its avowed purpose the expediting of an *inactivated* poliovirus vaccine, which up to this time had been on Jonas Salk's drawing board', the sniping that had previously been confined to scientific gatherings came out into the open. 'There seemed to be little use to stand on ceremony from then on.'[93]

The crucial meeting, which was intended to 'acquaint members of the Immunization Committee with the present status of the program for active immunization and to invite their collaboration in this very important undertaking', took place in Detroit on 23 October 1953. Though Thomas Francis had yet to be approached, the field trial was much further advanced than many of the committee had realised. This meeting, 'more than any other single factor', according to John Paul, who was there, 'was responsible for the belief and the accusation that the foundation had been secretive about its plans and had withheld information from its own scientific advisers'. If minutes were taken, they have not survived (which may in itself be significant). John Paul remembers:

> At the beginning of the meeting Albert Sabin and Joseph Melnick [of Yale] both presented individual work on attenuated strains of poliovirus with the idea that this approach might also be considered for developing a vaccine of an entirely different and more lasting nature . . . But when the question was asked whether a choice was going to be available as to the kind of immunization or the approach to be used against poliomyelitis, Mr O'Connor remarked sharply: 'That's not the function of this Committee,' a statement that was taken up and enlarged upon by Hart Van Riper.[94]

At this point it became obvious that 'the foundation had already thrown in its lot with the inactivated vaccine and the newly appointed Vaccine Advisory Committee'. Subsequent Cassandran discussion centred on the dangers of the virulent Type I Mahoney strain of virus Salk was using in his vaccine, and there are divergent views as to whether Salk made an undertaking to replace that strain. Paul concludes that 'it might have been better for the Immunization Committee to resign . . . But such an action would have been ascribed to pique, and this was not what the cause of poliomyelitis deserved.'[95] It would also have been (though Paul does not say this) tantamount to cutting off its nose to spite its face. Some committee members had already been looking for alternative sources of funding, such as the government's National Institutes of Health, but without success. It looked as if the scientists were stuck with the foundation, and the foundation was stuck with them.

For Salk it was an agonising, as well as an enviable, situation. The evening before the meeting he had visited John Paul in his hotel room and revealed his anxiety as to 'how certain members would react'.[96] This was the moment of truth in his relationship with his peers. If the foundation had 'thrown in its lot' with him, he had just as decisively thrown in his lot with it; his association with O'Connor had become much more intimate following the departure of Weaver, who had been the go-between (and might have prevented the *impasse* caused by O'Connor's jarring intervention at the meeting), and he was constantly calling him or calling upon him. As far as his fellow scientists were concerned, he had finally put himself beyond the pale; he was now O'Connor's creature – something very clearly revealed in the group photography commemorating the dedication of the Georgia Warm Springs Polio Hall of Fame in 1958, in which all the (living) scientists line up but make it plain with their body language that they are in two distinct groups: O'Connor and Salk, with Eleanor Roosevelt between them, form one, and the rest another.

Albert Sabin was, as ever, Salk's most redoubtable adversary. In June 1953, Sabin had used the occasion of the annual meeting of the American Medical Association, held in New York, to get his message across:

> Since there is an impression that a practicable vaccine for poliomyelitis is either at hand or immediately round the corner, it may be best to start this discussion with the statement that such a vaccine is not now at hand and that one can only guess as to what is around the corner . . .[97]

Sabin's dual purpose was to denigrate Salk's work and to demonstrate the infinite superiority of live vaccine:

> The objective is clear: to imitate what nature does to 99 to 99.9 percent of the population but without incurring the one in a hundred to one in a thousand risk of paralysis which in many parts of the world is the price for acquiring immunity to poliomyelitis . . . Unquestionably the ultimate goal for the prevention of poliomyelitis is immunization with 'living' avirulent virus which will confer immunity for many years or for life.[98]

Five or six months later, when leading virologists were canvassed, in what Salk regarded as a 'plot by an antivaccination clique', as to their willingness to let their children be vaccinated with Salk vaccine, Sabin not only endorsed the proposition 'that the formalin-inactivated vaccine is insufficiently tested for mass trial, potentially unsafe, of undetermined potency, and of undetermined stability', but compounded the offence by sending a copy of his letter to Salk with a note, thoroughly rubbing his nose in it: 'Dear Jonas, This is for your information – so that you'll know what I'm saying behind your back. This incidentally is also the opinion of many others whose judgment you respect. "Love and Kisses" are being saved up. Albert.'[99] Most of the scientific advisers to the National Foundation were docile enough, or too well bred, to make a fuss when they felt shabbily treated, but Sabin became a thorn in the foundation's side in much the same way as Sister Kenny had been a decade earlier. Both were grantees with powerful lobbies and strongly held convictions, and neither was prepared to acknowledge the foundation as the ultimate arbiter on matters pertaining to polio.

In both cases the foundation exercised restraint in responding to provocation. It could not, of course, openly attack Sabin or come to Salk's defence without compromising its neutrality as provider of research grants. But while maintaining the illusion of Olympian detachment it could always fight dirty, as it had when it excluded Sister Kenny from the first of its international conferences (she retaliated by attending as a journalist, but was prevented from participating even when her own treatment was under discussion). Albert Sabin could not be silenced thus, but the impact of what he was saying might be lessened if a more enticing counter-attraction could be scheduled for the same day – as happened on 11 March 1954, when Sabin was due to speak at an annual gathering of physicians in Michigan. It suddenly became imperative for Jonas Salk to drop everything and go to New Orleans and address the graduate medical assembly there.[100]

In the end, however, Sabin may have done the foundation less harm than good: 'If the field trial of 1954 was superbly designed and executed, its scientific *bona fides* beyond reproach, some credit is due Sabin for nipping at the organization's heels.'[101]

The drip-by-drip effect of Sabin's relentless sniping at the Salk vaccine was boosted at the eleventh hour by a frontal attack from a most unexpected quarter. Walter Winchell – like Earl Wilson – was a syndicated gossip columnist; he also did a weekly fifteen-minute radio programme (sponsored by the manufacturer of a hand lotion), which opened with the famously portentous greeting: 'Good evening, Mr and Mrs America, and all ships at sea . . .'[102] In Roosevelt's time, Winchell had been an influential columnist, but his influence was waning though his Sunday broadcasts were still very popular. In the early days he had been a champion of the National Foundation, and it still regarded him as an ally.[103] So his opening announcement, on 4 April 1954, 'In a few moments I will report on a new polio vaccine – it may be a killer!' was a blow to O'Connor and his henchmen, especially as Winchell had not seen fit to consult them before sounding the alarm.

> Attention all doctors and families: The National Foundation for Infantile Paralysis plans to inoculate one million children . . . with a new vaccine this month . . . The US Public Health Service tested ten batches of this new vaccine . . . They found (I am told) that seven of the ten contained live (not dead) polio virus . . . that it killed several monkeys. The name of the vaccine is the Salk vaccine, named for Dr Jonas Salk of the University of Pittsburgh.[104]

The National Foundation was swift with denials: any vaccine with live polio virus was not Dr Salk's; there were rigorous testing requirements; all batches of commercially produced vaccine found to contain live virus were discarded. Even Sabin called the broadcast 'irresponsible', though he went on to say that the field trial was premature.[105] Most editorials in the press supported the foundation by condemning 'the hysterics of Walter Winchell', who, according to the *New York World-Telegram and the Sun*, 'could do with an injection of truth serum, with or without live virus'.[106] The *Washington Post* was kinder but no less critical: 'We hope that Mr Winchell, whose good will has been made evident enough in the contributions he has made to the fight against cancer, will use his voice now to allay the anxieties which his own mistaken zeal has aroused.'[107] Far from following this advice, Winchell returned to the attack on 25 April:

> Now this is something serious and they can all pan me again but I've to report the news. Attention, parents of little children. The editorial pages from coast to coast, from border to border, recently scolded and spanked me and kicked me all over the lot for talking about the polio vaccine test because I warned you, the parents, that the new vaccine may be a killer. That still goes. It may be a killer . . .[108]

This time he identified his source, who turned out to be another old friend-turned-enemy of the foundation, who had good reason not to have forgotten Kolmer and Brodie:

> Holland, Michigan: Paul de Kruif sends me this urgent telegram – came a few minutes ago. De Kruif is a roving editor for *Reader's Digest* magazine. He is also a consultant to the Chicago Health Department. Here is what he sent me. Quotes. 'Walter, I have a letter dated April 22 from an unimpeachable scientific source. The quotes are. "Six lots of Salk vaccine, all supposed to be free of living polio virus on evidence of tissue culture tests, were then tested on monkeys, 54 monkeys to each lot. One or more monkeys in each group of 54 monkeys showed definite polio infection after being vaccinated. It is therefore believed there was living virus in each lot."' That's what I tried to tell you weeks ago.[109]

There were some withdrawals from the field trial as a result of these broadcasts, but they did not seriously affect its outcome. Indeed, the Michigan State Medical Society, which began by endorsing Winchell's scaremongering, reversed its decision and allowed the twelve counties in its jurisdiction which were involved to go ahead with the inoculations.[110]

There was another last-minute panic when the biologists at the NIH discovered that some mice which had been injected with commercial vaccine had developed paralysis. Once again David Bodian was called in. O'Connor recalled vividly that 'when asked what he thought of *that*, Dave studied a while and said, "I don't know." If he had stopped there, we'd have been ruined. But he waited a moment. Then he said, "But I know it's not polio." It was what they call Theiler's disease, a mouse paralysis unconnected with human polio.'[111]

Of all the scientists caught up in this drama, David Bodian emerges with most credit; his fairness and open-mindedness, his refusal to let his scientific judgment be swayed by issues of personality, were exemplary and, it has to be said, just about unique. One other person whose reputation was enhanced by his scrupulousness and – as it were – coolness under fire was the man of the moment, upon whose shoulders the responsibility for the field trial, and therefore the success or failure of the vaccine, rested: Dr Thomas Francis Jr.

Since the early crusading days of the Georgia Warm Springs Foundation, before the NFIP came into existence, first Roosevelt and then O'Connor had looked upon the conquest of polio as an adventure; and the field trial was, in a sense, its culmination. Those who took part in it and stayed the course of all three injections were rewarded with 'Polio Pioneer' buttons, a publicity gimmick, of course, but one with symbolic appeal to the children themselves: 'They were pioneers. Pioneers are special people who go first. They were very lucky to be able to get vaccine.'[112]

The field trial began on 26 April and a six-year-old by the name of Randy Kerr of McLean, Virginia, was officially designated the first child to be vaccinated. (More than quarter of a century later, Kerr, now a husky thirty-something wearing a check jacket and a tie that looks as though it is throttling him, and holding a gigantic replica of a Polio Pioneer button, was photographed standing beside a benignly smiling Jonas Salk, whose baldness on top is offset by the long grey locks curling over his collar.) Jane Smith recalls her experience as a first-grader in New York City's Public School 61, and 'the exciting sense that we were doing something brave and important'; she remembers what she was wearing and how, after the doctor had stuck a needle in her arm, 'my friend Nikki's mother gave me a lollipop, and I returned to my classroom in time for the afternoon snack of milk and cookies'. She had to go back for two more injections – only to discover that it had all been in vain: she had been given placebo and would have to go through the whole procedure again.[113]

These were the children of the postwar baby boom, reared according to the precepts of another famous Pittsburgh doctor whose name began with an 'S' ('Historians of the Baby Boom', Smith writes, 'humorously cite the three great doctors of that generation – Dr Salk, Dr Spock, and Dr Seuss . . .'[114]) whose *Baby and Child Care* gave comfort and reassurance to innumerable mums and dads on both sides of the Atlantic.

> Born in peacetime and growing up in an era of rising prosperity, these first beneficiaries of new schools, new subdivisions, new highways, and new shopping centers were now the first to receive a new medicine, the first to be protected against a nightmare that had haunted the past two generations. No effort, it seemed, was too great to improve the lives of these special children.[115]

No expense either. In the summer of 1954, when Salk, the members of the vaccine advisory committee and the staff of the National Foundation could do little more than hold their breath and hope or pray that no disaster would overtake the largest mass experiment in the history of medicine, O'Connor took an amazing gamble with money he did not have but was confident the could raise, and ordered twenty-seven million doses of vaccine at a cost of $9,000,000 so that the pharmaceutical companies would continue to manufacture it pending the outcome of the field trial. If the vaccine failed to obtain a licence, it would be nine million down the drain; but if, on the other hand, the field trial was a success and no vaccine was available, there would be hell to pay. O'Connor was putting his money – or the American people's money – where his mouth was.

'We have every reason to hope and believe that the vaccine will be effective,' he said, and he proposed to distribute the vaccine he had ordered, free, during 1955, with the children injected with placebo during the field trial having first priority.[116] According to John Rowan Wilson, 'O'Connor, to whom action and the taking of decisions were the breath of life, hardly hesitated.' But Wilson questions whether it was the right decision:

> This is not to argue that the vaccine should not have been bought, but that it should have been bought by the US Government [as a smaller amount was by the government in Canada, which also participated in the field trial], to whom nine million dollars is a relatively trivial sum. If this had been the case, the episode would have attracted little attention and the gamble would have been placed in its proper perspective, viewed, not in terms of dimes from the sticky palms of infants, but [in terms] of the testing of guided ballistic missiles at several million dollars a throw.[117]

The Third International Poliomyelitis Conference sponsored by the National Foundation was held in Rome in September 1954, six months or more before the results of the trial of the Salk vaccine were to be announced. All the big guns were there – firing from both barrels. In the words of one commentator, 'every panel discussion was a skirmish, every meal a reconnaissance'.[118] Among the Americans, only Tom Rivers, Thomas Francis and David Bodian stood above the fray; the remainder were *parti pris*. At one heated session chaired by the admirable Francis, Koprowski put forward his side of the argument:

> I would like, Mr Chairman, to make one closing remark – that we are living in the era of live virus vaccine. What we want is to elicit as nearly as possible all the latent capacities of human talent to apply principles, established by Jenner, Pasteur and Theiler, to the field of poliomyelitis.

Francis could not let this pass unchallenged:

> I think it is fine that you called forth the principles, but I think that we can also look forward in immunology to the idea that perhaps we can work through the lines of Avery, Heidelberger and others in terms of purified antigens. At least, it is a point of view for discussion.[119]

But they did not discuss it; Francis gently but firmly steered the meeting back to practical considerations.

By the mid-Sixties – when live-virus vaccine had become the vaccine of choice both in Britain and in the United States, where it had been endorsed even by the American Medical Association – Koprowski had completely changed his mind:

> This is not the era of live-virus vaccine [he told Richard Carter]. Now that it is becoming possible to isolate and purify the antigenic component of a virus, infectious vaccines are going the way of the dodo. To the extent that Salk realized this and attempted with his inactivation process to implement it, he was right and the rest of us were wrong.[120]

A remarkable admission and certainly not one that he or anyone else on his side of the dispute would have dreamed of making in Rome, where the voices of reason were drowned in a sea of scientific dogma.

The manner in which the results of the Salk vaccine field trial were presented to the world has been the subject of considerable criticism, if not outright condemnation. Thomas Francis, who succeeded in fending off all attempts to prise information out of him in advance of the official announcement of his report, was unable to prevent that occasion turning into a jamboree. Instead of quietly publishing the results in the *Journal of the American Medical Association*, he allowed the drug company Eli Lilly, one of the main manufacturers of the vaccine, to broadcast a special meeting held in the University of Michigan to 54,000 doctors in sixty-one cities by closed-circuit television, at a cost of a quarter of a million dollars.

But it was part of O'Connor's tacit agreement with the American citizenry that the National Foundation never passed up an opportunity to publicise its achievements – why shouldn't it give the people who provided the money the chance to celebrate the results? And the American appetite for this particular success had been stimulated by so many years of expectation that it required a big event to satisfy it.

Serendipity came into it, too. When the Ann Arbor announcement was finally scheduled for 12 April 1955, the last of three or four possible dates, Basil O'Connor asked a staff meeting at the foundation, 'Do any of you know the significance of that day?' Everyone looked blank, and O'Connor enlightened them: 'It will be the tenth anniversary of Roosevelt's death.' As far as he was concerned that was a mixed blessing. 'I had no more to do with holding that meeting on Roosevelt's anniversary than you did, ' he told Richard Carter. 'But nobody believes me.'[121] Yet criticism of the choice of date 'as "cheap exploitation"' – Greer Williams writes – 'seemed petty to millions of Americans who remembered Roosevelt as a friend of the common people. There was no cause, they felt, for apology on that score'.[122]

By the time Francis announced the results of the field trial the joyful news of success had been broadcast by NBC, breaking the 10.20 a.m. embargo on it, and church bells were already pealing in some American towns. The unseemly scramble by reporters from all over the world for press releases, abstracts of the report and the report itself so shocked the unsuspecting messengers that they 'backed off, pitching packets into the crowd like oceanarium keepers throwing fish to leaping porpoises'.[123] The message was that the Salk vaccine was 'safe, effective and potent'. That was all anybody wanted to hear; the qualifications in the small print were neither here nor there – except to the embattled Salk and his opponents. Celebrated already, Salk was now sanctified, yet on this day of days he managed to alienate what friends he still had in the scientific community and puzzle many others. 'I was not unscathed by Ann Arbor,' he said in retrospect.[124]

What upset Salk was that the results showed his vaccine to be least effective – only sixty to seventy per cent – against Type I, by far the commonest polio virus. He knew the reason for this and waited for Francis to give it, but that was not part of Francis's brief. The trouble was that the antiseptic merthiolate, which had been added to the vaccine to prevent contaminating bacteria on the instructions of the Laboratory of Biologics Control and against Salk's wishes, had seriously reduced the potency of certain batches. So when Salk's turn came to speak, he – in the view of his critics, particularly Tom Rivers, who never forgave him – upstaged the man whose day it should have been by saying how his new improved (merthiolate-free) vaccine might well be one hundred per cent effective, especially if the third dose was delayed six months, as he now recommended.

As if it were not enough to have offended Francis, by casting doubt on the significance of all his hard work before it could even be appreciated, Salk compounded his error by failing to mention his laboratory staff when he gave thanks to all and sundry for making the vaccine possible. The fact that he signed his paper, 'From the Virus Research Laboratory, School of Medicine, University of Pittsburgh', rather than with his own name, and paid generalised tribute to their work in a preface, might explain – though it hardly excused – this lack of public acknowledgement of their important contributions to 'his' vaccine.[125] Salk's assistants were not going to be interviewed by Ed Murrow on NBC's path-breaking live broadcast from Ann Arbor of *See It Now*; *they* were not going to be taken aside by an interviewer made famous by his wartime despatches from blitzed London and told that a great tragedy had now befallen them – they had just lost their anonymity. On the contrary, they might have welcomed a moment's elevation from their customary anonymity.

Yet Salk and O'Connor carried all before them. Even the licensing of the vaccine for commercial use by the Laboratory of Biologics Control was a foregone conclusion. The Secretary for Health, Education and Welfare in the Eisenhower administration, Mrs Oveta Culp Hobby, and her public health adviser, Surgeon-General Leonard A. Scheele, were waiting in the wings, or rather in Washington, to go through a special licence-issuing ceremony as soon as a panel of experts on the spot – including Salk, Sabin, Francis and Bodian (but not Enders, who declined the invitation to Ann Arbor) – had given the go-ahead. John Rowan Wilson is incredulous: 'It was plain to everybody, from the arrangements that had been made, that the Government had made up its mind beforehand, and was simply hoping to use the panel of experts as a rubber stamp.'[126] Some of the experts resisted this interpretation of their role, recalling the difficulties the drug companies had experienced in ensuring that every batch of vaccine was free of live virus and demanding more time for discussion. Sabin restated his objections to the dangerous Mahoney strain of Type I virus and argued that it should be replaced with a less virulent alternative before the vaccine was licensed for commercial distribution. But in the pressured atmosphere of Ann Arbor, crucial issues were either glossed over or overlooked. The most important of these was that the licensed vaccine would not be identical with the one that had been field-tested; it might be more effective, but they only had Salk's word for it.

Other problems were logistical: the Laboratory of Biologics Control was neither financed nor staffed on a scale that enabled it to test a vaccine once it had passed the experimental stage and gone into mass production; it relied on protocols and samples from those batches which had met the manufacturer's safety standards; yet the pharmaceutical companies were not obliged to say how many batches had failed to meet these standards, and would obviously be reluctant to do so for fear of being considered incompetent. With the benefit of hindsight, Greer Williams reveals the logical absurdity of the situation:

> The assumption that the 1955 vaccine would be entirely safe because the 1954 vaccine had been safe would have been justified only if the same safety-testing procedures had been observed. Actually, the conclusion on April 12 was that, because great care had resulted in great success, it no longer was necessary to be so careful.[127]

But the church bells continued to ring out in celebration of the 'conquest' of polio; the experts duly agreed that the vaccine deserved to be licensed; and Mrs Hobby, a little later than planned, signed the official documents, saying what a great and wonderful and history-making day it was for the whole world.

As for Salk, he had certainly lost his anonymity. For the next two weeks his feet scarcely touched the ground.

> Jonas Salk now belonged to the public, to humanity, more than to science. We loved this young man in white – a hero of test-tube magic, a saviour of little children, yet a modest person who asked nothing but to be left alone with his cherished research.[128]

Everyone wanted to shake or kiss his hand or touch the hem of his garment, write him letters of gratitude, shower him with cash and presents. Film stars wanted to visit him, film companies to immortalise him in Hollywood fashion; organisations voted him money, medals and honorary degrees; and finally Eisenhower himself summoned him to the White House to receive a presidential citation.

> The 1950s were bland times; the level of public excitement was geared to news of the 'cold war' or the convictions of such figures as Alger Hiss and the Rosenbergs, with the sordid hysteria of the McCarthy hearings as an underlying theme. People were weary of looking in closets for lurking Communists. They wanted a hero, and now they had one in Salk.[129]

The scientific community was less welcoming and less open-handed with honours: the coveted Nobel Prize went to Gentleman John Enders and his colleagues; and the members of the prestigious National Academy of Sciences repeatedly blocked Salk's election. If, as one member claimed, election to the academy had become 'a popularity contest', then Salk had no hope.[130] O'Connor called it a scandal – 'He shows the world how to eliminate paralytic polio and you'd think he had halitosis or had committed a felony.'[131] But the O'Connor connection was part of the problem. Other scientists took money from the foundation, but they used long spoons when they supped with the devil; Salk, in their view, had sold his soul, if not to the devil, at least to Mammon; in the words of an editorial in the *New England Journal of Medicine*: 'In the final analysis it is physicians who must assume some of the responsibility for allowing themselves to be drafted by methods of the modern impresario into a scientific version of grand opera . . .'[132] Individual scientists like Rivers might argue that Salk's exclusion from the club was a matter of scientific judgment, that there was nothing 'original' in his contribution, that he was no more than a glorified technician or 'biological engineer', as he was called in the *New York Herald Tribune*, but

> the real reasons for refusal were the press conferences, radio and TV appearances, the *Life* magazine spreads, the mad scenes at Ann Arbor, Salk's alleged slurs at Francis, the outpouring of public adoration, and all the other violations of the scientists' unofficial code of behaviour.[133]

As surely as night follows day, Nemesis lurks in the shadow of hubris. Scarcely had the church bells ceased to ring when worrying rumours of cases of polio unrelated to any natural outbreak began to circulate. Reports came from as far apart as California and Chicago, but a common thread was soon established: all the affected people were either children who had received doses of vaccine manufactured by the Cutter Laboratories of Berkeley, California, or members of their families or communities. For those who had been around in the Thirties this began to look like a sickening replica of the Kolmer/Brodie fiasco; for Walter Winchell it was the fulfilment of his prophecy, and he threatened further exposés.

As the cases mounted – there were over 200 in all, three quarters of them paralytic, with eleven fatalities – so did the scapegoating. No one accepted responsibility for what became known as 'the Cutter incident'; everyone blamed someone else. O'Connor blamed the government, saying that there had been no incidents of this sort during the field trial, while the National Foundation was in control; the Public Health Service denied that its responsibility extended beyond licensing the product and accused O'Connor and Salk of being in too much of a hurry to introduce the vaccine; the American Medical Association proclaimed its innocence of any part in the whole sorry saga; and Salk said that the manufacturers could not have followed his formula correctly, though they maintained that was exactly what they had done – followed to the letter the protocols approved by the Laboratory of Biologics Control.

But the buck had to stop somewhere, and it came to rest in front of the Surgeon-General – 'the calm and smiling Scheele . . . a man of colossal restraint and shock-proof benignity'.[134] His words, which could serve as a textbook illustration of Newspeak and doublethink, were belied by his swift action in requesting (which was all he had the authority to do) the withdrawal of the Cutter vaccine. It was, according to one account, 'one of the fastest governmental decisions in the history of the United States. It took almost three times as long to declare war on Japan following the attack on Pearl Harbor.'[135] What Scheele was not prepared to do, could not afford to do – though O'Connor in particular was constantly goading him to do it – was take the public into his confidence. Hence the bewildering sequence of contradictory statements that emanated from him over the safety of the vaccine.

The Public Health Service now found itself in an identical situation to the National Foundation prior to the setting-up of the vaccine advisory committee: it was a prey to conflicting advice from scientists on opposite sides of the dispute. At one particularly bitter meeting in May, John Enders leaned across the table and told Salk, 'It is quack medicine to pretend that this is a killed vaccine when you know it has live virus in it. Every batch has live virus in it.' Salk was mortified. 'This was the first and only time in my life', he recalled, 'that I felt suicidal.'[136] Under pressure from the live-vaccine lobby, Scheele suspended the entire vaccination programme pending investigation, particularly into the use of the Mahoney strain of virus, against which Sabin and Enders had repeatedly warned.

O'Connor, who was tremendous in adversity, fought back with every means at his disposal: rational argument, ridicule, threats. 'So long as the Salk vaccine and its research were in the hands of the National Foundation', he told a luncheon audience composed largely of science writers and publishers, 'you had some intelligence, intellectual integrity, and total courage and you had no politics whatsoever.' And he insisted that 'nothing that has been said affects the safety of the vaccine'.[137]

O'Connor may have won the propaganda battle:

> In general, the story the press told and the public believed was that Salk was a hero and Sabin a villain – that the Salk vaccine would conquer polio, and it was too bad the Cutter production mishap had to happen – that if the Public Health Service had lived up to its responsibilities, the Foundation's programme would have been a complete success.[138]

But though immunisation was resumed after new safety tests had been agreed by a committee predominantly composed of members of the NFIP's vaccine advisory committee, the damage was done, the momentum lost and the take-up of vaccine significantly reduced.

Scheele's report on the Cutter incident was nothing if not political. The critical issue of Cutter Laboratories' responsibility for passing as fit for use substandard vaccine was fudged (though this did not save the company from having to pay compensation as a result of legal actions brought against it) because Cutter was by no means the only commercial producer to have had difficulty in making the vaccine safe. Salk's 'margin of safety', which had already been questioned by the Nobel-prizewinning chemist

Wendell Stanley during congressional hearings in June, was the subject of further critical comment: 'The total experience of the manufacturers now reveals that the process of inactivation did not always follow the predicted course.'[139] O'Connor and the National Foundation, as well as Salk, were held responsible for pushing the vaccine from experimentation into large-scale production with 'unprecedented rapidity', if not unseemly haste.

That much was indisputable. But the most telling of Scheele's shafts was directed at the government itself:

> The records which manufacturers were required to submit did not include certain data which are essential for an adequate assessment of consistency in performance. The protocols submitted related only to lots of vaccine proposed for clearance, and gave no information concerning lots discarded in the course of manufacture.[140]

Scheele recommended reorganisation of the Laboratory of Biologics Control; it was renamed the Division of Biologics Standards, a new head was appointed and its staff increased more than fourfold. This belated recognition that 'the increased tempo of all medical research created problems in biologics control amenable to solution only with the accumulation of knowledge and experience' amounted to an admission that the government had been remiss in not making it its business to acquire such essential knowledge and experience.[141] Whatever the ostensible reason for the resignation of the Secretary for Health, Education and Welfare, Mrs Hobby – in time-honoured fashion she said it was in order to spend more time with her family, in this case an ailing husband – a sense of dereliction of duty in allowing Basil O'Connor to shoulder what should have been governmental responsibilities played a part. During the field trial O'Connor had visited Washington with the express purpose of alerting her to the need for a federal policy on vaccine allocation.

> He told her it was his personal judgment that the vaccine was going to work, and that if it worked the question of distributing a limited supply of vaccine to a clamoring world would be a major public health problem. Furthermore, he continued firmly, it was a problem the National Foundation couldn't and shouldn't handle, because public health was her responsibility, not theirs. Hobby's reply, in essence, was that the marketplace would control supply and demand, and that her responsibilities ended with licensing. As O'Connor was wont to say, he had gotten exactly nowhere.[142]

Mrs Hobby's *laissez-faire* attitude was entirely in keeping with the American approach to medical economics. But her response to a Senate committee which questioned her failure to anticipate the need for an adequate supply of vaccine – 'I think no one could have foreseen the public demand' – has gone down in history as a classically inept statement.

Had FDR still been around there can be little question that things would have been handled very differently. If for no other reason – and there were many other reasons – his friendship with Basil O'Connor would have ensured the closest co-operation over a matter so near to both their hearts.

Mrs Hobby was not the only casualty of the fallout following the Cutter incident. Dr W.H. Sebrell Jr stepped down as director of the National Institutes of Health and was replaced by his chief assistant, Dr James A. Shannon, who had been one of the leading critics of Salk's inactivation process and had insisted on further safety measures before he was ready to agree to a resumption of the vaccination programme.

Eleven years later, in 1966, Dr Shannon made a speech in Oklahoma City, in which he addressed the issue of social intervention in medical research. 'NIH is too frequently perceived', he said, 'as a science agency interested in biological and medical problems in themselves, rather than a health agency utilizing science and research for the solution of disease and health problems.' He took as an illustration of the dangers attendant upon intervention 'one of the triumphs of postwar medical research – the development of the polio vaccine' and argued, first, 'that the decision of the Foundation to throw its resources behind the development of an inactivated vaccine markedly increased the difficulties and greatly protracted the time required to develop the generally accepted [live] polio vaccine we have today', and secondly,

> that the development of Salk vaccine had outrun its scientific base, thus putting a substantial number of vaccine candidates at serious risk and necessitating a redesign of production methods and safety testing of Salk vaccine, which in turn required a redefinition of the fundamental concepts upon which both inactivation and testing were based.[143]

There remarks drew a spirited response from the historian of the National Foundation, Saul Benison, in a letter addressed to the Executive Secretary of the National Academy of Sciences' committee on science and public policy. Benison blamed the Cutter incident on the NIH for taking the drug companies at their word over the safety of their product. Various researchers, including Salk himself, 'discovered that when virus fluids were stored that virus particles tended to conglomerate, and that this conglomeration interfered with the process of inactivation'. This had not occurred during the Salk vaccine field trials because the virus fluids were used so rapidly in the manufacture of the vaccine that they were never stored. To cope with this problem of 'clumping', as it came to be known, an extra filtration requirement had to be written into the production process. Benison saw this more as a commentary on a breakdown in the administration and policing of the production process than as a lack of basic scientific knowledge of polio virus: 'To suggest, as Dr Shannon does, that the Cutter incident required a redefinition of the fundamental concepts upon which testing and inactivation were based is simply a misuse of the words fundamental and concept.'[144]

Benison then turned to Shannon's other point, that the 'production of Salk vaccine greatly delayed the development of the Sabin vaccine', and pointed out that it was not until the summer of 1952 that Sabin, 'under the auspices of the National Foundation', began working on a live-virus vaccine, and that when Salk's vaccine was ready to be tested in 1953, Sabin 'was truly at the beginning of his labors'. The foundation, far from standing in Sabin's way, facilitated his progress by organising a meeting of grantees to help with a particular problem he encountered and by supporting him throughout.

> To argue that it was a mistake on the part of The National Foundation to develop Salk vaccine instead of throwing all effort into the development of Sabin vaccine simply overlooks the fact that in 1954 there was no Sabin vaccine. An acceptable Sabin vaccine was not produced before 1961 [when the NIH licensed it]. In those seven years Salk vaccine proved itself to be an effective and safe immunizing agent and markedly reduced the incidence of polio in the US.[145]

(Even such a staunch supporter of live vaccine as John Paul concedes that 'once the painful episode of the Cutter incident had subsided, the triumph of the Salk-type vaccine became even more manifest', achieving a decline in the rate of paralytic poliomyelitis from 13.9 per 100,000 in 1954 to 0.5 in 1961.[146])

O'Connor was so pleased with Benison's letter that he circulated copies of it to the deans of several US medical schools. But it is not the last word on the subject. In promoting scientific research the National Foundation, for the first decade of its existence, was in some respects closer to being a government agency like the NIH, or an independent philanthropic body like the Rockefeller Foundation, than a social movement; it supported any and every line of investigation that its panel of scientific advisers thought potentially productive, however remote from its main objective. But once it got a sniff of a possible vaccine, first Weaver, than O'Connor himself entered the fray and started to influence the outcome directly; and this changed its image irrevocably. Just as Salk was, among scientists, forever tainted by his association with the foundation, so the NFIP never recovered its unchallenged pre-eminence in the field of research once it had thrown in its lot with Salk; other agencies, such as the World Health Organisation, would become increasingly important in polio prevention.

In the late Forties and Fifties there was one disease – cancer – that had an even higher public profile than polio. The Basil O'Connor of the American Cancer Society (ACS) was Mary Lasker, the wife of an advertising tycoon; she too used aggressive fund-raising tactics to accumulate huge amounts of money for the cancer crusade. The difference between the ACS and the National Foundation was that the former was never a significant agency for research; the National Cancer Institute (NCI) was the biggest of all the National Institutes of Health. The problem for the NCI was one that had plagued O'Connor from the founding of the NFIP – how to justify wide-ranging and long-term research when the public was clamouring for instant results. Cancer research failed to come up with any equivalent of the Salk vaccine, but the harmonious way in which the NCI and ACS worked together – with the ACS concentrating on public education and awareness and the NCI on research – could have served as a model for the NFIP if O'Connor had been less autocratic. (Despite Richard Nixon's identification with the 'War on Cancer' for political gain in 1971, results did not justify either the high hopes or the millions of dollars invested in it; and in the changed mood of the late Seventies its fortunes resembled those of another war. An expert commented at the time: 'By comparison with the fight against polio, the war on cancer is a medical Vietnam.'[147])

O'Connor and Salk's mutual esteem and loyalty invite admiration – and they surely did each other more good than harm – but after the glory days of 1954 and 1955 (up to the Cutter incident) neither would ever again reach such heights. O'Connor's wife died in that eventful summer of 1955, and he remarried two years later; his second wife had been a physical therapist at Warm Springs. He outlived both of his daughters, who died in the Sixties, and guided the foundation into the post-polio era. But the tenth anniversary of his old friend and patron's death remained his finest hour.

He continued to give Salk lavish support, helping him realise a dream of building a scientific utopia on the coast of California, the Salk Institute – as O'Connor insisted it had to be called – to which the world's most eminent scientists were invited to think great thoughts. But this must have seemed more of a curse than a blessing to Salk when it turned out that these 'great minds' had 'great egos to match' and looked down on their host as a 'mere technician'.[148]

Caught between science and celebrity, between active research and mystico-philosophical musing, Salk failed to find a marriage of great minds or to sustain his own. He was divorced in 1968 and he got married again in 1970 – to the painter Françoise Gilot, though she was better known as Picasso's former mistress. His institute became a mausoleum in which he was entombed alive; he resigned as Director but kept his suite of offices and has recently been working to develop a vaccine for AIDS.

7
The Quick and the Dead

. . . in the United Kingdom . . . there is no urgent reason why we should change our allegiance from Dr Salk to Dr Sabin, or, putting it another way, why we should rush to make the difficult choice between the quick and the dead! No personal allusion there, of course! . . .

From a speech by Dr W.H. Bradley, of the Ministry of Health, at the Fifth International Poliomyelitis Conference, Copenhagen, 1960

White-haired, dark-eyed, with a sharp nose, small, dark moustache, nubbin chin, and general mien of Reynard the Fox, Sabin is a sociable man and something of a gourmet; he also has a reputation in virological circles for asking the sharpest questions, in unctuous, Oxonian tones.

Greer Williams, *Virus Hunters* (1960)

Nowadays [Sabin's] face is familiar to millions – a spare, tough, sharp-eyed man with a vulpine countenance, a small moustache and a resemblance on occasions to Groucho Marx. Like Groucho, he is intelligent, serious and extremely businesslike.

John Rowan Wilson, *Margin of Safety* (1963)

Albert Sabin could be more than generous on occasions, and equally egotistic and possessive on others. It was in controversial situations with his reputation at stake that he really came into his own . . .

John R. Paul, *A History of Poliomyelitis* (1971)

. . . I am opposed to the idea that – in this atmosphere of a political convention as one of our audience has called it – one must be either pro-Salk vaccine or pro-Sabin vaccine. I belong to a third party . . .

From a speech by Dr Thomas Francis at the Fifth International Poliomyelitis Conference, Copenhagen, 1960

The National Foundation's support of the Salk vaccine, which ensured that it continued to be used in the United States despite the Cutter incident, meant that the protagonists of live-virus vaccine had to try out their wares elsewhere. 'By a reversal of Canning's famous dictum', John Rowan Wilson writes, 'they called in the Old World to redress the balance of the New.'[1] The Old World had watched events surrounding the introduction of the Salk vaccine with a mixture of fascination and scepticism, and different countries reacted in different ways. The Danes wasted no time in manufacturing their own Salk vaccine; the French produced a variation made by Dr Pierre Lépine of the Pasteur Institute; and the British, well, temporised.

Tom Rivers professed himself baffled by the British attitude: 'I hate to blame it on British conservatism – heck, vaccination as a procedure is a British medical innovation.'[2] But there was no other word for Sir Weldon Dalrymple-Champneys's ultra-cautious approach. This man from the Ministry of Health had visited the foundation early in 1955 and had seemed enthusiastic about the vaccine. But that was before the Cutter incident. A year later, the superb-sounding Dalrymple-Champneys replied to a letter from Rivers:

> I am sorry that you found my optimism with regard to vaccination against poliomyelitis unduly restrained, but I believe that it is useful to err, if anything, on this side when trying to make people understand the problems involved in a matter of this sort. This does not mean that I do not regard the great trial of the Salk vaccine as a very remarkable achievement, which is bound to have a profound influence on the history of poliomyelitis control, even if the vaccine eventually used may be different in type.
>
> In this regard, I am very interested to read your opinion of the probable attitude of your people to a live virus vaccine . . .[3]

John Rowan Wilson, a qualified surgeon whose job as Medical Director of the UK subsidiary of the American pharmaceutical company, Lederle Laboratories, involved him in the organisation of live poliovirus vaccine trials, is less reluctant than Rivers to castigate British conservatism:

> In the very nature of things, the Ministry advisers, members of an ultra-conservative caste in an ultra-conservative country, could not have been otherwise than shaken by the way in which the Foundation had handled its public relations. To them, planned miracles and real-life Dr Kildares meant less than nothing. They respected Salk and Francis, it was true, but they respected Sabin equally, and Enders possibly more than either.[4]

The British version of the inactivated vaccine, tried out in 1956, substituted Enders' modification of Brunhilde Type I poliovirus (known as

Brunenders Type I) for the more virulent Mahoney strain. In general, the 'stop-go progress' of the Salk vaccine 'repeated itself on a minor scale in the United Kingdom'.[5]

Everything to do with polio in Britain – not least the disease itself – was on a minor scale. Outside Scandinavia, no one in Europe had been unduly concerned about polio until after the Second World War. An American medical team sent to war-ravaged Berlin in 1947 to monitor an epidemic there and advise on the care of the acutely ill was appalled by both the lack of facilities and the postwar fatalism of the German people – 'their feeling that it was useless to exert a lot of effort to save just one life'.[6] The apathy of the defeated was in marked contrast to the incipient panic among the victors in Britain, which also experienced its first large epidemic in 1947. The *British Medical Journal* printed nervously nannyish editorials – on 26 July:

> A state of panic is rather easily produced by Press publicity, and it is to be hoped that the daily papers in this country will not draw undue attention to the present outbreak. In any year the cases and deaths due to poliomyelitis will be far fewer than the injuries and deaths caused by road accidents . . .[7]

and again on 16 August: 'Throughout history certain diseases have caused widespread panic . . . Today poliomyelitis is arousing widespread anxiety among the public out of proportion to the low morbidity and mortality rates among those infected . . .'[8] Before 1947 the largest number of notified cases of polio in a single year in Britain had been just under 1,500 (that was in 1938). The final tally for 1947 was 7,776.[9] As the Senior Medical Officer at the Ministry of Health, Dr W.H. Bradley, told delegates at the National Foundation's First International Poliomyelitis Conference in New York in 1948,

> During the 1947 epidemic in Great Britain when we had nearly five times as many cases as ever before in our history; when the disease killed four times as many people as diphtheria; and when we suddenly realized that poliomyelitis was our most important problem at the time, we had to do a great deal of letter writing and telephoning to experts all over the world to get advice . . .[10]

Bradley compared the attack rate in England and Wales in 1947–18 per 100,000 – with that of the 1916 epidemic in the United States. He summed up the problem of polio the world over in three words: 'ignorance, impotence and insecurity'.[11] But if the British were more phlegmatic in their response to the 1947 epidemic than New Yorkers had been in

1916, it probably had less to do with their famously stiff upper lips than with the more advanced state of knowledge available to them – thanks to predominantly American research.

Yet polio *was* a frightening disease. Dr Geoffrey Spencer (see Introduction) was a medical student at the beginning of the Fifties and recalls vividly the polio epidemics of that era:

> They were actually terrifying . . . There was an epidemic – it must have been, I suppose, in '51 in London – and several of us were asked if we would help out in some of the infectious diseases hospitals. The one I was drafted to was in Hendon, in North-West London. I was feeling terribly grand, you know, 'I'm really a proper doctor at last' sort of thing, and I was taken into a ward where there were a bunch of iron lungs and various other breathing machines, and hectic activity. It was . . . it was almost a slaughter house really. Because the bad ones came in, and they were bunged into various machines which didn't seem to work particularly well, and the death rate was fairly high . . .
>
> Anyway, after a little while it was quite obvious that a fair number of the staff and the people who were drafted in to help cope with this epidemic were going down with the disease themselves . . . A lot of them made a reasonable recovery, but the terrifying thing about it was this Sword of Damocles hanging over everybody's head, and however fired up one was with the enthusiasm of the medical student to sacrifice oneself heroically on the altar of medical care and all that stuff, there's no doubt it was absolutely bloody terrifying . . .[12]

Although Dr Spencer specialised in anaesthetics, his involvement in the after-care of mainly respiratory polios did not begin until the late Sixties. In retrospect, he sees the Copenhagen epidemic of 1952, in which an exceptionally large number of patients with bulbar involvement (causing breathing and swallowing difficulties), and a chronic shortage of iron lungs, obliged the Danish doctors to seek desperate remedies, as 'the "1066 and All That" of polio'. It revolutionised the treatment of respiratory polio.[13]

The Copenhagen outbreak occurred the year after the Second International Poliomyelitis Conference was held in the city, and some people still wonder whether that was entirely coincidental: might not the physicians attending the conference have been carriers of the disease? Whatever the cause, it was an epidemic of unprecedented severity: the attack rate of 238 per 100,000 was far greater than that of New York 1916 or any subsequent epidemic. Thirty to fifty patients were admitted to the Blegdam Hospital daily, of whom six to twelve were 'drowning in their own secretions'. The hospital had only one tank and six cuirass respirators, and the doctors were 'soon faced with the intolerable dilemma of having to choose which patient to treat in the few respirators at hand and which not to treat... The need for improvisation became imperative.'[14]

Given that twenty-seven of the thirty-one patients treated in respirators during the first three weeks of the epidemic died – nineteen of them within three days of admission – the acquisition of more respirators would hardly have solved the problem; it might simply have meant *more* deaths.

According to Dr H.C.A. Lassen, the hospital's Chief Physician, he and his fellow physicians acted swiftly and decisively, applying the principles of anaesthesia to the problem and inviting anaesthetists to join the staff, 'the first being Dr Bjorn Ibsen'.[15] In reality – the medical historian, Dr G.L. Wackers, claims – Lassen was initially sceptical of the ability of a mere anaesthetist to help. In 1952 anaesthesia was 'an emerging but low status medical discipline'. Most anaesthetists, including Ibsen, were not employed full-time but worked on a free lance basis, their 'range of action and acknowledged competence... restricted to the operating theatre where they delivered anaesthesia to patients' under the surgeon's direction. Dr Lassen 'had to overcome a certain degree of professional pride in seeking "outside" help'; and anaesthetists were just about bottom of his list of potential helpers. It was only when all else failed, and on the advice of another member of the hospital's medical staff, that he turned to Ibsen.

Ibsen suggested replacing the expensive, failing and cumbersome negative-pressure respirators by cheap, manual positive-pressure ventilation. He was not inventing a new method; he was merely adapting the anaesthetic technique already used in the operating theatre to the needs of the polio ward: he would pump air into paralysed lungs through a tube inserted directly into the trachea, bypassing the nose and mouth. He tried it out on a severely paralysed twelve-year-old girl; but to begin with things went badly. The girl seemed to be dying in his hands. The other physicians, who had come to watch the experiment, regarded it as a failure and soon lost interest. But Ibsen persisted. To ease the girl's struggles he gave her a shot of Pentothal, at which her own efforts to breathe ceased. This enabled him to ventilate her fully, without encountering resistance.

> When the other physicians returned... the girl's skin colour had returned to normal – without the supply of extra oxygen. Both body temperature and blood pressure were restored to normal too. This twelve year old girl was the first patient during this polio epidemic who survived as a result of a medical intervention.[16]

After this success – and despite his initial reluctance to seek an anaesthetist's help – Lassen 'energetically devoted himself to the implementation of the new, emergency therapy for all patients with respiratory insufficiency and swallowing impairment'. The difficulty was that the treatment had to be performed manually. Someone had to sit at the patient's bedside twenty-four hours a day squeezing an air-filled bladder; and at the height of the epidemic there were seventy patients requiring constant ventilation. Lassen appealed to medical – and dental – students at the university, who came and worked six-hour shifts in return for 'a moderate financial allowance' from the Copenhagen municipal authority. It was tough on the students: 'They were literally keeping the patient to whom they had been assigned alive with their hands.' For six hours at a stretch they had to sit beside a child (the majority were children) who was unable to speak because of respiratory paralysis and the tracheotomy, and learn to communicate by other means. Many quitted after a few weeks, suffering from emotional as well as physical exhaustion.[17]

Yet it was 'not a sad time', according to Anne Isberg, one of the epidemic children: it was 'like wartime, the spirit of the resistance'; everybody 'was doing their best'. And Anne remembers Lassen with particular affection:

> We were his children and he gave us life. He liked to follow us and kept in close contact. He was a very special person. When he retired, he retired – he left the country and nobody heard from him. He went to France, lived there close to Lyon, but when he got sick he came back to his old hospital and died there. That was it.[18]

Once the success of Ibsen's treatment was demonstrated – and at the end of three months, Dr Spencer points out, the mortality rate had been halved from eighty per cent among the 'lucky patients' in iron lungs to forty per cent among the 'unlucky ones' dependent on the ministrations of the medical students – it did not take long for European companies to design and market *mechanical* positive-pressure respirators. The iron lung became obsolete in Europe, though not in America, where so many had been manufactured and there was some resistance to the surgical aspect of tracheotomy – what Geoffrey Spencer cheerfully calls 'cutting throats'. In Britain 'we were sort of halfway between, because we were still making and developing iron lungs'.[19]

The Danes (largely due to the 1952 epidemic no doubt) were among the first to take advantage of Salk's vaccine, only awaiting the results of the 1954 US field trial before embarking on their own programme of vaccination.

In Britain, these were the early days of the welfare state, as Dr Bradley told his international audience in New York in 1948. He pointed to the 'comprehensive provisions by various Ministries in the Government for the care of paralyzed persons, their education and absorption into industry', while at the same time allowing that there was still a need 'for the more personal interest which . . . a voluntary body could provide'.

> The Infantile Paralysis Fellowship in England, equivalent to the great National Foundation of America, is a very small body indeed. Last year, its income was less than 10,000 dollars, mere chicken feed, and obviously this sum will provide very little in the way of special assistance to the several thousands of cases of poliomyelitis in 1947, quite apart from old cases.[20]

To mention the IPF (now BPF – British Polio Fellowship) in the same breath as the National Foundation was to highlight the astronomical difference in scale between British and American polio initiatives. O'Connor's powerful organisation, founded by a President of the United States, supported by a citizen army of volunteers and funding every significant research effort in polio and related fields, was dedicated to nothing less than the conquest of polio; whereas the IPF had the modest beginnings and limited ambitions of a society for self-help and mutual aid, ahead of its time only in the sense that it was an organisation of the disabled run by the disabled themselves.

Patricia Carey had contracted polio in India – after the First World War – at the age of eight. Her father was a brigadier. Years later in England, in the loneliness and frustration caused by her disability, she decided to try and learn French and German, and her cousin, who was a physiotherapist, put her in contact with another polio, Frederic Morena, an ex-actor now teaching languages for a living. Morena had abandoned law studies to take up acting and had been a member of the Old Vic Company. Polio struck him down when he was forty-two, and he was ever after confined to a wheelchair. When Carey first mentioned her dream of starting an association of polios, Morena was unenthusiastic: Europe was on the brink of war and this was no time for such an enterprise. But the more he thought about it, the more attractive the idea became; and he was instrumental in setting up the inaugural meeting held in a quiet street in London's Bloomsbury on 29 January 1939.

> Some arrived in wheelchairs, others were supported by calipers and walked with crutches or sticks. Some 30 polio disabled persons were meeting in a small art gallery in response to a press announcement inviting those interested in forming an organisation to help solve their problem of social and physical rehabilitation . . .[21]

The IPF did not really get going till after the war, when successive epidemic years underlined the need for such an organisation. By 1956 there were eighty-six branches and groups, a total membership of 16,000, of whom 10,000 were polios. Its aim was largely to improve the quality of life of its members. Welfare officers gave out more advice than money, their role being to ensure that everyone knew how to take advantage of the statutory services and provisions for the disabled. It might be a shoestring operation in comparison with the National Foundation for Infantile

Paralysis, but the British outfit had some advantages – according to a visiting American professor, who was a member of his local chapter of the NFIP in Michigan:

> In America we have been so intent on the purely medical aspects of polio, we have given but little attention to the problem of social rehabilitation. We have, so far as I know, nothing that compares to your IPF – a group of people, having a common handicap, who are banded together with the avowed purpose of doing something for each other . . .[22]

In 1950, the IPF opened its first holiday hotel by the sea, the Lantern Hotel in Worthing, which still provides holidays and residential care for the polio disabled. Five years later it held its first National Swimming Gala in London, and many young polios whose parents enrolled them in the fellowship now only remember the swimming – the late Dr David Widgery, for example, who contracted the disease in 1953 at the age of six:

> Then there was the thing called the Infantile Paralysis Fellowship, which was a group for disabled children to swim, and we used to go to the North Finchley heated bath on Sunday morning and you'd be chucked in the water with people with all sorts of muscular disability. I was much the least disabled of all those people and I learned to swim then, but it was a daunting experience. I won a prize at the IPF's annual swimming contest at the Marshall Street baths, just by Carnaby Street, and I remember getting presented with a *Superman* comic book. My parents were very keen on swimming anyway; they used to make me go to this. But I hated it. I remember the stink of the swimming costumes . . . I thought I'd much rather be reading a book. But it was done because swimming was thought to build up the muscles, that was the main thing . . .[23]

In the mid-Fifties the IPF became involved in the setting up of a residential home for polios who could not remain indefinitely in hospital and were either too disabled to go home or had no home to got to. An unconventional orthopaedic surgeon and disciple of Sister Kenny called Lancelot Walton got together with the President of the Walton-on-Thames Rotary Club, himself a polio and one of Walton's patients, and they found a pleasant house surrounded by woodland in Cobham, Surrey. They bought 'Silverwood' with the support of the Rotarians and the IPF and converted it into a residence for twenty or more people with varying degrees of disability. The ninety-year-old Mr Walton explains: 'I had no interest in somewhere where you'd go and sort of rot for the rest of your life'; this was to be 'a staging post' on the road, if not to recovery, to what would now be called independent living. Silverwood was formally opened by the actress Julie Andrews (Walton's daughter-in-law at the time) in the autumn of 1956.[24] It served its purpose in enabling a number of polios of both sexes, from different backgrounds and with different levels of disability 'to get their bearings in the world before getting integrated again in the community'.[25] (See pages 250–51).

But only a few could go to Silverwood. The IPF *News* pointed out that: 'Many young people, as well as older ones, who have no relatives to look after them, are perforce lodged in homes for the aged and dying, and homes for incurables, simply because they have nowhere else to go.'[26] Janet Keegan was one such young person. At seventeen years of age, she was working at a council-run day nursery in the London borough of Hammersmith, looking after pre-school children. This was in 1956, when children, but not young adults, were being vaccinated. Two weeks before Christmas, Janet fell ill. Can't be polio, said her father, wrong time of year. But it was polio (at the Fourth International Poliomyelitis Conference in 1957, Dr W.H. Bradley reported that in 1956 the seasonal curve in Britain was made unusual 'by the occurrence of 3 or 4 discrete "off-season" outbreaks which deserve special study . . .'[27]). After three months in an iron lung in an isolation hospital, Janet was moved to an orthopaedic hospital, where, despite the usual intensive physiotherapy, she made only limited progress. All four limbs, as well as her breathing, were affected; and though she can use her hands, her shoulders are as useless as her legs. She remained in the hospital two years: 'In my case, they didn't know quite what to do with me because home was out, so I think they were dragging it on with me.' Her mother had died when she was born; her father had remarried when she was eight and her stepmother never wanted anything to do with her, so she had been brought up largely by her paternal grandparents. As there was nowhere for her to go, she was sent to a home in Welwyn Garden City.

> That was another shock. They were old there . . . An awful lot of people seemed to have multiple sclerosis, so of course they were getting worse. A most unsuitable place to put a young nineteen-year-old, as I was then . . .[28]

But she was nothing if not adaptable. She admits to an inclination to let things drift: 'I didn't think much about anything, I wasn't a very forceful person.' And there she might have remained indefinitely if Cupid had not intervened in the shape of a young Irishman called John. John had been persuaded by his sister to give up a perfectly good job in his native land and come to England to be an orderly at the home in Welwyn, where she was working. He came, hated the job, which was not his sort of thing, and left after four months – but not before he had been smitten with love for Janet – to go and work for the Post Office. That Christmas, with his sister's help, he took Janet to Ireland where her mother-in-law (to be) welcomed her into her already enormous family. It was assumed that she and John would get married; and John's mother invited Janet to stay until they could find a place of their own in England. For Janet,

> it was a wonderful sort of intermediate place because the going was easy . . . John's father had died some years before, and mother-in-law had to deal with all these children. So it was quite handy if you could just sort of sit there and be useful . . .

While John looked for somewhere for them to live, Janet listened to his siblings read and helped them with their homework. They did get married and, after living in rooms to begin with, moved into a prefab, where they stayed three years; their son was born while they were living there. After the home in Welwyn even a prefab was luxury – 'your own little place, and quite roomy inside'.[29]

Duncan Guthrie's interest in polio began in 1949, when his elder daughter, who was not yet two, contracted the disease. Guthrie had had 'a good war'; he had served behind enemy lines in France and then in Burma, where he managed to evade capture by the Japanese despite having broken a leg when he was parachuted in (he wrote a book called *Jungle Diary* about his Burma experiences). After the war he worked for the Arts Council on the 1951 Festival of Britain – that 'rainbow', in Michael Frayn's words, which 'marked the ending of the hungry forties, and the beginning of an altogether easier decade'.[30] The festival that threw up such marvels as the Skylon and the Dome of Discovery was, as Frayn describes it in a brilliant essay, the last triumph of the 'Herbivores' before the 'Carnivores' took over.

> Festival Britain was the Britain of the radical middle-classes – the do-gooders; the readers of the *News Chronicle*, the *Guardian*, and the *Observer*; the signers of petitions; the backbone of the BBC. In short, the Herbivores, or gentle ruminants, who look out from the lush pastures which are their natural station in life with eyes full of sorrow for less fortunate creatures, guiltily conscious of their advantages, though not usually ceasing to eat the grass . . .[31]

After the Festival of Britain, Guthrie did not want to go back to the Arts Council because his department was now redundant and he would inevitably have been demoted. So he did a very Herbivorish thing: 'I thought, well, I will set up an organisation to raise money for polio research.'[32] Through his experience with his daughter, he had been shocked to find how little was known about the disease. Initially he worked through the IPF, for whom he had organised an exhibition of humorous art at the big London store, Gorringe's, in June 1951, while he was still Arts Council Representative for the Festival of Britain.

One of the original aims of the IPF was to encourage 'research into the causes, cure and prevention of poliomyelitis', but according to an IPF pamphlet from the early Fifties,

> there appears to be less enthusiasm on the part of medical men for research on Infantile Paralysis than on some other subjects. Certainly the Ministry of Health has expressed its willingness, almost its eagerness, to support more research on poliomyelitis than is at the moment being carried on. It may be that researchers who, quite reasonably, wish to undertake work which offers good prospects of eventually successful results, do not, on the basis of their expert knowledge, see good grounds for hope of success in poliomyelitis research.[33]

It may also be that they thought poliovirus was too dangerous to handle, the risk not worth taking.

In 1952 Guthrie was appointed director of the IPF research campaign and given the go-ahead to establish a fund for research. 'One thing is certain: the Fellowship must be the initiator,' he told readers of the IPF *Bulletin*. 'The Research Fund will be the child of the IPF and the IPF will, before very long, have good reason to be a very proud parent.'[34]

Alas, it did not work out that way; Guthrie found the IPF a 'not very inspiring organisation' and Morena, in particular, a 'strange character'. So he went off and set up an independent Polio Research Fund. (When Morena was ousted from power in the IPF in 1957 in what seems to have been a very British and bloodless coup, there was renewed co-operation between that body and the National Fund, as Guthrie's organisation was now called.)

> I say, with a certain amount of truth, but perhaps embroidering a little, that I started the National Fund with 7s 6d – three half-crowns. I'd always wanted a roll-top desk, and in this one I bought there was an old tobacco tin with three half-crowns in it. So I say, that was my working capital.[35]

In fact, there was also a gift of £1,000 from an eccentric oil man who lived in an enormous mansion in a Miss Havisham-like state of decay and claimed to be the oldest surviving polio. As a layman, Guthrie turned to the professionals for expert advice and initially used the Medical Research Council as his scientific advisory panel. 'A year or two later I was against this, because it gave the impression that we were just paying for what the government should have been paying for and wasn't.' So then he assembled an independent advisory panel – made up of experts on research rather than on polio, to avoid the pitfalls of professional jealousy. His first Chairman was an admiral and he filled the board with 'duchesses and people like that . . . a window-dressing committee of VIPs'.[36]

Like the National Foundation, which now operates under the name of March of Dimes Birth Defects Foundation, Guthrie's National Fund is still in business as Action Research for the Crippled Child. John Rowan Wilson called the Fund 'a very, very pale British counterpart of the Foundation', but it played its part in subsidising important polio research –

none more so than that undertaken by 'a young and energetic Professor of Microbiology named George Dick' at Queen's University, Belfast.[37]

Dick had done research on Yellow Fever in Africa and was very interested in live-virus vaccines. He knew what was happening in America because he had worked at Johns Hopkins with David Bodian, Howard Howe and Isabel Morgan. (They had done the early work on evaluating methods of inactivating virus and established that formalin was the most effective agent.) Dick's interest in polio dated from the war years, which he had spent in Africa in the Royal Army Medical Corps. When there was an epidemic in Mauritius, Major Dick was sent to investigate. The army was concerned about the troops on the island; it was not interested in the local population. But Dick, after he had satisfied himself that army personnel were unaffected, turned his attention to the Mauritians. 'How do you know there is an epidemic?' he asked the medical officer there. 'Well', the man replied, 'I've seen so many funerals going past my office.' But nothing was being done to combat it. On his return to the mainland, Dick planned to make a report; but before he could do so he encountered Dr H.J. Seddon from Oxford, an acknowledged polio expert, who had been sent out by the Colonial Office. Seddon persuaded Dick to return to Mauritius with him and, together with a third doctor who joined them there, they did 'what was probably the best epidemiological study ever done at that time'.[38]

When he was offered the Chair in microbiology at Queen's in Belfast, Dick took the attitude that 'before you could receive any support you had to show that you'd been useful to the community'; and what better way of doing that than by organising a trial of the only existing live poliovirus vaccine at that time – the one developed by Hilary Koprowski of Lederle Laboratories in America? Northern Ireland, with its own parliament at Stormont, was in a unique position to dispense with the red tape governing medical experimentation in the rest of the United Kingdom.

> We couldn't have done *any* of this in England. None. In Northern Ireland you could phone up the Medical Officer, and go and see him. And tell him what you wanted to do. And show him the data. He said, 'You'd better come and say hello to the Prime Minister and tell Brookie what you're going to do.' And I did. I went and saw him, and he said, 'Well, if you think it's safe, that's all right.' There was great *trust* at that time – and great co-operation. The town and the gown were very close together . . .[39]

Koprowski was delighted with the prospect of a trial in Europe and sent over a technician with his attenuated polioviruses of Types I and II (he had no Type III) to help out. Dick pursued his study throughout 1956, beginning by vaccinating his own family and those of colleagues in the university, then a number of children in a home and in the wider community. He gathered his data meticulously, checking the children's stool samples for virus daily and injecting excreted virus into monkeys to test its virulence.

The results, which came as a shock to Dick himself, were disastrous for Koprowski and damaging for all protagonists of live poliovirus vaccine. Dick's report revealed that excreted Type II virus caused severe paralysis in monkeys, while Type I virus had been transmitted to an unvaccinated adult and a child. He concluded: 'we do not consider that [these] vaccines should be used at the moment on a large scale'.[40]

The measured language masks a fierce indignation which is still apparent when Professor Dick addresses the subject: 'I felt incredibly let down by Koprowski; I felt that his data was inaccurate. At that time he was doing a big study in Africa. How could he evaluate it? I mean, what if a few African children became paralysed? I became very anti.'[41] Dick and his colleagues established the criteria they felt were necessary for live-virus vaccines. The pro-live vaccine lobby regarded these requirements as impossibly stringent and Koprowski retaliated in an impassioned peroration to the paper he gave at the Fourth International Poliomyelitis Conference in Geneva in 1957:

> The advocates of 'safety' do not want to pay any price for immunization; yet, exactly what are the costs one might have to pay for a method of immunization which would not only protect the vaccinated subject against the disease but also may lead to elimination of poliomyelitis?
>
> From all the data accumulated to date with recently studied attenuated strains in two or three different laboratories, there is no evidence that the degree of mutation of these polio viruses in the human intestinal tract . . . is such that it will ever lead to the production of a strain which will be highly paralytogenic for man. Experimental evidence supporting such ideas should be furnished by those who propound them, not by us . . .[42]

It was good fighting talk, more rhetorical perhaps than scientific. But there were many who agreed with Koprowski – the level-headed Dr John Paul for one, who grumbles in his book that 'Dr Dick was loud in his denunciation of the live virus principle, and as a result the whole group of attenuated poliovirus strains suffered a brief but undeserved eclipse.'[43] Now in retirement, George Dick is still ready to fly to America when called on by US attorneys to give evidence in cases of vaccine-damaged children.

Koprowski was down but not out. Dick's adverse report on his vaccine strains led to the break-up of his partnership with Herald Cox, who was his senior at Lederle but had been content – or perhaps obliged – to let him go his own way so long as Koprowski was coming up with the goods. According to John Rowan Wilson, who knew all the principals, Cox was 'unquestionably the simplest, least sophisticated character in the polio story'. He lived for his work and his family, and his favourite relaxation was a few weeks' fishing in his home state of Montana. A stocky, bald-headed man, he 'had a curious habit of shaking hands with his friends in the course of conversation, as if in search of reassurance from physical contact'.[44]

Early in 1957 Koprowski left Lederle Laboratories, taking several of the staff with him, to become Director of the Wistar Research Institute in Philadelphia. He had no intention of giving up work on the live vaccine, but both he and Cox had more or less to start again – Cox without his energetic and resourceful partner, and Koprowski without Lederle's (or its parent company, American Cyanamid's) millions. And there was another competitor in the field, who had now caught up with them and was better equipped than either to walk off with the spoils.

The wily Albert Sabin was as conscientious as Cox and as clever as Koprowski, and more ambitious than either; his approach, in Wilson's words, 'deserves study as an example of scientific generalship of the very highest order'.[45] While Cox and Koprowski were marginalised, carrying out trials in South America and the Congo respectively, Sabin found the perfect place to try out his vaccine, a place where there would be maximum impact but no problem in finding 'volunteers' and no press interference of the sort that had plagued Salk – the huge and largely virgin territory of Russia.

'In the Soviet Union', Dr Valentin Soloviev reported to the delegates of the Fourth International Poliomyelitis Conference in Geneva, 'the marked increase of poliomyelitis was first noted in 1955 when the morbidity rate reached a figure never previously observed, namely, 8 cases per 100,000 inhabitants.'[46]

Sabin's Russian origins and his ability to speak the language gave him an entrée into the world of Soviet science; and since President Eisenhower himself had stressed the United States' willingness to make available to all the world the fruits of Salk's research, he could hardly be accused of a lack of patriotism, or of selling secrets to the enemy. Indeed, the Russians were already attempting to manufacture Salk vaccine, but had encountered the familiar problems with the inactivation process and faced even greater ones in 'injecting three shots of polio vaccine into more than 200 million people'.[47] The logistics of distributing Sabin's oral vaccine were far less daunting, and the attraction of stealing a march on their great adversary irresistible.

Nevertheless, they took their time, starting small and then vaccinating ever larger numbers of children as their confidence in Sabin's vaccine grew. They began in 1957, but since they published no reports nothing was known for certain until mid-1959, though rumours abounded that Sabin 'was due to sweep the board . . . mainly on the strength of the Russian work, which, so it was claimed, would surprise everyone with its spectacular success'.[48]

At the International Scientific Congress on Live Virus Vaccines held in Washington in June 1959 under the auspices of both the World Health Organisation and what was to become the Pan-American Health Organisation (and financed, incidentally, by the National Foundation's rival, the Sister Kenny Foundation), Sabin announced that his vaccine had now been fed to four and a half million people, mainly in the Soviet Union and Eastern Europe, but also in Mexico and Singapore, without mishap. The Russian virologists, Mikhail Chumakov from Moscow and Anatoli Smorodintsev from Leningrad, substantiated Sabin's claim with detailed statistics. In addition, Smorodintsev addressed the question of reversion to virulence, admitting that it could happen but denying that it could ever reach a level which would hold any danger for man – whatever it might do when injected into monkeys. The Russians were confidently looking forward to the day when polio would be eliminated in the Soviet Union.

Sceptics and believers alike acknowledge the profound impression made by the Russians in 1959 and 1960. Richard Carter, a sceptic, writes: 'the Russians moved in phalanx from one international polio conference to another, reporting one enormous, absolutely unqualified scientific success after another and standing shoulder to shoulder in support of Sabin's contention that all fundamental questions were now answered'.[49] And according to John Rowan Wilson, a believer, 'the appearance of Sabin and the Russians at meeting after meeting was becoming a triumphal progress'.[50]

Cox and Koprowski were left trailing in their wake. But these triumphs were no more than preliminary skirmishes; the real contest was between live and killed vaccine, Sabin and Salk – or rather, Sabin versus the alliance of Salk and Basil O'Connor. And the showdown would not be in Europe but in the United States of America.

In Britain, meanwhile, despite the continuing prevalence of polio, a campaign launched in 1958 to encourage the take-up of Salk vaccine made little headway against public apathy. It was not until the spring of 1959, when the Birmingham City and England international footballer, Jeff Hall, died of polio, that the message got through. At twenty-nine, Hall himself was outside the age group then eligible for vaccination (fifteen to twenty-six), but his death did more to persuade young people to have their injections than all the appeals made by the Rt Hon. Derek Walker-Smith, Minister of Health, during the previous six months.

Queues formed outside special poliomyelitis clinics in London and elsewhere; football clubs broadcast a recorded message from the Minister of Health over their public address systems at half-time during Saturday league matches; and Bristol's Medical Officer of Health took advantage of this sudden interest to initiate a 'jabs while you jive' campaign, in conjunction with the Mecca Dance Hall Company, in the city's youth clubs and dance halls.[51] On 20 April, the *Daily Telegraph* reported that

> medical officers of health, with depleted stocks of poliomyelitis vaccine, face an even bigger rush for immunisation than has occurred in the last fortnight. It will come when thousands of young people inoculated recently seek second injections. There was Sunday working yesterday at the Kent factories of two producers of the vaccine.[52]

A week later the *Telegraph* announced that 'the first 300,000 of the million doses of Salk poliomyelitis vaccine ordered last week by the Ministry of Health arrived by air from the United States yesterday . . .'.[53]

As a result of this rush for immunisation the incidence of polio decreased sharply. By 1961, cases were being counted in hundreds rather than thousands, and it seemed that polio would soon become a thing of the past in Britain. But an outbreak of the disease in Hull, starting rather later in the season than usual (the first case, a girl of nine, was admitted to hospital on 20 September 1961), put an end to such complacency.

Rosalind Murray, a native of Stroud in Gloucestershire who had moved north when she married a man from Hull, had had two children in two years but had lost one of them. In the summer of 1961, at the age of twenty-four, she was pregnant for the third time.

> That summer was particularly hot, heatwaves and everything, and I used to go down, when I was pregnant, on the pier to cool off with my little lad, who was two. And there were all kinds of notices up, you know, warning about polio: would we go to this mass place and be vaccinated. So along I toddled and they said, 'No, we'll do you after the baby; not right now.' They refused. I had Robert at the end of July and I wasn't very well afterwards, so it would have been six weeks after that before I went down again. Still boiling hot, by then Septemberish, and they did some tests and asked a lot of questions – do you get asthma, eczema, hay fever . . . ? Well, I get hay fever really badly, so I admitted it. 'Oh well, you'll have to come back,' they said . . . They were going away, this mobile thing, but they'd be back in a month.
>
> In the meantime I started to feel a bit rough, like flu. Still I walked two miles to my friend's, with the two children in the pram, and nearly crawled back. The next day, the 12th of October '61, I got out of bed but somehow my legs were stiff. My head was pounding and I thought, 'Heck, I think I'd better have the doctor.' Well, the little lad said to me, 'I'll go to the shop; she'll ring' – we didn't have a phone in those days; it was a little one-up and one-down and a loo in the back yard. So the doctor came, and he had me moving my arms, moving my legs, walking around. 'Hmm, you walk a bit stiff for a twenty-four-year-old, that last child must have really done for

> you,' he said. 'I don't like it, your temperature's rising.' There'd been a scare apparently – we have this fair once a year, Hull fair, and there'd been some isolated cases, two or three children in the local isolation hospital, suspected polio, but they hadn't proved it yet. He said, 'I don't like the sound of it. I don't want to say it's polio. I'll be back after tea. If it's any worse, I'll know better then.' So he came back after tea with a specialist from Leeds, a Mr James, polio expert. By then I was much worse. The back of my neck – I felt so sick, I was stiff. In fact I was crawling on my hands and knees to the tap in the yard to cool Robert's bottle. I couldn't stand up . . . He said, 'I think it's pretty obvious . . .' So the ambulance came and I remember looking in the pram, with the baby there, having one last look. I was one of the first grown-ups to get it. Even the ambulance men hadn't been vaccinated, and they were really scared . . .[54]

Rosalind Murray (who bounced back to have two more children, despite residual paralysis of her lower limbs) went into hospital on Friday the 13th, an inauspicious date. By then there were forty-six suspected cases in the Infectious Diseases Hospital, and only the day before the health authorities had made an historic decision. In an attempt to put a stop to the epidemic they sought permission from the Ministry of Health to use live vaccine for the first time in Britain.

The action of the Salk vaccine – generating antibodies in the blood and, as it were, waylaying the virus before it could do any damage in the central nervous system – did not prevent the spread of infection and was therefore powerless to halt an epidemic. Sabin's oral vaccine, by contrast, worked in the same way as the disease itself, spreading it in a harmless form, even to those who had not themselves been vaccinated but were in contact with people, generally children,who had, and so might effectively impede the spread of 'wild' virus.

The Ministry of Health readily agreed to Hull's request. Live vaccine was already being manufactured by Pfizer Ltd of Folkestone, a British subsidiary of the American company, and was being held in reserve for just such an emergency. So 'elaborate plans were laid for the vaccination campaign, the chief aim being to vaccinate in the shortest possible time, the entire population of Hull, and any other persons who presented themselves'.[55]

Though the epidemic was caused by Type I poliovirus, the inhabitants of Hull were given Type II vaccine – just as the inhabitants of Singapore had been, nearly three years earlier, in a similar attempt to curtail an outbreak. This was primarily a safety precaution, so that the vaccine could not be held responsible for causing cases of polio, unless they happened to be Type II; and it was reckoned to be just as effective a way of preventing polio, since the viruses were known to interfere with one another – which was why the three types of vaccine could not be given simultaneously (a problem Cox had been trying to overcome with his three-in-one, or trivalent, vaccine).

By 17 October, when the oral vaccination began, seventy-six cases of suspected polio had been admitted to hospital, making this the most serious outbreak in Britain for two years.

> On the first evening of the vaccination campaign, 76,000 people presented themselves for vaccination within four hours of the clinics having been opened. The following day the weather was wet and stormy, yet a further 157,000 were vaccinated, and such was the tremendous demand for vaccine that some of the centres had to be closed temporarily in the late morning through shortage of supply. Fortunately, the RAF gave timely assistance and flew in additional supplies . . . and all vaccination centres were in full operation again in the same afternoon.[56]

In just one week, 358,000 people – more than the entire population of Hull – were vaccinated, most of them within the first two days; and in a fortnight the epidemic was over. Was this cause and effect, or would the outbreak have anyway ended then? The authorities were convinced that the vaccine was responsible for bringing the epidemic to so abrupt a conclusion, and they were pleased to note that it was safe as well – there were no cases of Type II polio. As John Rowan Wilson writes,

> Dick's cautious attitude was now being progressively overtaken by events. Ritchie Russell, the most consistent and staunch supporter of live vaccine in Britain, criticized him for suggesting the vaccine might be dangerous and then recommending its gradual introduction. For even Dick was now beginning to accept that live vaccine must come . . .[57]

In Britain, at least, the battle was won. The Sabin oral polio vaccine [OPV] was officially endorsed early in 1962 – even by George Dick, who writes, 'As soon as I was satisfied that the tests for safety and surveillance were good I supported the use of OPV.'[58]

'Picture a Cheshire cat riding a jet-propelled steamroller and you have the Albert Sabin of 1959 and 1960.' So writes Richard Carter. 'International polio conferences were as frequent as the seasons during those years, and Sabin was at each, with his triumphant Russians. By late 1960 they and others had fed his vaccine to upwards of 100 million persons.'[59] John Rowan Wilson sees events 'moving irresistibly' in favour of Sabin.

> Opinion was swinging over to his side at a rate that even his admittedly impressive results hardly seemed to justify . . . Everybody likes to back a winner, and as soon as Sabin began to radiate an aura of success he began

> to gather support at astonishing speed. Just as Salk, in the years of his triumph, had seemed to acquire an almost magical power, so did Sabin in 1959 and 1960.[60]

Even the US Surgeon-General recognised that polio vaccination 'had become a matter of cold war prestige, like bombs', and people were beginning to 'talk of the "polio gap" between Russia and the United States – like the "missile gap"'.[61] But 'some of those attending [the Fifth International Poliomyelitis Conference in Copenhagen in 1960] grew nauseated with the atmosphere of spite and partisanship for which polio vaccination was now famous'.[62] And Salk himself said that 'the vying for position was absolutely brutal'.

> . . . long before a particular report was read, you heard a dozen whispered allegations about the lies contained in it and the number of deaths unmentioned by it. It was like Lisbon during the war – everybody hawking secrets to everybody else.[63]

The unfortunate Herald Cox was the object of much of the whispering in Copenhagen. He had made valiant efforts to keep Lederle in the vaccine race, but the tide was running against him. His obsessive attachment to chick embryos as the medium in which to grow virus (pioneered by Ernest Goodpasture in the early Thirties) did not help; following Enders, both Sabin and – latterly – Koprowski cultivated their virus strains in the more versatile monkey kidney tissue culture. And he suffered a serious setback when his and Sabin's strains were tested by an independent investigator, Joseph Melnick, who found both wanting in terms of safety, but all three of Cox's types more virulent – and therefore more dangerous – than Sabin's. Cox trudged wearily round the international conference circuit, a homespun American looking hopelessly out of place and unwilling or unable to tout his wares in the requisite manner. Yet in 1960 he had two heaven-sent opportunities to gain some ground on the all-conquering Sabin.

The first was in the USA itself, in Dade County, Miami, Florida, where despite the fact that people had had one or more – sometimes all three – injections of Salk vaccine there had been nearly fifty cases of paralytic polio in 1959. Dade County wanted to try out live vaccine and turned to Cox and his one-off, trivalent vaccine served in cherry-flavoured syrup (Sabin's came in a sweet or a sugar-lump). The second chance came in May, in West Berlin, where 280,000 children were vaccinated. Inevitably, politics played a part in this. Like Russia, East Germany was Sabin territory.

> It became suddenly apparent that this would leave a small unvaccinated pocket behind the Iron Curtain, in the shape of West Berlin. If West Berlin

> were to show a much higher polio rate than in the so-called Free Democratic Republic, it would obviously be most deplorable . . . Lederle, it was thought, had saved the day for free Berlin.[64]

On this occasion, alas, the US Cavalry – or American Cyanamid – rescue operation went badly wrong. Forty-eight cases of polio were recorded before the end of the year, twenty-five of them occurring within four weeks of vaccination. The following reasons have been given for linking many of these cases, directly or indirectly, with the vaccine:

> (a) that most of the cases occurred at a time of the year – May and June – two to three months earlier than the usual seasonal epidemic; (b) that the cases were randomly distributed over the whole of West Berlin and not aggregated into groups as they so often are in a natural outbreak; (c) that two-thirds of the cases were in adults; and (d) that, of the 25 cases occurring in the 4-week period following vaccination, 17 had received the vaccine, giving an attack rate fifteen times higher among the vaccinated than among the unvaccinated population. Of the eight unvaccinated cases, it may be remarked, six had been in contact with vaccinated persons.[65]

All the evidence suggests that these twenty-five cases were vaccine-associated, but there has been some dispute over the precise causation. The British microbiologist H.V. Wyatt, for instance, argues that since the virus recovered from these cases was not markedly more virulent than the vaccine virus, there has to be some additional reason why these persons and not others fell victim to the disease; for him the culprit is the drug thalidomide, which was being sold over the counter and widely used in West Berlin in 1960: in association with the oral vaccine, it could have caused 'provocation poliomyelitis' in susceptible individuals.[66] (Provocation poliomyelitis is the name given to paralytic attacks of the disease in persons who have been exposed to poliovirus but might not have been paralysed by it had they not been recently inoculated against other diseases such as diphtheria and pertussis. An 'association between the site of the recent injection and the site of the paralysis' was first noted by British and Australian doctors in the late Forties and early Fifties.[67] The paralysis is not necessarily limited to the particular limb which has been jabbed, but that limb – be it an arm or a leg – is invariably affected in cases of provocation poliomyelitis.)

The Cox-Lederle vaccine was also implicated in causing polio in Dade County, Florida, where there had been seven cases, five of them adults, within three weeks of the vaccination of 412,000 people.

At the Copenhagen conference, no accusations were made publicly. But 'it was generally felt that Cox would never get over this, whatever the final verdict might turn out to be'.[68] Within two months of the Copenhagen conference the US Surgeon-General announced that his office would recommend the licensing of Sabin's vaccine strains, and Lederle took the hint: it abandoned its own man's vaccine in favour of Sabin's, which it now undertook to manufacture.

With Cox out of the running, Sabin's only remaining rival in the field of live vaccine was Koprowski, who was not yet ready to concede defeat. His mass trial in what was then the Belgian Congo predated Sabin's Russian trials, but it was largely confined to Type I vaccine, which was sprayed into the mouths of more than 240,000 people (2,500 also got Type III vaccine, but no Type II was used). The virus that caused most paralytic polio in the Congo, as elsewhere, was also Type I; and for that reason Koprowski would have had difficulty monitoring his experiment even if he had had adequate laboratory facilities to do so and the trial had not been overtaken by civil war.

At the time, the main interest in this trial was in whether or not it was supported by the World Health Organisation: with his references to 'following the recommendation of the Expert Committee of the World Health Organisation' in administering his Type I vaccine in the Congo,[69] Koprowski gave the impression that it was, while the WHO itself denied it. (John Rowan Wilson points out that the WHO became involved in the whole question of vaccines only reluctantly: 'The polio issue was a political hot potato, with endless possibilities for trouble and recrimination.'[70])

In recent years, however, following the publication of an article in *Rolling Stone* in March 1992, Koprowski's Congo experiment has been the focus of renewed attention for a quite different reason. If the AIDS pandemic originated in Africa in the Fifties and started out as a monkey virus lethal to man, what triggered it off? Could the vaccine, in which live poliovirus was cultivated in minced monkey kidneys, have been contaminated by other unrecognised viruses? Monkey kidneys were known to harbour innumerable monkey viruses, and Sabin himself claimed he had found an unidentified virus in Koprowski's poliovirus strains.

The place was right, the time was right, and the suggested link between the Koprowski polio vaccine and HIV plausible. But there are significant flaws in the theory. For a start, as Sabin pointed out, the AIDS virus will not survive swallowing. Ah, but the process of squirting the vaccine into people's mouths could mean that some of it got up their noses and into their lungs, enabling the virus to enter the bloodstream that way; or though a cut or sore in the mouth. But HIV and the related monkey virus, SIV, do not grow in monkey kidney cells; nor would they be likely to survive the cycles of freezing and thawing the vaccine went through before it was used. Then there is the question of the time it would take for SIV to mutate into HIV-1, the virus known to have been present in Zaire in 1959.[71]

Despite these objections, the Wistar Institute took the claims about Koprowski's vaccine seriously enough to set up a scientific committee to investigate them. The committee report, published in October 1992, did not entirely discount the possibility that the vaccine contained a monkey virus related to HIV, but found it extremely unlikely. One member of the panel admitted, 'We are never going to be completely conclusive on this issue.' But the panel's 'almost complete certainty' that the trial was not the origin of AIDS was supported by the discovery of HIV-1 in a Manchester sailor who died in 1959. He had 'visited North Africa and Gibraltar in the mid-1950s but had left by the "first half" of 1957, before the Central African trial started'.[72] Koprowski himself, whose vaccine was also tried out in Poland, Switzerland and Croatia, is scathing about the hypothesis:

> The argument for the safety of polio vaccination lies in the absence of any AIDS-related disease among the hundreds of millions of people vaccinated throughout the world; the fact that AIDS is rampant in subequatorial Africa can only be attributed to the polio vaccine by the wildest of lay speculation.[73]

In 1960, Koprowski fought a rearguard action against Sabin, arguing that the safety standards for live vaccine were 'picked to fit a particular strain and not the strain to fit the standards'.[74] John Paul admits that it was 'in some ways a pity that Sabin happened to be a member of the WHO committee because it was subsequently charged that there was a bias in establishing criteria for the selection of live attenuated strains', but claims that the impartiality of Macfarlane Burnet, the authoritative Australian chairman of the WHO's expert committee on poliomyelitis, ensured 'wise ground rules' for live vaccine trials.[75]

'Clearly, the control of poliomyelitis is unfinished business,' Dr Alexander D. Langmuir of the US Public Health Service's Communicable Disease Center in Atlanta, Georgia, announced at the Fifth International Poliomyelitis Conference in 1960. The previous year had seen 5,000 paralytic cases in the United States, which was fifty per cent up on the year before and more than double the record low figure of the year before that.[76]

It was not the effectiveness of the Salk vaccine that was in doubt; that was estimated to be 'in the range of 80 per cent for 3 or more doses and 90 per cent for 4 or more doses'. With the exception of an outbreak of Type III poliomyelitis among the well-vaccinated population of Massachusetts – which could possibly be explained by a lack of potency in the Type III component of particular early batches of Salk vaccine – there was a consistent pattern to the epidemics that did still occur: they broke out in those ethnic, racial or religious groups with the poorest vaccination levels.

The new pattern was almost a reversal of the old: before the Salk vaccine, 'if class distinction did exist, almost invariably the higher attack rates occurred among the upper socioeconomic groups', but now it was the lower socio-economic groups, and blacks in particular, who were subject to paralytic attacks. In Des Moines, Iowa, in 1952, in a 'Negro' population of just under 10,000 there had been no reported cases of polio; by 1959, the Black population had gone up to just over 10,000 and there were 33 cases, a rate of 321 per 100,000 – whereas the attack rate among the best-off Whites had dropped from 224 per 100,000 to a mere 16.[77]

The obvious solution was to intensify the immunisation programme so that no group, however poor or inaccessible, was left out. The problem with killed vaccine, however, was that *everyone* had to be vaccinated individually with at least three separate injections, and the logistics of this exercise were more daunting in the slums than in the suburbs – which was why there was a problem in the first place. With live vaccine, not only was distribution far simpler – quite literally as easy as giving candy to kids – but incomplete coverage mattered less, since the vaccine-induced polio would spread of its own accord, and that would be beneficial so long as there was no danger of a reversion to virulence.

Some virologists refused to be partisan. Thomas Francis did not want 'to be listed as pro one or pro the other' type of vaccine,[78] and Macfarlane Burnet 'persisted in the view that the desirable procedure was Salk vaccine in infancy with Sabin vaccine 'boosters' in early school age and perhaps at eighteen to twenty years'.[79] But these voices of sanity were drowned out by the hardliners on both sides; and there were increasing commercial and political pressures.

At a US Public Health Service meeting in February 1961, Basil O'Connor and Tom Rivers 'almost came to a parting of the ways'. It was largely through Rivers's unwavering support that Sabin had continued to receive National Foundation funding throughout what might be called the Salk years, and he was not about to desert his champion now. O'Connor disapproved of Rivers making public statements to the effect that the Sabin vaccine was ready for licensing, and that it would eliminate epidemics. He supported Salk's desire that the inactivated vaccine be allowed a further year to establish its efficacy without the statistically confusing introduction of another product. O'Connor also annoyed PHS officials by intimating that they were 'acting as advance men for a commercial promotion and that they were doing so at the expense of polio prevention'.[80]

The counter-allegation – that American vaccine manufacturers were holding back on the oral vaccine because they were making so much money out of Salk vaccine – was put forward by the Russian newspaper *Izvestia* and investigated in an article in the April 1961 issue of *Redbook*, 'the magazine for young adults'.

In answer to the question, 'What's delaying the new polio vaccine?' the authors exonerated the National Foundation (which then distributed reprints of the article) on the grounds that it had been 'continuously supporting Dr Sabin's oral vaccine research since 1953 with more than $1 million in March of Dimes grants'. They cleared the manufacturers, too. Out of four companies planning to manufacture the Sabin vaccine, only two were involved in the production of Salk vaccine. These two might therefore 'have some slight incentive to go slow. *But the other two companies have no Salk vaccine to sell*' (italics in the original). Lederle Laboratories, for instance, had 'already spent $11 million on oral vaccine research and an additional $2 million for a new oral vaccine plant and equipment'. And since it would get no return on its investment until it could market the Sabin vaccine, Lederle obviously had more to lose than gain by the delay. In this respect, the manufacturing situation supported O'Connor's contention rather than *Izvestia*'s. As to the government's role, the Division of Biologics Standards had learned to be extra cautious back in 1955, when it had licensed the Salk vaccine in a hurry and paid the price. Now DBS officials and scientific advisers 'decided to proceed with what Dr Roderick Murray, head of DBS, describes as "all deliberate speed"'.[81]

This caution was justified by the results of the Dade County trial of the Cox-Lederle vaccine – had the vaccine been licensed there would have been an outcry. But Sabin had also held a field trial in the United States in 1960; he had vaccinated 180,000 children in the town of Cincinnati, where he had been a Professor of Research Paediatrics in the University College of Medicine for a number of years, and there had not been 'a single case of paralytic polio traceable to the vaccine'. The DBS, however, was determined to avoid a repetition of 1955, and insisted on the strictest licensing requirements. 'No manufacturer will be permitted to sell his first batch until he has proved by tests on five successive batches that he is able to turn out a safe vaccine *consistently* [italics in original].'[82] This meant that the Sabin vaccine would not be generally available until the end of 1961 at the earliest. Sabin might protest that it had already proved its safety, but the DBS was taking no chances. If the DBS erred at all, the *Redbook* article concluded, 'it erred in the direction of greater safety for the millions of people who will take the new vaccine'.[83]

It is clear from this that there was no more substance to O'Connor's innuendo than to *Izvestia*'s allegation. John Paul recalls that the deliberations of the US Public Health Service ad hoc committee on live polio vaccine, of which he was a member, 'lasted almost four years'.[84] If they were indeed 'acting as advance men for a commercial promotion', they were taking an unconscionable time about it.

In the final act of 'this scientific tragi-comedy', as John Rowan Wilson calls it, 'all the roles became reversed'.[85]

Salk's tetchy response to a persistent questioner on the safety of his vaccine – 'It is safe, and you can't get safer than safe'[86] – which came back to haunt him after the Cutter incident, stands as a classic instance of the perils of giving such hostages to fortune. But Sabin's definition of the moral issue, following that same incident, is less often remembered. Speaking at the Congressional hearings on the vaccine in June 1955, he had said:

> I cannot ask anyone to come forward and take the chance this year of getting paralytic polio which he otherwise might not have because in so doing he would be doing a great service for perhaps 2,000 others who would be prevented from getting polio this year.[87]

When *his* vaccine came under the same intense scrutiny as he had given Salk's, Sabin took a rather different line.

Though no one questioned the overall success of Sabin's mission in Russia, it was, as Smorodintsev belatedly admitted during a visit to the United States in 1964, 'a public-health measure, not a field trial'.[88] Sabin's live vaccine was never subjected to the kind of rigorous field trial that Salk's killed vaccine had undergone in 1954.

Writing in 1962, David Bodian argues that, given the 'sharp division of opinion' in the scientific community between virologists 'heavily in favor of the live virus approach to immunization' and immunologists whose 'candidate, as it were, is the multiple killed-virus antigen', the need for 'an intermediate stage of oral vaccine introduction and evaluation should have been especially apparent'.[89] John Paul disagrees:

> . . . a gradual turnover did not hold the promise of bringing out the full potential value of the oral vaccine in terms of community-wide protection. This called for intensive mass vaccination campaigns conducted over a limited period of time and directed primarily to preschool and school-age children . . .[90]

Despite the fact that – in Paul's words – 'the question of strain reversion to previous virulence within the body of the vaccine remained the most important of the unsolved problems', by 1960 it was generally accepted that the danger was 'either absent or insignificant'.[91] On this assumption, the American Medical Association took the unprecedented step of endorsing Sabin's vaccine in June 1961 – before it had even been licensed for public use. 'Quite by chance', Richard Carter writes ironically, 'a former medical director of Pfizer's Laboratories Division, Dr William C. Spring Jr, was Secretary of the AMA Council on Drugs when it presented to the House of Delegates its extravagantly biased review of the virtues of the Sabin vaccine. . . .'[92]

Less than two months after the AMA's action, in mid-August, the DBS licensed Sabin's Type I vaccine. In October, it licensed his Type II. But it held back on his Type III strain, which was as controversial as Salk's Type I Mahoney strain had been, until the end of the following March. And that was the one that caused the trouble.

By the end of August 1962 sixty-two cases of polio occurring within thirty days of vaccination had been reported from different non-epidemic areas of the United States; the majority of the stricken were infected with the Type III strain, and a high proportion of them were adults, who – especially since mass immunisation with Salk vaccine had reversed the upward trend in age susceptibility – were unlikely to have been infected naturally. 'The old tragedy was with us once more,' John Paul concedes. 'Some felt that the whole program should be called off immediately. A few opponents of the live virus program were loud in condemnation, calling it a national scandal.'[93] No one was louder than Basil O'Connor. In a National Foundation press release he argued that this mass vaccination campaign had come 'too late'; the Salk vaccine had already done the job: by the end of 1961 polio had been reduced by ninety-seven per cent. He maintained there was 'no scientific evidence' that Sabin live poliovirus vaccine was more effective than the Salk killed poliovirus vaccine, or that it produced a more lasting immunity. His most provocative point was a neat reversal of the claim made on behalf of the live vaccine, that it spread immunity even to those who did not take it: 'A killed poliovirus vaccine can never put the live poliovirus back into the community. A live poliovirus vaccine could do that.'[94] More damaging than O'Connor's comments, perhaps, were those of a polio expert of comparable stature to Sabin himself, David Bodian:

> . . . widespread publicity has been given to the idea that poliomyelitis and poliovirus could be eradicated from the USA by means of a nation-wide mass campaign with oral vaccine. This provocative idea was coupled with the notion that this was the only way of accomplishing the goal of eradication. Yet, there is hardly any doubt that killed-virus vaccine has exhibited the capacity to eradicate poliomyelitis in this country and in others.[95]

The main problem, as Bodian saw it, was not which vaccine was superior, or even whether the oral vaccine was entirely safe, but of achieving universal acceptance, and take-up, of either vaccine: 'It is really an age-old political problem, which comes under the heading of "getting out the votes".'[96]

The Public Health Service played down the – admittedly minimal – risk associated with live vaccine for that very reason: for fear of putting people off. But it did gently discourage older people from taking the Sabin vaccine, and set up a special advisory committee to determine the number of cases attributable to it. Out of a final tally of eighty-seven cases in non-epidemic areas, '57 were regarded by the committee as being "compatible", that is, as possibly due to the vaccine'.[97] After the publication of the committee's report in 1964, the PHS lowered the age over which it advised people *not* to take live vaccine to eighteen. 'Like the historic "full-recognition-of-very-small-risk" document of 1962, which had recommended conservatism among persons older than thirty, this new press release was couched in language of the greatest delicacy....'[98] Richard Carter is contemptuous. But John Paul, an insider and committee member, explains that while some members of the committee reiterated their plea that live vaccine be restricted to pre-school and school-age children, 'when it came to a vote it was repeatedly rejected with the argument that such a limitation might stop the oral vaccine program altogether'.[99]

Sabin inevitably disputed many of the committee's findings. But he agreed with it that 'if any risk did exist, it was very small, and should not be allowed to interfere with the practice of routine immunization of all infants followed by a reinforcing dose to children on entry to elementary school'. The immunologist Sir Graham Wilson, whose words these are, concludes that 'if the Sabin vaccine is to be incriminated on epidemiological grounds, then so also must the Salk', since 'both in the United States and Great Britain numerous cases of paralysis have occurred within 4 weeks of inoculation with apparently satisfactory batches of Salk vaccine'.[100]

John Paul points out that it is 'a rare biological product with a high degree of effectiveness that exhibits complete safety'.[101] But Sabin's acceptance of a minimal risk in his own product contrasts sharply with the moral line he laid down in relation to Salk's nearly a decade earlier. He may have triumphed, in that his vaccine superseded Salk's as 'the vaccine of choice' in the United States as well as in most other countries throughout the world, but the live versus killed vaccine controversy refused to die.

On 20 December 1976, for example, the *Daily News* reported that, that year, there had been six cases of polio in the United States, four of which might be attributable, directly or indirectly, to the Sabin vaccine.[102] As a result, the Canadian firm, Connaught Laboratories, the only North American company still manufacturing Salk vaccine, stepped up its production in the hopes of increasing its US sales. Salk had long argued that the changeover to live vaccine was 'both unnecessary and ill-advised',[103] and even though in 1977 the National Academy of Sciences' Institute of Medicine 'recommended sticking with live vaccine' the sharp increase in medical litigation – which Sabin called 'a plague upon American society' – has brought the Salk vaccine back into contention.[104] The death of Sabin at the age of eighty-six in March 1993 may have brought to an end a bitter and long-running scientific feud, but the issue has yet to be finally resolved.

The Fifth International Poliomyelitis Conference held in Copenhagen in 1960 was the last in the four-yearly sequence of meetings organised by the National Foundation. Since the mid-Fifties, when the foundation had effectively thrown in its lot with Salk, its political influence had been waning and many people assumed that, once its objective had been achieved, it would gracefully go out of business. They were reckoning without Basil O'Connor, whose commitment to the eradication of polio was equalled only by his passionate espousal of the 'volunteer principle' in relation to health organisations. He spelled out his 'new philosophy . . . – namely, the free and voluntary engagement of a whole people in a joint enterprise for its own better health' – in the organisation's 1957 annual report.[105]

John Rowan Wilson, an English observer writing in the Sixties and imbued with notions of the welfare state, remains unconvinced. He recognises O'Connor's own dynamism and capability, but regards the foundation as rather old-fashioned:

> Flag-days, birthday balls, Polio Pioneer buttons, are a little like political party conventions – they are extroverted, wild, wasteful, exciting, entertaining – a part of America's rumbustious past. But not of her future. The collection of voluntary contributions nowadays has little to do with real charity . . . People are cajoled into giving, bludgeoned into giving, shamed into giving. It is not easy to see what moral advantage these processes have over the more simple method of taxation.[106]

Wilson may argue that 'voluntary organizations are not necessarily dynamic and Government departments are not by definition sluggish',[107] but the very formulation of the sentence suggests what the norm is, however regrettable that may be. The American medical journalist Victor Cohn rejects an 'either-or' approach in favour of pluralism, pointing out that from a research point of view, 'where the scientist can . . . turn to more than one treasury, he is freer', and citing the case of the 1954 Nobel chemistry prizewinner, Linus Pauling, who was once turned down for a government grant 'because of his unorthodox, left-wing politics' but received March of Dimes grants to the tune of $340,000.[108]

Voluntary funding of research does not necessarily supplant state support; it may even stimulate statutory funding.

> When FDR and 'Doc' O'Connor started the drive against polio in the 1930s, the role of government in supporting biomedical research was modest. By 1955, it was becoming important. Today it is overwhelming. Yet the money you and I give to voluntary organizations often funds projects that are crucial to medical progress and health.[109]

The problem is, what proportion of the money donated actually goes to support research? In the two decades from 1938 to 1958, according to one article, the National Foundation 'spent twice as much on fund raising as on research'. O'Connor drew no salary during that period (from 1959, however, he received $50,000 a year – increased to $60,000 in 1967 – plus expenses, a fact not widely publicised), but one year, according to the same source, his 'combined expense and special travel account . . . was about $72,000'.[110]

Another criticism is that 'the National Foundation was always first among the major health agencies in fund raising and last in the number of victims needing care'.[111] Taking the year 1957, the National Foundation raised $44,000,000 to care for 15,000 patients, while at the other end of the scale the National Association for Mental Health raised only $817,000 to care for 16,000,000 patients. But severely paralysed polios are expensive to look after and require long-term care stretching over a number of years, so the figures for any one year are deceptive; and the benefits of money spent on polio research were by no means restricted to polio. Still, the argument that individual drives for causes like polio, backed with heartrending posters of crippled children, distort the real balance of needs in a society has weight. The move towards federated fund-raising, united funds and community chests, was an attempt to restore that balance and reduce the 'mad scramble for volunteers'.[112]

But such an argument cut no ice with Basil O'Connor, who resisted all overtures from the Arthritis and Rheumatism Foundation when the National Foundation entered that field. He remained a staunch opponent of federated fund-raising for three main reasons:

1. Federated drives cannot raise nearly as much money as several individual drives.
2. Forcing causes into federated drives (as industries and unions have done in some cities) 'has no business in a democracy', and amounts to 'communization' because 'it will ultimately ruin all private health and welfare activity and lead to complete government control'.
3. 'Most important', he says, destroying the country's great volunteer agencies will destroy one way Americans still have of 'participating', of 'having important personal relations with each other', in an age in which bigness has destroyed many old ways.[113]

But even O'Connor was unable to drum up the same kind of popular enthusiasm for the cause of arthritis and birth defects as he had for polio. The annual contributions to the March of Dimes in 1963 amounted to $21,000,000, which represents a considerable decline from the mid-Fifties peak of $67,000,000.[114] There was some truth in the allegation that O'Connor had become 'a man without a disease'.[115]

When he died in 1972, at the age of eighty, the *New York Times* paid this 'Polio Crusader' second only to Franklin D. Roosevelt a suitably double-edged tribute:

> Imperious, vainglorious, stubborn and blunt, Mr O'Connor generated a host of critics in the health field . . . But for all his foibles, he created what Gerard Piel, publisher of *Scientific American*, recently described as a 'unique social invention: a permanently self-sustaining source of funds for the support of research – the voluntary health organization'.[116]

8

Born Too Soon

. . . Today [National] foundation funds supply . . . care for some 20,000 victims who, as O'Connor says, 'got born too soon'.

Medical World News, 31 January 1964

. . . Once lionized as heroic examples of human fortitude, the thousands of polio survivors who continued to need medical and financial help were suddenly ignored as embarrassing emblems of their own poor timing, clumsy enough to get polio before the vaccine that could have protected them was found. As veterans of other wars would continue to discover, the same civilians who pray for them in the heat of battle don't like to be reminded of the wounded and the dead after the war is over.

Jane S. Smith: *Patenting the Sun* (1990)

I have to go it alone out of the basket world. I have to protect myself from the cripples all around me, who are my mirror image and who, at all times, are falling, puking, muling, slipping, spinning, wheedling, pulling, hanging on, all about me, trying to pull me down with them.

My brothers! I stand apart from you: you, and your wasted limbs, your blighted breath, your palsy and your lordosis. I could care less about your sob stories . . .

Lorenzo Wilson Milam: *The Cripple Liberation Front Marching Band Blues* (1984)

Persistence in strenuous activity was *the* great rehabilitative virtue and the pursuit of ambulation second only to the quest for the Holy Grail.

Alice Mailhot, 'Age and the Old Polio: Do the Virtuous Fade First?' *Rehabilitation Gazette*, Vol. xxiii, 1980

At the bottom of the hill on which the Roosevelt Warm Springs Institute for Rehabilitation stands, the cracked, now empty, set of swimming pools bears witness to a bygone era; the trickle of spring water leaves a forlorn and solitary old treatment table high and dry. There is talk of renovating the site, but it would be expensive both to do and to maintain. Why bother? Probably only a few of the thousands who annually troop through the Little White House to pay their respects to FDR find their way down here. Anyway, its state of decay is appropriate. There is the kind of sadness about the spot that affects all abandoned places once vibrant with laughter and purpose. But who wants to turn back the clock and see the pools full of crippled children waving their flail limbs in the water? Better let it rot, consigned to history along with the disease that brought it into being.

Like its other half, the National Foundation, Warm Springs tried to change direction. In January 1964, in the first issue of a magazine for 'former patients and friends' called *Contact*, Dr Robert Bennett quotes Basil O'Connor's words – 'Today we are seeking to create for all types of physical disability the same philosophy of total care that we have, in years gone by, attempted to develop for poliomyelitis.'[1] But this 'difficult transition', as Bennett called it in a letter, was only partially successful, and depended on the continued support of the March of Dimes.[2] After O'Connor's death, Warm Springs was threatened with closure. In 1974, the State of Georgia, in the person of Governor Jimmy Carter, bought the whole caboodle for *one dollar*. Now the Roosevelt Warm Springs Institute for Rehabilitation is just another state-run medical rehabilitation unit, with all that that implies in terms of institutionalisation and bureaucracy.

At least there are a number of black people to be seen on campus, not just behind the lunch counter but in wheelchairs too. When Emelie Simha, a French national who had contracted polio in Egypt, arrived soon after the war she was enchanted by the place, which reminded her of the TB sanitorium in Thomas Mann's great novel, *The Magic Mountain*, but shocked by the racism still in evidence there.[3] There were some black patients, but they were confined to a basement and ate their meals separately from the whites, an arrangement which was justified on the grounds that it suited both parties, the blacks supposedly preferring the informality of wandering in and out of the kitchen, spooning the food directly out of the pots, to the awkwardness of being waited upon by their own kind amid the crystal and linen. The Administrator Woodall Bussey once told Viva Erickson, a physical therapist who came from Iowa, 'Oh, but they like it that way.'[4]

The mother of Elaine Strauss, a polio quadriplegic who was 'both a Jew and a Yankee', was so incensed by the treatment of black patients that she complained to Lester Granger of The Urban League in New York. But his inquiries into conditions brought 'a dreadful solution'. Black patients were simply transferred from Warm Springs to a general hospital in Columbus, Georgia, with no rehabilitation facilities for polio. Strauss found it ironic that while being a Jew did not prevent her attending a memorial service for FDR in the Warm Springs chapel on the anniversary of his death, being black barred her Christian attendant – despite the latter's eagerness to pay her respects.[5] Given his own evasiveness on this score, FDR might have found it less than ironic.

Diane Smith, who had polio in 1949 at the age of ten, went to Warm Springs after three months, but remained so severely disabled that she had to return there later in her teens for 'functional training'. The coal-mining area of southern Illinois, where she had lived for the first ten years of her life, excluded blacks, so she had had no relations with them until she went to Warm Springs. The only black patients she remembers seeing were *out*-patients, and all the black workers had to be off the foundation grounds by dark. 'I was *appalled* by that. Because I found that they were much more compassionate and had much more feeling towards the patients. They were just better to you. They would look at you as if they understood.'[6] The polio centre for black people at the Tuskegee Institute nearby in Alabama was still functioning, however, and Warm Springs established a consultative service there in April 1954. For a brief period Dr C.E. Irwin and Dr Robert Bennett made weekly visits; Dr Irwin performed several operations and both doctors held clinics. But Irwin left Warm Springs in 1955 and Bennett discontinued his visits, pleading lack of time. At this point Dr John Gorrell, the National Foundation's Director of Medical Services, suggested that members of the Tuskegee team might visit Warm Springs rather than the other way round. But this did not work out either, as the Medical Director of the polio centre reported to the President of the Tuskegee Institute:

> The problems encountered in this endeavour can be exemplified by the experience of the Director of Orthopedics, Dr J.F. Hume, who in 1956 visited the Warm Springs Center . . . There was an embarrassing incident which occurred when the Foundation requested that the orthopedist eat in a separate facility. As a result of this, it is felt that until race relations improve to the point where visiting personnel can be equally accommodated . . . it would be unfitting to attempt this form of liaison between these two institutions.[7]

Eventually, when the foundation had to seek support from sources other than the March of Dimes, it was obliged to integrate in order to qualify for funding. Warm Springs, it might be said, was dragged kicking and screaming into the second half of the twentieth century. Emelie Simha recalls wheeling along the corridor of the surgical wing one day and finding the maintenance crew busy erecting a door frame.

> I said, 'What are you doing?' And they said, 'Well, this is where the blacks are going to be' – only they didn't call them blacks then, they called them

> Negroes, and sometimes they didn't even call them Negroes. So I said, 'Why?' 'Well, they're *separate* from us.' Now this was just a psychological barrier they had to have because they were so insecure within themselves that they had to have someone beneath them. To me, this is the epitome of stupidity and ignorance.[8]

The illogicality of it all became even plainer when the same people who resisted integration were confronted by black patients from El Salvador and Guatemala. 'They would say, "Oh yes, he's kind of dark, isn't he?" And I'd say, "Yes, he's dark. He's just as dark as Raymond down there in the kitchen." "Oh well, that's not the same thing! He's Spanish".'[9] A polio who arrived at Warm Springs a decade after Emelie Simha sums it up by saying that most of the professional staff 'were themselves very kind and not overtly racist at all – they were trapped in a system'.[10] Just as FDR himself had been.

Another adult polio of the late Forties, Turnley Walker, who would write the semi-official history of Warm Springs, recalls in a memoir 'the side-by-side pressures of my longing for such a haven and my recognition of the necessity to leave it as quickly as possible and get back, somewhere, to a life with my family and work'. His wife's restlessness – 'For her, no destination had been reached' – prevented Walker from identifying too closely with the place, but there were no such constraints on many other patients, particularly the children whose presence was largely responsible for its cheerful atmosphere.[11] The physical therapist Viva Erickson remembers that the foundation became so concerned with patients not wanting to go home that it sent a psychiatrist, Dr Morton Seidenfeld, from New York to find out why.

> The obvious thing was they were in an environment in which they were like everybody else, number one. But number two, they – the able-bodied people and the patients themselves – they didn't . . . we didn't consider them handicapped or unusual, or anything. We just said, 'They can't do this and we're going to have to figure out a way to show 'em'. And by working together we did that.[12]

Many were reluctant to leave Warm Springs, but Hugh Gregory Gallagher did not want to go there, because that was where *crippled* people went. He had never liked crippled people and he certainly did not want to think of himself as a crippled person. But Franklin Roosevelt Jr lived across the street from Hugh's parents; his secretary had worked for Hugh's father; and when he called Hugh's mother and offered his help it would have seemed churlish to refuse. So Hugh went to Warm Springs.

> And I got down there, really frightened, and went into this four-bedded room. There was only one other person in this room and he was in a hospital bed, rolled up. A huge, six-foot-four long guy, with glasses on the end of his nose, doing the *New York Times* crossword. He was about my age. And he looked up and said, 'What is seven letters for . . . whatever?' And God help me, I knew. And I just said it, and we've been very close friends ever since.[13]

This was Lorenzo Milam, and he and Hugh became prominent members of what he might have called the class of '53. In his polio memoir, Milam is quite scathing about Turnley Walker's sentimentalisation of the Warm Springs story – 'a dingbat book', he calls Walker's history. Yet no one writes more ecstatically about Warm Springs – 'Paradise of Meriwether County, Georgia' – than Milam himself:

> The tranquility of the nights. Those warm nights out of the American South. Voices on the soft wind. Soft lights shining. The shadow of a disease banished. Laughter, joking. Humanity. People coming to life again. People like me, who have been buried in ignorant drab little hospitals all over the country; being brought by bus, by car, by train, by airplane to this paradise: this white, clean, alive, human, deliciously hopeful environment.[14]

There are pages of this in Milam's *The Cripple Liberation Front Marching Band Blues*. But there is a complex sensibility informing Milam's rant, a person at odds with himself behind the hyperactive persona, doubly challenged by homosexuality as well as disability. No wonder he looks back longingly to the simplicities of an innocent boy-girl love with the beautiful but crippled Francine under the mythical 'Bush Thirteen', famous for lovers' trysts at Warm Springs.

> I think there is no way that I can again have the happiness that came to me that summer, that swollen summer of joy in 1953. I can never ever be normal among the normals who make up the world, and, for the last time that summer, I am among my own people, the polios of 1952–53, joined together in the last summer of our youth, together in the heady air of . . . Georgia.[15]

When he was writing his study of Roosevelt's disability, Hugh Gallagher revisited Warm Springs and met Viva Erickson, who had retired by then. She told him, 'You talk about this Independent Living Movement. Well, it turns out that's what we were doing. We just didn't have a name for it.'[16]

Erickson emphasises the familial aspect of Warm Springs and attributes it not just to the fact that every patient was a polio, but also to its geographical isolation. This meant that, among the staff, only dedicated people stayed, the ones who 'would sit around at night and talk about the patients and say, "Now I have this patient with this problem: do you have any idea what we can do about it?"'[17]

But even in its heyday Warm Springs had its critics. Dr Richard Owen of the Sister Kenny Institute had a chance to go and work there early in his career, but he turned it down. As a specialist in rehabilitation medicine and a polio himself, he might have been expected to leap at the opportunity, but he reacted against the intensely personal atmosphere and felt that it created too great a dependency among patients;[18] and it could certainly be argued that the relationship between the polios and the physios sometimes bordered on the unhealthy. In the end it hardly mattered; once the success of the vaccines was assured, Warm Springs' days as a *polio* rehabilitation centre were numbered.

Despite the ever-mounting number of cases of paralytic polio in the United States during the pre-vaccine period between 1949 and 1956, the fatality rate almost halved, while the age of susceptibility continued to rise; and these trends combined to cause a sharp increase in the number of 'catastrophically disabled survivors, hundreds [of them] chronically dependent upon respirators for maintenance of life itself'. By the late Fifties, two thirds of respiratory polios – 'respos' or 'responauts', as they were sometimes called – were young adults.[19]

The National Foundation – with its resources already stretched to breaking point – set up fifteen regional respiratory centres after a pilot project in Baltimore had established that the advantages of pooling resources and skilled staff outweighed the inconvenience of having to move some patients long distances from their homes and families. One of these centres was the Toomey Pavilion (named after Dr John A. Toomey), which had previously been the contagious ward of the City Hospital in Cleveland, Ohio.

There had been 'a ghastly polio epidemic' in Cleveland in 1949, and the newspapers had appealed for volunteers to learn the Kenny method of preparing and applying hot packs. Gini Laurie had been one of the first to respond: 'Naturally, I took the course, and I've never stopped volunteering with people who are disabled since.'[20] Why naturally? Because Virginia Wilson Laurie came from a family in which all four of her siblings had gone down with polio in 1912. Two sisters died, another was mildly disabled, and a brother was very severely disabled. Gini was born the following summer and named after a deceased sister. She recalls: 'From the time I was mobile, I was my brother's auxiliary arms and legs. I didn't notice how disabled he looked. I thought he was supergreat and followed him like a shadow.'[21]

This dearly loved brother died of pneumonia at the age of twenty-one, when Gini was sixteen. She never forgot him. So when she and her husband Joe moved to Cleveland from St Louis for business reasons in 1949, she was ready to drop her plans for creating a herb garden and renovating the old house they had bought in the nearby village of Chagrin Falls to answer the call for volunteers on the polio wards.

For ten years she frequented the Toomey Pavilion, both before and after it was converted into a respiratory centre.

> I loved being with the patients. I came to know their families, their problems, and their joys. I shopped for them and ran errands. I fed, bathed, dressed, toileted, shampooed, shaved, manicured, read to, and wrote for. I did whatever needed doing and I tried to relieve the boredom and to make life as much fun as possible by giving birthday parties and decorating the wards for every holiday.

For Gini Laurie, 'fun' was the key word. Fun and togetherness.

> The patients watched over each other. If someone made the major alarm sound – a 'cork-popping', clicking sound (a sound that you can make by clicking your tongue against the roof of your mouth even if your trachea is unplugged and you are voiceless) – then everyone with a voice yelled, 'Nurse!' I still love the sounds of the bellows of an iron lung and the pulsation of a portable respirator because I have had so many close friends attached to the swooshing and pulsing sounds.[22]

Many respiratory centres produced mimeographed newssheets in the Fifties. They had punning titles like *The Croakers' Chronicle, The Res Parader, The Rocking Reporter, The Rock 'n Roll, The Vital Capacitator, The Weakly Breather,* and *Gulpers' Gazette*. The one that outstayed them all was originally known as the *Toomeyville Gazette*. By 1957 it had been 'shelved and forgotten', but it was revived under a new name, *Toomey j Gazette,* in 1958 (providing Dr John Toomey with a measure of the posthumous fame he deserves for being right about polio in the Thirties – even if he could not prove it – when almost everyone else was wrong).[23]

By then the wards were beginning to empty. The respiratory centre in the Rancho Los Amigos Hospital in Los Angeles had pioneered home care for ventilator-dependent patients early in the decade. It was felt that, once patients' condition had been stabilised and nothing was to be gained by keeping them in hospital, they would be better off at home: 'Basic to this plan, naturally, was a family willing to accept such a responsibility.'[24] But even when a family was ready to accept responsibility for a chronically disabled member, the community might take a different view – as the following story told by a medical social worker from Ann Arbor, Michigan, illustrates:

> An adolescent from a sparsely populated area was near the end of his treatment program and we planned to send him home. Many things

> needed to be considered. Continuing medical care, facilities for emergency hospitalization, schooling, periodic nursing supervision and completion of the vocational plan had to be arranged. The family attempted to interpret these needs in their community but met with little success. When the medical social worker arranged to spend several days in this small city, she learned quickly that the community generally was horrified that the hospital considered sending a young man home when he could not breathe without mechanical assistance and was totally dependent on others to meet his personal needs. First, the local doctor had not had experience in caring for such a patient. Next, the community hospital flatly refused to admit such a patient or even to have a tank respirator in the building. The school superintendent was irritated that the hospital would so much as think of risking this young man's life by sending him home. The public health nurse refused to take any responsibility for supervising home care. However, after two days of interpretation and discussion, the community became most interested and co-operative. When the young man's condition was really understood, everyone was eager to help him again become a useful citizen and they did so.[25]

Gwendolyn Shepherd, a physician on a visit from Argentina – which had recently experienced an epidemic reminiscent of Copenhagen 1952 with doctors having to 'tracheotomize and hand-breathe the children' – was amazed at the attitude of the staff at the respiratory centre she went to:

> The whole time I was there I didn't once hear the '*¿Vale la pena? ¿Para que? ¡No hay nada que hacer!*' ('Is it worthwhile? What for? They haven't a chance'), that we had had thrown at us from the beginning of the epidemic, even from the lips of those who should have known better. The thought didn't even seem to occur to them. On the contrary, their immediate reaction was, 'Well, now what has to be done to get this patient back into meaningful living?'[26]

This positive approach to rehabilitation had the additional advantage of relative cheapness, the price of home care 'ranging from one tenth to one quarter of what hospital costs would be'.[27] As a result, the system was soon being adopted all over the United States; and that, in conjunction with the Salk vaccine, was responsible for the emptying and eventual closure of the regional respiratory centres.

In 1958, when Dr Robert Eiben, Medical Director of the Cleveland centre, asked Gini Laurie and the only other surviving volunteer in the Toomey Pavilion to revive their newsletter – 'because each polio outpatient who came in for a checkup or a bout with pneumonia plied him with questions about all the other outpatients' – they wrote to the medical directors of the other fourteen respiratory centres and drew in polios from all over the country. But even that did not satisfy their ambition, which was 'to reach and advance respiratory polios all over the world and to share the problems, experiences, thoughts and adventures that it would be of value to know about each other'.[28]

An immediate problem for all American polio survivors was the changing policy of the National Foundation (no longer 'for Infantile Paralysis'), following the success of the vaccine and the consequent drop in March of Dimes contributions. Future financial support could no longer be guaranteed. The NF's 'new theme song' was: 'we have no money.'[29] For respirator-dependent polios, in particular, the writing was on the wall, however often the foundation denied that it was turning its back on polio and focusing attention on its new fields of interest.

In 1961 the *Toomey j Gazette* initiated a survey of 'people with long-term post-polio disabilities'. Its findings were published in a research monograph in May 1962. Out of 1,400 people on *TjG*'s mailing list who were sent the questionnaire, 806 replied – 453 women and 353 men. The severity of their disablement is clear from the details of the report:

> Twenty-nine per cent have no ability to feed themselves. About 40% are able to do so without assistance, and nearly 31% can or could feed themselves with assistive devices. Inability to dress was reported by 83%. The ability to write unassisted was indicated by 48% and the ability to type by 27%. Writing and typing with assistive devices was specified by 18% and 35%, respectively. About 32% are unable to write and 28% cannot type . . . The ability to transfer from bed to chair with no one assisting was reported by about 40%. An additional 40% can transfer with help. Almost half of the respondents cannot propel a wheelchair . . .
>
> The need for attendants is also a measure of the degree of disability. Over 90% of the respondents indicated this need, the majority depending upon family members and about a third upon volunteers, hired help or a combination.[30]

The vast majority (eighty-four per cent) of the respondents lived in either their own or their parents' home. The number of parents in professional and managerial occupations revealed the 'high socio-economic status of respondents' families as compared to the general population'. About half the sample were married before polio, and a majority of those who were subsequently divorced or separated attributed the breakdown of their marriage to their disability. The educational levels of the respondents were 'much higher than those in the average population', and 'a comparison of the educational levels before paralysis and at the time of the survey shows remarkable achievements on the part of many of the individuals'. Earnings were commensurate with levels of education, but bore little relation to the degree of impairment; and dependence on breathing aids had little effect on working status. The report concluded that 'many need but cannot afford more care. It is obvious that long-term disabilities require long-term aid.'[31]

As a result of this survey and hundreds of letters from *TjG* readers spelling out the consequences of the withdrawal of March of Dimes financial support, the government called a meeting in April 1963, to which the National Foundation and other interested public and private agencies were invited – but not *TjG*. The meeting was 'inconclusive' and the attitude of the National Foundation was to wash its hands of the whole business, leaving it up to the volunteer membership of individual Chapters to take local initiatives. Yet *TjG* found that the majority of these volunteers, 'the men and women of good will in our communities', were '*unaware* of the curtailment of aid to polios'.[32]

Donna McGwinn, one of the *TjG* team of 'horizontal' editors, had parents who were both supportive but lived apart. When she came out of hospital she lived first with one and then with the other, helped out by attendants, whose selection could be a hazardous business. 'To make a mistake', she writes, 'is possibly to endanger your life.'

> Once one of my helpers was late in returning from the weekend. As she came in the front door my mother went out the back on her way to work. It didn't take me long to realize the woman was beyond comprehension with too much alcohol. She barely knew what she was doing and was so groggy that I was afraid I would lose her for the rest of the night if she once fell asleep. Somehow with repeated instructions I guided her into hooking up the telephone I could work myself. (You can see why it's important to do as much as you can by yourself!) I called my mother who promptly returned home. For many months after came the recurring thought, what if she had tripped over and disconnected the rocking bed cord before I could call for help?[33]

(The utility of rocking for artificial respiration was discovered by an Englishman, Dr F.G. Eve, in 1932. But it was not until 1947, when Dr Jessie Wright of Pittsburgh 'studied the principle of forcing air into the lungs by the diaphragm which was pushed up and down by the viscera as a result of the see-saw motion', that the rocking method was applied to poliomyelitis patients. Another crucial discovery, by Dr Clarence Dail of Rancho Los Amigos, was glossopharyngeal breathing, more commonly known as 'frog-breathing'. Dr Dail 'observed a patient using his tongue and throat muscles to push air into his lungs'; he studied this technique of 'frog-like gulping', and, with the help of some physical therapists, taught it to other respiratory polios.[34])

Donna McGwinn was fortunate in having a father who left her enough money to enable her to remain at home and employ attendants after his death. For other respos, Gini Laurie writes in 1963, the 'most important need is immediate planning for the future, when, with the deaths of their parents who now care for them, they will be faced with the soul-killing prospect of vegetating in county homes with the senile'.

> We envision small, homelike, residence-care centers throughout the country . . . In such homes . . . the residents would participate in the management, and would share their remaining abilities to engage in profitable business and creative work, and thus pay all or part of their own expenses and have the opportunity to live – not merely to exist.[35]

Gini Laurie may have been influenced in her thinking about homes for the disabled by a letter she received from an English polio, who was also a writer. Peter Marshall lived for two years – his 'National Service', as he calls it in his autobiography, *Two Lives* – in a hospital. He coins an 'ugly hybrid' of a word, 'humangosophy', a composite of humour, anger and philosophy – 'the three bastards of the rapist, "disability"' – to sum up his acerbic, nonconformist, but robust attitude to life in a wheelchair: 'Of course, I miss walking, miss my independence, but I don't envy other people . . . If I can't be Marshall walking, then I'd rather be Marshall crippled than be anybody else . . .'[36] In what is plainly an autobiographical novel, *The Raging Moon*, published two years later, Marshall charts his crippled protagonist's unsuccessful attempt to live with family – in this case, a brother and sister-in-law – and his subsequent incarceration (not too strong a word in this context) in a home run by the Cheshire Foundation (set up by the late Group Captain Leonard Cheshire of *Dambuster* fame): 'I was going . . . to a sheltered, plastic world where people in wheelchairs talked to people in wheelchairs all day long; a world where I would never want for anything except everything.'[37] His account of life in this home for incurables is so bitter that he feels obliged to preface the novel with a disclaimer: 'I would like to make it clear that this [fictitious home for the disabled] does not resemble Heatherley, the Cheshire Home with which I am associated and where an urgent need is being met by a handful of selfless people. . . .'[38] But in a letter to Gini Laurie, he tells a different story:

> It is a sad but undeniable fact that Cheshire Homes have failed on all but the strictly practical level . . . basically, they asked me to leave because I was 'anti-social'. This means that I spent a large part of the day by myself, working – I was writing *Raging Moon*. It seems I was expected to sit all day in the communal lounge being 'jolly'. If one lives in a Cheshire Home, one has to give up everything worth living and dying for – self-respect, one's identity as an individual, personal liberty (no one is allowed out after 9.30 in the evening), any form of work other than occupational therapy, the right simply to be alone, the right to choose one's entertainment (one is expected

to watch everything provided as a mark of 'gratitude'), the right to choose one's friends, the right to freedom of speech . . . Power only retains morality for as long as it is balanced by an equal power, and in Cheshire Homes the residents are powerless, completely at the mercy or otherwise of the Management. Living in one is a degrading and humiliating experience: one gradually becomes a zombie in a strictly regimented society . . .[39]

Marshall begs Gini Laurie 'not to let similar organizations develop in your society'. In an ideal home, he argues, the 'basic requisite' is privacy. Numbers should be kept small – 'any number above one is an excess'. There should be space for possessions and, if possible, for entertaining friends privately. There should also be 'the fullest possible contact with the outside world'. Charities should be avoided 'like the plague'.

In this country we have Women's Institutes and Rotary Clubs who 'visit' because they feel it is their duty. They ask personal questions about one's body – from me they get the information that I have movement, yes, below the waist, no, not in my right leg, no, not in my left leg. When realization dawns they are shocked. There is a subversive belief that polios are sexually neuter – can't think why. In this category of charitable organizations, I would place the Church.[40]

The later Fifties and early Sixties was the era of the 'Angry Young Man' in Britain. In America, too, there were dissenting voices among young polios at this time. Because paralytic polio was no respecter of persons and prostrated an unusually high proportion of articulate, young middle-class achievers in the decade following the Second World War, assertion of rights replaced gratitude for charity among the polio disabled and paved the way for future disability legislation.

The first Center for Independent Living created and staffed by severely disabled people – polios among them – did not come into being until the early Seventies, in Berkeley, California. But the concept of independent living for the severely disabled was at least twenty years older, dating from the Rancho Los Amigos home care experiment in the early Fifties.

In England in 1963, there was a survey of polio survivors which was not confined to the very severely disabled but included varying degrees of disability and all ages. It was conducted by the Deputy Director of the Office of Health Economics, Michael Lee, in collaboration with the British Polio Fellowship (as the Infantile Paralysis Fellowship now called itself). Out of an estimated total polio-disabled population in Britain of between 11,500 and 12,500 people, 1,710 BPF members – roughly one seventh – participated in the survey.[41]

Lee found that 'the social class distribution of the households in which members live, determined by the occupation of the principal breadwinner, who may or may not be the disabled person, shows a greater proportion in the higher social classes than in the population at large'. This was particularly marked in 'households where the disabled person contracted poliomyelitis as an adult'. And since 'the severity of disability is greater with those who contracted poliomyelitis as adults', the upper classes 'include a higher proportion of severe disability'. (The economist John Vaizey, who suffered from osteomyelitis as a boy, writes in a bitter memoir of hospital life during the war that 'the social class of the boys with polio was on average considerably higher than that of the boys with osteomyelitis or tuberculosis . . . [they] carried an air of aristocracy about them . . . their bodies were lean and hard and clean, and did not fester and smell; quite fortuitously, too, they carried with them the glory of President Roosevelt'.[42])

The figure for the break-up of marriages which predated the onset of polio – approximately one in twelve – is similar to that of the *TjG* survey, with the disabled partner most commonly being the wife. 'Divorces are associated with severe disability. Both legs were affected in most cases, a large proportion of which involved total paralysis. Most of the divorced or separated females had their arms affected.' Yet sixty-seven per cent of disabled men were married, and forty-three per cent of women, and Lee is careful not to paint too gloomy a picture: 'Over nine in ten marriages survived the disability of one partner, and nearly four in ten of those disabled have subsequently made a lasting marriage.'

As in the American survey, Lee notes the high levels of education attained, but here it is particularly among 'those who completed their secondary education before they contracted poliomyelitis'. Employment figures were eighty-four per cent for men and seventy per cent for single women, as against ninety-five and ninety-two per cent respectively for the non-disabled working population in that far-off era of almost full employment. Lee concludes the summary of his findings with a description of the two groups in which the degree of rehabilitation achieved contrasted most sharply:

> The first is men and women who contracted poliomyelitis as adults. With these, although disability tends to be severe, they are on the whole more active and successful members of the community. When they were disabled, they had accumulated a substantial amount of 'social capital' in their education, training, family life and experience on which they could fall back. The problems they face appear to be more sophisticated than those normally associated with handicapped people . . . The second group are single women who were disabled in childhood. With these, the problems of rehabilitation are more extensive. In childhood, their education and training appear to be relatively neglected, while their chances of marriage are

> low. They live in their parents' households until late in life; the proportion in employment is low. Among the aged a high proportion live alone where the problems of isolation are aggravated by disability . . .[43]

The 1965 annual report of the British Polio Fellowship drew attention to 'the plight of the respiratory polio'. The BPF's National Welfare Officer had done a survey which estimated that there were 'between sixty and seventy' respiratory polios living either at home or in hospital.

> Those in hospital are there not because they are ill, but in many cases because home conditions do not permit a totally paralysed person to be in permanent residence, very often this is a financial problem. The Fellowship is convinced that many at present in hospital could, subject to medical approval, live at home, and those at present living at home could live in less straitened circumstances if the government were willing to acknowledge their total disability and award a pension, as in the case of those permanently disabled by polio contracted during active service in the forces.[44]

Britain's best-known responaut (along with the ex-dancer and successful mouth painter Elizabeth Twistington Higgins, if an appearance on Eamonn Andrews's *This Is Your Life* television show is taken as a measure of celebrity) was Paul Bates. He was lucky in the sense that he had contracted polio as a young National Service army officer in Malaya and therefore received a war disabled pension. Once it became clear that he could expect little physical improvement, he was encouraged to go home. Yet it took a fire in the polio unit at the Western Hospital in Fulham to prise 'this limpet from its rock', as he puts it: 'frantic visions of getting movement back in this extreme emergency flashed before me but with all the will in the world I couldn't move an inch to save myself'.[45] Eventually he acquired a home of his own, but even for him it was a struggle.

Paul Bates went on to achieve his fifteen minutes' worth of fame when he participated in the *Daily Mail* London to Paris air race in 1960. He was greatly assisted in leading an active life by a remarkable English invention known as 'Possum' – coming from the Latin for 'I am able' and standing for Patient Operated Selector Mechanism. This 'suck and blow' mouth-operated remote control for a selection of electrical appliances – toaster, television, light, typewriter, telephone – was the brainchild of an industrial chemist by the name of Reg Maling, who was a member of the 'Friends of Stoke Mandeville', the spinal injuries centre, and his colleague Deryck Clarkson. They were supported by Duncan Guthrie's National Fund and Bates was invited by Guthrie to make a second trip to Paris to demonstrate their enabling invention there.

But it is the less public moments in the life of a 'horizontal' polio, which he describes with unflinching candour in his co-authored memoir, that linger in the mind:

> Having polio is to be forced to do nothing for there seems to be nothing you can do; to have polio is to stare at the ceiling; to have polio is to do nothing for yourself; to have polio is to fight with a disease which would, if it could, rob you of the ability to do anything, presenting you with each twenty-four hours as a barren waste with no choice but to endure them . . . [46]

The anaesthetist Dr Geoffrey Spencer came to be interested in polio in a roundabout way. As a Senior Registrar at St Thomas's Hospital, he had the opportunity to set up the first purpose-built intensive care unit in a London teaching hospital in 1968 – 'the era of Harold Wilson and the "white heat of technology" and all that.'[47] He soon found that 'if one jacked up the quality of artificial respiration to the limit of what was technically feasible in those days, straightaway one created a problem – a lot more people survived who were relatively fit apart from the fact that they couldn't breathe'. The question was, 'what the hell to do with them'. In an intensive care unit of ten beds, 'we had two beds permanently blocked, as the jargon has it, by people who were stuck on respirators'. Dr Spencer admits he was 'stumped'.

> Then I heard that Thomas's was taking over a slum hospital in the back streets of Brixton, because it couldn't function; and we were going to be lumbered with it. And there was an old polio unit in the Western Hospital at Fulham, which the North West Thames regional health authority wanted to shut down but couldn't. Polio tends to be a middle-class disease; a disproportionate number of sufferers are, relatively speaking, educated, and they have the capacity to make a noise. And one or two of the people who had been looked after and were maintained by the Fulham hospital started making a very large noise . . . And it suddenly occurred to me that if we took on this unit, these people – these funny people with polio – had actually had some experience of long-term breathing, and some of them, the financially fortunate, and those with a spouse, had actually made it home.
>
> So I thought, if we took on that polio unit, maybe we could learn something from their experience which we could apply to the new generation of respirator-dependent people that we have created as a result of being too damn clever at keeping alive people who would previously have died.

Geoffrey Spencer had his way and, after some negotiation between the two hospital boards, St Thomas's took over the Fulham polio unit and moved it to the old South Western Hospital in Brixton. Without being fully aware of the importance of what he was proposing, Dr Spencer said,

> We must have a home maintenance system, so that people at home with breathing machines will be visited regularly by chaps from this unit. And

> we must have a couple of technicians who will go round and see the patients at home and look after the equipment.

An anaesthetics technician called Frank Kelly had put this idea in his head when Spencer discovered that Kelly spent his annual two-week holiday 'pushing some girl who was paralysed following polio round the Lake District in an iron lung'. Kelly was clearly the man for the new unit. Spencer made an even more surprising discovery about Kelly when he overheard him talking to some Russians who were visiting St Thomas's in their own language. 'And eventually the story came out . . .'[48]

The son of a First World War American GI whom he never knew, Kelly had joined the Royal Army Medical Corps in 1944 and volunteered as a paratrooper. He was parachuted into Arnhem, where the British troops were surrounded and captured by the Germans. Here he came into his own as a medical orderly, first looking after the wounded, then caring for his fellow prisoners in Stalag 4B at Mühlberg. When the Russians liberated the camp, Private Kelly did not wait around to be repatriated; he commandeered a bicycle and took off into the German countryside, where he worked hard, as he puts it, 'doing a round of villages on the bicycle, dressing wounds and acting like a District Nurse'.[49]

He admits that 'you may think it odd that I didn't think of myself as a deserter, but I didn't'. He was, he says, 'a sort of "displaced person" by nature': 'I was born a drifter and drifting was my way.' Besides, he was doing good business, in a Harry Lime-ish way, as 'a kind of underground VD station for the Russians in those parts'. It was an ideal life for Frank Kelly, but it was too good to last. When the Soviet authorities caught up with him, they put him on a train to the border; but Kelly preferred to stay in the Russian sector. The next time they arrested him they began to suspect him of spying – why else should he remain? He was incarcerated in Leipsig Prison for five months. Private Kelly's ignorance, far from suggesting innocence, appeared to the Russians as the cunning deviousness of a master spy. They let him go to see where he would lead them, and then picked him up again. This time, they took him to Potsdam, trussed up like a chicken in the back of a car.

There he was brutally beaten up, not just once but over and over again for a period of weeks, until he cut his wrist with a sharp chip of enamel he had secreted away. He was in trouble, but he did not care.

> That was a turning point for me. I had a purpose. I would fight all the time. I would open up my arm a dozen times. I would hunger strike. I would complain about everything or nothing. I would keep on protesting until I died or they let me go.

Private Kelly, the alleged spy, was flown to Moscow, where he was sentenced to ten years in the political prison of Vladimir, 200 kilometres to the north – at least it was not in Siberia. It was the end of the prison road for Kelly and he soon adapted to it. 'No more interrogations – no more beatings – no more treatment. And the only torture the mental torture of caged men.'

Once again, his services were required 'as a bit of a Doctor and nurse': 'it always did me good when others were sick and I could help. That suited me best.' And the men he felt called upon to help were of a type new to him – bishops, priests, professors, political prisoners. A Lithuanian Archbishop who persuaded Kelly to teach him English taught him Russian in return, gently overcoming Kelly's resistance to learning the language of the hated enemy. They played chess, and the Archbishop and others opened up for Kelly the world of books and ideas. Kelly did not let up on his protests, his periodic wrist-cuttings and hunger strikes, but the influence of these fine men – 'the only real and true friends I have ever known' – worked a subtle change in him: 'Because of them I shall never be as I was before I went to prison.'[50]

In the brief and partial thaw following the death of Stalin, Kelly was released and taken to Berlin, where he was handed over to the British military authorities – who promptly placed him under arrest again. But though he was held in the detention wing of the British Military Hospital, he was treated kindly. A court martial decided against any disciplinary action and gave him a good conduct sheet for the seven years he had been absent. He was also given time to recover before being thrown to the wolves – as the battery of press photographers and reporters impatient to get at him seemed, in his confusion.

In one newspaper he was described as being 'near collapse as he spoke, his face contorted as if a devil dentist was at work'. The report continues: 'But rallying to full strength for one half-minute he cried out, "I was innocent. These barbarians, with their brutalities, are blinding and paralysing innocent people."' In a rather different vein, Kelly was quoted as saying, 'I had no knowledge of England. I never knew about the Coronation . . .'[51] And he describes himself as 'a kind of Rip Van Winkle', requiring 'a long process of rehabilitation' in the Royal Military Hospital at Woolwich before he was ready to face the world again.[52]

Perhaps this goes some way towards explaining his extraordinary empathy with polios, who also required a long process of rehabilitation – at the end of which there was no guarantee of recovery. But mostly he had a gift, something that came naturally to him and gave him strength. In prison in Russia, he had saved a German doctor's life, nursing him through epileptic fits and then, when they were separated, going on hunger strike until he was allowed back in the doctor's cell to resume his vigil.

Geoffrey Spencer says, 'I had enormous trouble with him, when it came to filling in forms or any sort of bureaucracy. But he was *so* good with the patients, who *loved* him, that it was a minor matter.'[53]

Kelly was 'the instigator of the home maintenance service' – news of which spread rapidly through the 'polio grapevine' and by the end of 1968 there were forty people with respiratory polio on the unit's books. Dr Spencer was impressed with the calibre of his new patients: 'You've only to give them a shove in the right direction and they're off. And they don't expect miracles or cures, because they have come to terms with their state.'[54]

Richard Crossman, Labour Minister of Health in the late Sixties, had the impossible dream of creating an 'integrated health service', cutting across the entrenched bureaucracies of public health, social services, local authorities, general practitioners and all the other interested bodies. Disabled activists saw the opportunity of getting the remaining respiratory polios – those lacking accommodating spouses or the means to set up homes of their own – out of hospital. Crossman listened to the argument that it would be not only better, but also cheaper, to support patients at home rather than have them block expensive hospital beds. This could be a model for his integrated health service, and he instructed his officials to look into it. They were less keen, as civil servants are wont to be when change is in the air; they saw it as the thin end of the wedge: if polios received special care at home, what was to stop 'every poor old thing with a bad hip' demanding it? So the polios staged a protest, in which a slightly embarrassed Geoffrey Spencer took part, disguised in an old sweater.

Crossman and the protesters had their way. A research project was set up to compare the cost of home care for respiratory polios with that of hospitalisation, and the volunteers were canny enough to insist that the care provided should not be withdrawn once the three-year period of research was over. The scheme was successful in that home maintenance freed hospital beds and gave the respos greater freedom.[55]

Some years later, when the respiratory unit was threatened with closure, the polios once again demonstrated their propensity for direct action by forming a patients' association. Under the chairmanship of Felicity Lane Fox, who was subsequently made a life peer, they set up an appeal and raised one and a quarter million pounds to pay for the re-siting of the Phipps unit, as it was then called, at St Thomas's itself. When the new unit opened in 1982, it was named after Lady Lane Fox, who had died the year before. More than a decade later, the Lane-Fox unit still plays an indispensable part in the lives of British polio survivors.

In 1981 – the International Year of Disabled People – Geoffrey Spencer was one of a number of specialists invited by 'that *great* woman, Gini Laurie', to an international symposium held in Chicago, to discuss the question: 'What Ever Happened to the Polio Patient?' This was the first of many 'post-polio' conferences, and it attracted some 200 people from all over the world, seventy of them in wheelchairs and thirty of those respirator-users, to the Americana-Congress Hotel on Chicago's waterfront – 'a beautiful building in that rather stolid 1930s style of architecture, very impressive, with lots of room for wheelchairs'.[56]

Over the years *Toomey j Gazette* had outgrown its parochial origins and changed its name to *Rehabilitation Gazette*, the 'International Journal for Independent Living by Disabled People'. It was based in St Louis, following Gini's husband Joe Laurie's retirement from business and their return there. Gini, aptly described as 'the glue that holds the polios of the world together', saw herself as 'archivist' of the movement for the disabled – also as 'the catalyst, the catcher of straws in the wind, the gatherer and giver of facts'.[57]

In an article in the 1979 *Gazette*, Larry Schneider, a polio survivor from Tucsan, Arizona, drew attention both to the deleterious effects of aging on polios and to the lack of doctors who knew anything about the disease any more. Regretting that 'there is no longer one center like the old Warm Springs Foundation which has answers to polio after-effects and aging', he suggested that *Rehabilitation Gazette*, 'which seems to be the last polio link', draw up a 'national directory' of experienced and 'simpatico' doctors for the benefit of surviving polios in need of advice and treatment.[58]

This short article 'struck a responsive chord with many others who are post-polio experiencing similar physical limitations that are increasing with age'.[59] Schneider and the *Gazette* received a quantity of letters, and on 5 July 1980 NBC aired the problem on the television programme, *Prime Time Saturday*. The producer told Schneider 'that they wanted to make the program "as mild as possible" so as not to overly alarm any post-polios'.[60] But it had the opposite effect; and Gini introduced a special feature on 'Respiratory Rehabilitation and Post-Polio Aging Problems' in the 1980 *Rehabilitation Gazette*, with details of the forthcoming Chicago conference and a post-polio questionnaire for readers to fill in. One article – by a woman who described herself at twenty, the age at which she contracted polio, as 'either too foresighted or too slothful to engage in the strenuous exertion that would have put me back on my feet' – points to the paradox confronting middle-aged polios with declining strength: 'It isn't their vices that are catching up with them. It's their virtues, good old-fashioned rehabilitative virtues: exercise, effort, and physical achievement.'[61] In the Forties and Fifties, when most of the surviving polios were getting over the acute phase of the disease, they had been pushed to the limits of their strength and endurance in the hope of recovery, not just by doctors and physiotherapists, but by family and friends as well. It was really a joint conspiracy, since they themselves were for the most part willing collaborators. Now rehabilitation specialists were telling them to slow down, take it easy, perhaps all that exercising had not been such a good idea after all.

The basic tenet of the disease, the core of the optimistic outlook characteristic of polios – that what you gained, you held – was being called into question.

The importance of this article of faith can hardly be exaggerated: *onward and upward* is the message of all the inspirational literature spawned by the disease. Turnley Walker is a fine example (even his commonplace surname acquires significance in the context of paralytic polio): just as he sentimentalises FDR in his history of Warm Springs, so he brings to the story of his own rehabilitation a Hollywood sense of the dramatic. At lunchtime on the day after leaving Warm Springs *en route* for California in a car laden with family, pets and luggage, he pulls into a restaurant parking lot. His wife Flora takes the children inside and comes back to escort him in on his crutches – though he would much rather she got out the wheelchair from under the luggage.

> There were several things I was going to say, violent things, cruel things. But when I looked down, her eyes were on mine, and I knew I would never forget what was in her eyes. I took first one crutch and then the other. She turned and walked a little ahead of me . . .[62]

They get into the restaurant without mishap, and he slumps gratefully into a chair. But at the end of the meal, as he struggles to his feet again, Walker has 'a foreboding of disaster', exacerbated by the feeling that everyone's eyes are upon him.

> Halfway up everything was fine, but then I lurched and put the wrong angle of pressure on the table. It slid a little, then tipped, then crashed. And I followed it to the floor, immediately . . .
>
> I went blank with rage and terror. I didn't want to move. I listened to the quiet of the room, and it seemed to press me tighter to the floor.
>
> But slowly I turned around and hoisted myself to a sitting position. I did it because I knew Flora was watching and waiting for me to do it. She expected me to do it, quickly, and with no nonsense. Sitting there, I looked up at her. She made it clear that she intended for me to get up . . .

So it goes on, another couple of pages of gruelling attempts to rise from the floor amid the broken crockery, the other customers stunned into silence. Eventually, Turnley makes it back to the car with minimal help from his wife. 'The legs of my gray flannel trousers were black with sweat.' Once out on the highway, 'I began to feel all right.' Soon they are laughing about it. He tells her, 'That was quite a performance'; and she replies, 'You're such a dramatic man.' That night in a motel, he absorbs the lesson: 'I began to see that Flora would never falter in her belief that I was still strong enough and capable enough to be the head of that family. This regard was softening the thick wall my fear had raised.'[63] Walker calls the first of his two polio memoirs *Rise Up and Walk*, and that just about sums it up. Other titles from that era convey a similar message of painful striving: *The Long Road Back, The Long Walk Home, Through the Valley, Through the Storm, Keep Trying . . .*

A historian who has studied thirty-five biographical and autobiographical accounts of polio finds their 'themes of recovery and redemption . . . reminiscent of the Puritan Covenants of Works and of Grace'.[64] An American sociologist writing in the early Sixties makes much the same point: 'The physiotherapy regime, which in its very design faithfully captures the essence of the Protestant ideology of achievement in our culture – namely, slow, patient, and regularly applied effort in pursuit of a long-range goal – has built into it, as it were, its own prophecy of success'; and adds a revealing footnote:

> While many orthopedists privately doubt that physiotherapy *per se* contributes much beyond what natural recuperative processes themselves can accomplish, few press the point in actual treatment situations because they are aware, implicitly at least, of the valuable psychological functions it serves.[65]

Yet even in the Fifties the work ethic was not universally or uncritically accepted. Dissenting voices were raised within the very citadel of the triumphalist approach to polio, Warm Springs itself. At a staff medical meeting on Wednesday 27 January 1954, Dr Bennett introduced a guest speaker, Dr Clint Knowlton.

> Dr Knowlton discussed neuromuscular fatigue and presented evidence to support his feeling that excessive exercise produced definite damage to muscles that have been injured by poliomyelitis. His reasoning is that fatigue is a safety valve to warn against overworking a muscle and a sensation of fatigue is lacking in muscles weakened or damaged by polio, hence the danger of over-exercising a non-normal muscle. This apparently is a completely new theory, according to Dr Bennett, and one that deserves much consideration.[66]

In a paper they published four years later, Bennett and Knowlton concede that this new theory may not be quite so new as they originally thought:

> Overwork weakness is certainly not a new and original finding. Lovett, in his survey of his experience with the victims of the 1913 Vermont polio epidemic, repeatedly found that activity caused deterioration rather than the expected and usual improvement in muscle strength.[67]

They admit that their overwork theory is controversial: 'Many highly capable men in the field of medicine and its supporting sciences do not agree that muscle can be overworked through voluntary effort.' And

The shadow of the crippler: a still from the March of Dimes propaganda film, *The Daily Battle*, known colloquially as 'The Crippler'.

'Fight Infantile Paralysis' was the ubiquitous message of March of Dimes propaganda in broadcasts and on billboards in the Forties and Fifties.

Hooray for Hollywood: the 1954 March of Dimes poster child is flanked by Danny Kaye and Bing Crosby.

Charismatic leader: though FDR had been dead four years by 1949, his face still adorns the dimes in this March of Dimes poster design by Erik Nitsche.

In 1949 the March of Dimes poster child was Linda Brown.

The military metaphor was made explicit in this 1952 March of Dimes poster during the Korean War. (The poster child was Larry Jim Gross.)

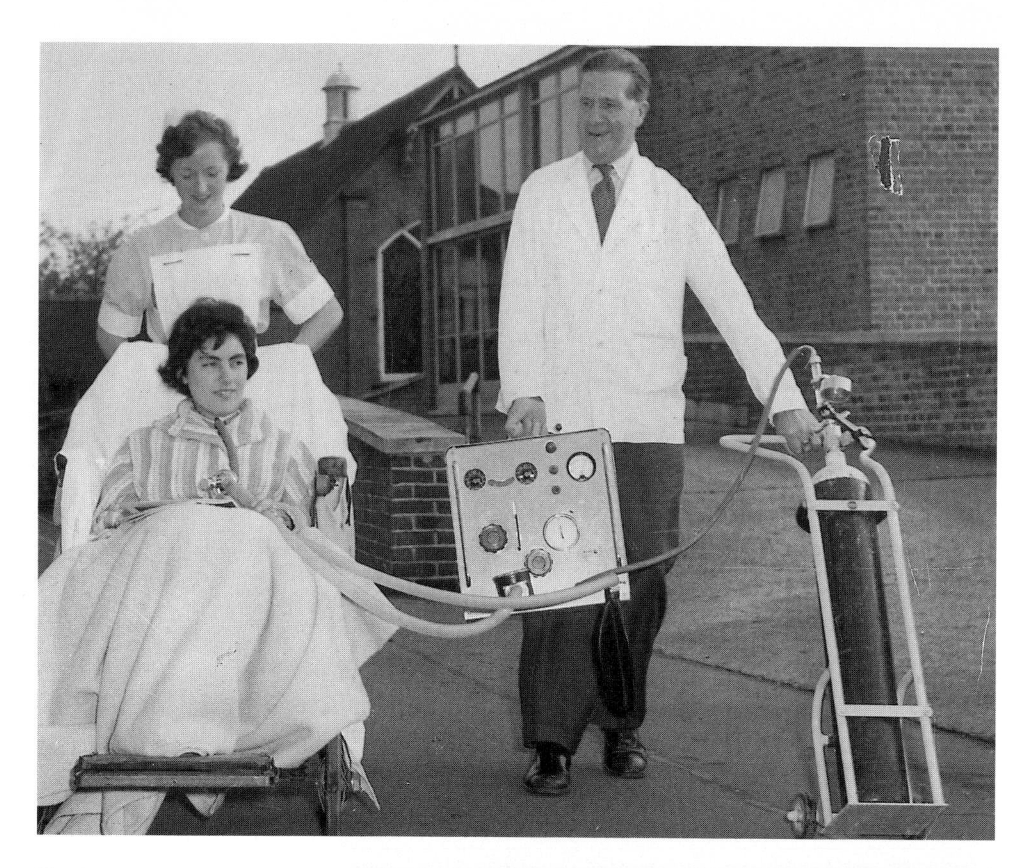

The Barnet Ventilator: one of the portable, positive-pressure respirators which replaced the more cumbersome iron lungs following the 1952 epidemic in Copenhagen – note the tracheotomy.

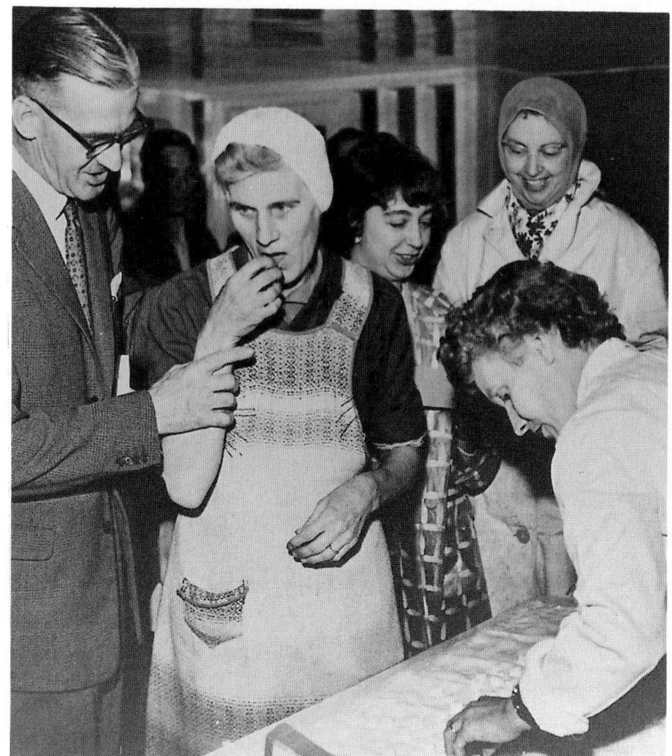

Hull 1961: in the last significant outbreak of polio in the UK, Sabin's oral polio vaccine was used for the first time in Britain.

Taking their medicine: these Scottish children prepare to digest vaccine-impregnated sugarlumps.

The first British National Swimming Gala for polio survivors – organised by the Infantile Paralysis Fellowship at the Seymour Hall Baths, Marylebone, London.

Sister Kenny, wearing the hat that made her 'look like Admiral Nelson'.

A cartoon from the Washington *Times-Herald*, 31 March 1945, dramatising Sister Kenny's predicament: loved by her patients, spurned by many doctors, she threatens – once again – to depart.

Sister Kenny demonstrating her controversial method of treatment.

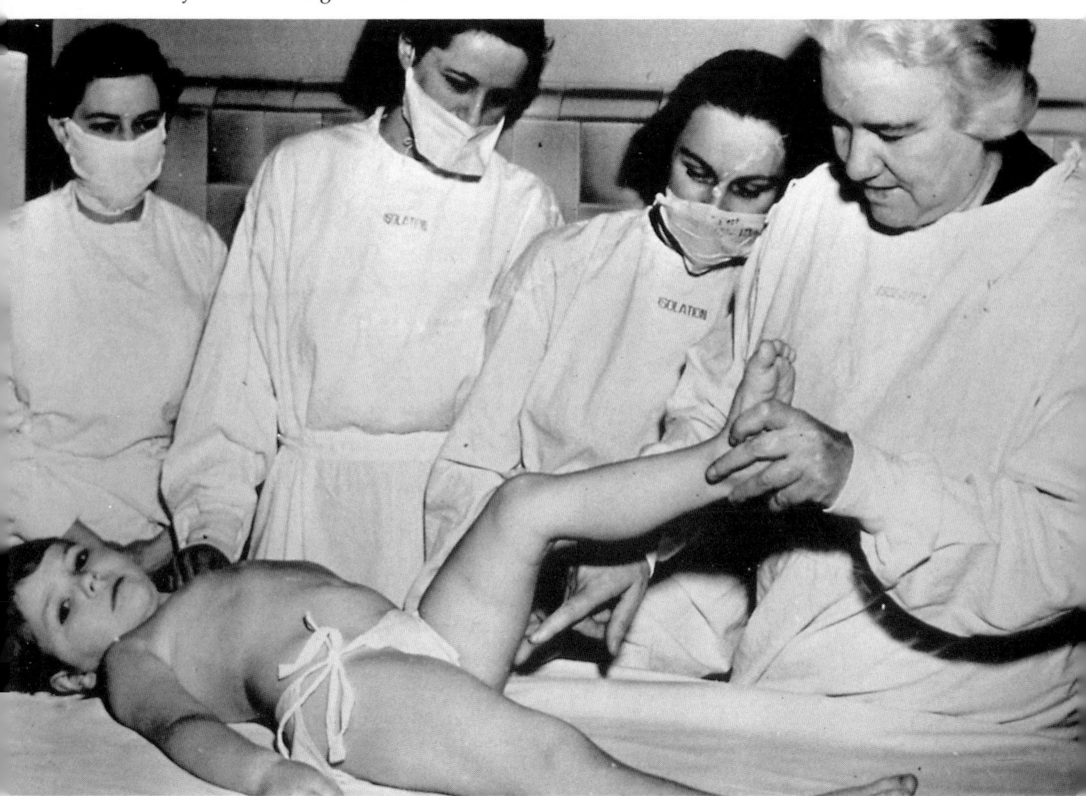

The men in long white coats: Drs Jonas Salk (*left*) and Albert Sabin (*right*), whose championship of their different kinds of vaccine led to a bitter rivalry.

Polio pioneers: schoolchildren prepare to participate in the field trial of the Salk vaccine – the biggest in the history of medicine – in 1955.

Handle with care: boxes of polio vaccine and a reminder of what they are designed to prevent.

they are careful to distinguish between different kinds of fatigue, emphasising that 'exercise carried to a feeling of tiredness has but little relationship to overwork'. The art is to gauge precisely the right amount of exercise. Deterioration following overwork of muscle 'may be irreversible'.[68]

A decade after that initial staff meeting, the Warm Springs newsletter, *Contact*, reported that 'increasing numbers of our old polio patients who return to Warm Springs for re-evaluation are showing evidence of muscle weakness from over-activity'.[69] And in 1967, Dr Bennett announced in *Contact* that Warm Springs was conducting a survey of former polio patients 'in order to determine whether, in their opinion, they have experienced any unusual loss of muscle strength or atrophy of muscle'.[70] In this, as in so many other things, Warm Springs was the trailblazer; but its days as an independent entity were numbered and the initiative passed elsewhere.

There were other studies of 'the problem of functional deterioration in the middle-aged patient with a history of acute anterior poliomyelitis many years previously' – as three doctors from the Department of Rehabilitation, Montefiore Hospital, New York, describe it in an article published in the *Lancet* in 1972.[71] But the Chicago conference of 1981 was the first concerted attempt to address, not just the problem of functional deterioration, but the increasing medical ignorance about polio (people troubled by its after-effects were dismissed by doctors either as hypochondriacs or, worse, neurotics in need of psychiatric treatment) and the neglect of polio survivors following the virtual elimination of the disease in the West as a result of the vaccines.

Many important disability issues were addressed during the three days of the conference, but its enduring value lies in the fact that it provided both an impetus and a focus for future action. Deserted by the March of Dimes (and ignored by the march of time), the children of Warm Springs and the National Foundation no longer had any need to feel orphaned: 'If you ask Gini', a journalist wrote in a profile of Gini Laurie, 'if she and Joe had any children of their own, she will say, "yes, 50,000" – the readership of the *Gazette*'.[72]

Lauro Halstead contracted polio when he was a college student travelling in Europe. Left with a frail right arm, he had no great difficulty in adapting to his disability; a keen squash player, he simply switched to playing left-handed. On a travelling scholarship to Japan in 1957, he climbed Mount Fuji, and it was only when he reached the summit that he recalled that it was 1 August, three years to the day after he went down with polio. It seemed to him that the third anniversary marked the completion of a very satisfactory recovery.

His grandfather was a physician, and so was his brother. 'I always said that medicine was my second choice; I didn't know what my first choice was.' Consciously at least, becoming a doctor had little to do with his disability, and for twenty years or more Dr Halstead pursued his career in rehabilitation medicine without giving a thought to polio, his own or anybody else's. With a good job at the Baylor College of Medicine in Houston, Texas, he was 'seeing patients, and doing research and writing and travelling and raising a family'. He did not feel any special need to conserve his energy.

> And then in the fall of '82 I became aware that I was getting quite tired in the afternoon. And the thing that particularly caught my attention was that on some occasions I was driving home from work and I stopped at a red light and I actually fell asleep. Which had never occurred before. I would just doze; and then to make sure it wouldn't happen again I actually pulled off the road several times going home. This wasn't a long drive – fifteen, twenty minutes. And I wasn't working particularly hard – 5.30, 6.30 in the afternoon. But I just felt this *extraordinary* sense of fatigue.[73]

The next thing Lauro Halstead noticed was that it was getting harder for him to walk the long distances from hospital to hospital on Baylor's huge campus. 'And I was taking the elevator rather than climbing steps. So I became aware that I was now developing weakness in my legs.' But it did not occur to him that these alarming developments might be in any way related to polio. It was only after undergoing a battery of tests which failed to provide an answer that he began to wonder if they might not be connected.

At that time, another specialist in rehabilitation medicine, Dr David Wiechers, came to Baylor from Columbus, Ohio, to give a lecture on electromyography (the recording of electrical activity in the muscles and nerves useful in the diagnosis of neuromuscular disorders). Halstead had heard that Wiechers had just published an article on post-polio, 'so I sort of had the assumption that maybe he had done a lot of study and knew a very great deal, and I made a long list of about twenty questions I wanted to ask him'.

> I got to about the third question and he said, 'Look, put your list away.' He said, 'You can tell me about polio.' He said, 'I studied ten folks.' He said, 'It looks like there are some interesting things going on in the older polio group, but I haven't studied them . . .'

Before they parted, Wiechers and Halstead agreed to collaborate in organising a conference. Their original intention was to get together 'a bunch of elder statesmen in our field, who had seen a lot of polio, and find out what they have to say'. The obvious place for such a gathering was Warm Springs.

Lauro Halstead had had no dealings with Gini Laurie or *Rehabilitation Gazette* – 'I wasn't on the mailing list and didn't know anything about it; I was doing fine.' Now he heard that a second post-polio conference was to be held in St Louis in the spring of 1983. He designed a questionnaire, applying the lessons he had learned about epidemiology during a year's course at the Harvard School of Public Health, and, in addition to sending it out to polios whose names he got from the mailing lists of what had once been the big polio hospitals, he obtained Gini's permission to distribute it at the conference.

The pre-conference response to the questionnaire was disappointing. From the hundreds he had sent out, Halstead received scarcely a dozen replies. So he went to St Louis in a mood of frustration. What he heard there only increased his irritation; the post-polio discussion was so unfocused, and the panel of experts so *in*expert in their responses to the myriad questions raised by the 150 or so polios among the 400 people attending the conference. The role of rabble-rouser had no appeal for him, but for once he could not contain himself.

> So I got up, and made an extempore impassioned speech through the microphone they had there. I said, 'I've been sitting here for the last half an hour and I'm getting very frustrated and upset. I'm speaking now as a person who's wearing my physician's cap, and also as someone wearing my post-polio cap.' . . . I said, 'My feeling is that there aren't any answers.' I said, 'I know of several published articles in recent years, but they report one or two cases, they're of a very anecdotal nature.' And I said, 'Unless we look at this as a legitimate problem that needs to be studied in a scientific way, we're going to be here a year from now, or two years from now, and all bitching and moaning and complaining – but no further ahead.'

The speech had a great impact. People clapped and cheered and came up to Halstead afterwards and told him, 'You really pulled the whole conference together.'

This response, combined with the fact that when he got home he found his mailbox stuffed with replies to his questionnaire – with more coming in by every post – convinced Halstead that he must go ahead and organise the scientific conference he and Wiechers had discussed. But he was no longer so keen on the idea of a gathering of elder statesmen, 'because I figured, hell, if they knew all of this stuff, why weren't they writing about it?' There was nothing on this in the textbooks. He was also afraid that the conference would be given over to reminiscence and argument among eminent medical scientists about who should take credit for what; and at the end of it all nobody would be any the wiser. So instead of going for acknowledged polio experts, he sought out the best and brightest in neurology, pathology, virus research and related fields.

A transatlantic visit in the summer of 1983 'suddenly became a crusade to pick the best minds I could find in Europe'; and everywhere he went Halstead found medical scientists more than willing to participate, such was the lingering fascination of polio. Warm Springs, which was seeking to attract some of its old patients back, jumped at the opportunity to stage the conference, and took on the task of publicising it. In due course the necessary money was raised; and one elder statesman, the great neuropathologist David Bodian, was persuaded to come out of retirement and synthesise everything he had ever written on polio 'in the light of this new problem'.[74]

By setting the scientific agenda and, at the same time, attracting considerable lay and media interest, the conference helped restore polio to its historical eminence in the American consciousness. Such phrases as 'the Late Effects of Polio' and 'Post-Polio Syndrome' were bandied about in newspapers and on television. (Dr Richard Owen of the Sister Kenny Institute recalls that 'there were at least eleven names for the condition', and states his preference for Dr Bodian's 'things that happen to people who had polio years earlier'.[75] (But greater familiarity did not lead to greater understanding, which was hardly surprising considering the experts themselves were at loggerheads over what these 'late effects' were and what exactly constituted 'post-polio syndrome', or if indeed it could be called a syndrome.

The English doctor, Geoffrey Spencer, for example, was scathing about it in a speech he gave at the annual conference of the British Polio Fellowship in 1989:

> People are trying to lump together the many things which can happen and are calling them a syndrome. A syndrome is a group of symptoms caused by a condition or disease. What affects people who have had polio is many different complications. I would not like you to think that the troubles you are experiencing are in any way unreal or that I deny their existence, far from it. A paper published by my unit last year showed that some 78% of a group of 250 who had had polio had experienced some sort of functional deterioration between six and fifty-four years after the onset of the acute polio. It is a real thing but it is by no means a *single condition* and it is destructive to think of it as a single condition [italics in original].[76]

Despite the disagreement over nomenclature, everyone accepts that the functional deterioration experienced by so large a number of polios – usually thirty to forty years after the onset of the disease – amounts to a serious medical problem. But what causes it is the subject of considerable speculation and debate. The most alarming theory – propounded by Wiechers and Hubbell, among others, in the paper Lauro Halstead had heard about before he met Wiechers – links it with the rapidly degenerative amyotrophic lateral sclerosis (the commonest form of motor neurone disease).[77] But this theory is now generally discredited.

Among other theories, one – according to Halstead – 'suggests that there may be persistence of the polio virus or viral fragments that have lain dormant and then are reactivated by some unknown trigger mechanism'; another, 'that an immunologic mechanism may play a role'; a third 'concerns changes in the spinal cord that might compromise motor neuron function'; a fourth hypothesis 'is that the new neuromuscular changes are caused by premature aging of the polio patient'; a fifth, 'and the most plausible in light of the limited information available at this time, suggests that the new clinical changes are a result of motor neuron overwork that eventually produces neuronal dysfunction'. In this last scenario, the 'giant motor units characteristic of postpolio reinnervated muscles . . . may be unable to sustain the metabolic demands of all their sprouts indefinitely'. Progressive weakness results from 'a slow deterioration of individual terminals and a dropoff of reinnervated muscle fibers'.[78]

Geoffrey Spencer employs the analogy of athletic performance, which peaks in people in their twenties – or even earlier if one thinks of tennis, for example – to explain functional deterioration in layman's terms:

> Somebody who is using all their remaining muscle power to perform an everyday procedure like, say, getting their food to their mouth, maintains their muscle fibres in a state of training. If you look at them down a microscope, there are very few fibres but they're huge. They're the same size as a weight-lifter's fibres. And of course a weight-lifter, who trains himself into some sort of magnificent muscle-bound creature, is not growing extra muscle fibres, he's simply making the ones he's got *bigger*. And people who're rehabilitating themselves after polio are enlarging their fibres. So what happens is that when they reach an age when their athletic performance would begin to fall off, they can no longer do something which previously they could just manage.[79]

This is not to be confused with the normal process of aging. 'Everybody starts to lose anterior horn cells when they're over sixty, which is why the elderly get weak.' There is no evidence to support the theory 'that the anterior horn cells damaged but not killed by polio die early'.[80]

But the idea of premature aging persists. Just as the years of retirement are labelled 'the third age', Lauro Halstead calls the onset of post-polio weakness 'the third phase of polio' the first phase being the initial acute illness and the second the period of recovery and stability. In his own case, he conserves energy by resting every afternoon for an hour, using a little scooter to get around the hospital (he now works in the National Rehabilitation Hospital in Washington, DC) where previously he would have walked, and by wearing a lightweight brace on one leg. But he is busier than he was before, having added post-polio clinical work to his already considerable involvement with spinal cord injuries; and he now has a national, quasi-political role as a spokesman on post-polio issues.

After the 1983 St Louis post-polio conference, which also marked the twenty-fifth anniversary of *Rehabilitation* (plus *Toomey j) Gazette*, Gini and Joe Laurie went on a weekend 'retreat' with their board of directors to look back over the past and plan the future. They decided to give themselves a new name, choosing the unwieldy Gazette International Networking Institute for its acronymic aptness: the volunteer Gini had grown into the professional GINI, with 'a salaried executive director and secretary to assist Joe and me'.[81] Although it is such a mouthful, the name accurately reflects the function of the organisation, with the emphasis on networking: 'a new word', as Gini writes, 'for the ancient system of support services supplied by the family and community'. Gini saw GINI's role as the coordination of 'information and networking on the late effects'.[82]

Through the Eighties, post-polio clinics and support groups sprang up all over the United States, often alongside one another: 'The clinics referred people to support groups and the support groups referred their members to clinics.'[83] The archetypal support group, in a sense, was Roosevelt's original Warm Springs, which had at least two of the three characteristics Gini Laurie ascribes to support groups:

> They are voluntary, nonprofit, and charge little or no fee for the involvement of their members.
> Their members are peers or equals who help each other by sharing common problems or predicaments through mutual aid and self-help.
> Their members have a sense of ownership.[84]

The 1981 and 1983 post-polio conferences were 'the catalyst' for the support groups: 'They brought survivors of all degrees of disability together and revived their former togetherness and sharing of problems and information.'[85]

For years most polios had regarded themselves as *sui generis*, unique. A seriously disabled respiratory polio recalls.

> When I first had polio, the counsellors at the hospital said, 'There, look at these little magazines – *Toomey j Gazette* you need to get in touch with these people.' And I was like, you know: No, that ain't me at all. They're polio victims, and they have different things. I'm not at all like that. For a long time, I didn't want to even see or hear anything about that.[86]

Once free of hospitals and rehabilitation centres, polios had tended to shun their own kind. An attorney in Washington DC, who had had the disease at the age of six and had successfully suppressed his feelings about it until he began to suffer the late effects, remarks: 'I don't think I had ever met or spoken to anyone who'd had polio for thirty-some years since I'd had polio.'[87]

These late effects, and the publicity about them, changed many people's attitude. New vulnerabilities made for a new openness. Some groups, like the Washington DC Polio Society, advertised their existence. According to one Washington woman who went to the meetings, their fascination was in the storytelling – the myths about 'how we thought we'd caught polio', for example:

> One person thought she got polio because she pulled her dog's ears when she was told not to. Another person believed he caught polio because he ate blueberries he wasn't supposed to eat. And I mowed the lawn when it was too hot out . . . We associated what happened with misbehaviour. We must have picked that up from the family – some of these people were four years old . . .[88]

That children – and their parents – felt guilty about their illness is borne out in a sociological study undertaken at the time when most of these polios now sharing their memories first had the disease: 'they frequently gave as the cause of their paralysis "messing around too much", "playing when my mother said I shouldn't", or "falling down when I played too rough"'. Among the parents, some 'expressed guilt for having been harsh with the sick child prior to his illness'.[89]

Dr Jennine Speier of the Sister Kenny Institute in Minneapolis always asks the people who come to her post-polio clinics questions about their original polio.

> Many of them have talked about the traumatic experience of having polio as a child. Not sure if it had to do with separation from their parents. The attitude to when children should or should not be separated – there was a tendency to leave the child, because that was better for them, and a tendency not to let them cry and ventilate feelings. Almost a punitive nature to that. And many people, as they told me about it, would actually break down in tears and say, 'This is the first time I've talked about it for thirty years.' It was such a trauma. So I can see where support groups, in particular, can be very helpful . . . It's quite a contrast to how we would treat a child with a disability today.[90]

Different people had different ideas of how to organise a support group: some wanted a proper structure with a charter and membership; others preferred something less formal, more intense. The latter kind worked more like therapy and had a limited lifespan:

> It just seemed like a phase to go through . . . you know, the stages of grief. Once you've passed denial, you have a chance to – I don't like to use the

> word 'grief' because I never really grieved. But it's sort of allowing yourself to think about what happened. And I do sometimes feel that something's missing because I haven't gotten terribly angry. People say, why aren't you angry? Now that people are dealing with post-polio stuff, some anger is coming out . . . you know, why do I have to deal with this all over again? But still, the basic anger of 'why did this happen to me?' – I don't feel that. Sometimes I get scolded for not being angry![91]

Support groups have encouraged this introspection – the psychologising of polio. This same Washington woman who worries about her lack of anger recalls how, when she was young and 'it was all the rage to see a shrink', she went to a psychotherapist and discussed her 'romance problems and career problems and stuff like that'. She mentioned she had had polio as a child, but 'in about three years of therapy we never talked about it – it never came up'.[92]

Through talking to each other for the first time in support groups, polios have discovered what two anthropologists call 'a shared polio pride'.[93] Professor Jack Genskow, a respiratory polio and psychologist who coined the phrase 'the glue that held the polios of the world together' to describe Gini Laurie, is also aware of a new solidarity:

> There's almost a disability pride that's developing among people with disabilities throughout the country, but also in the polios group. I guess maybe at these conferences you see so many people who've led extremely effective and impressive lives. So there's something to be proud of there . . .[94]

This has led to speculation as to whether there is, or is not, a 'polio personality' and, if so, 'whether this was a function of social circumstances, the individual's response to the disease, or some kind of natural selection', as Lauro Halstead puts it.[95] Though the question of a genetic predisposition to polio has been debated at least since the New York epidemic of 1916, when Dr George Draper started to develop his bizarre theories on the subject, the new focus is on the characteristics of polio *survivors*. According to Richard Bruno and Nancy Frick, who conducted a survey on 'Stress and "Type A" Behavior as Precipitants of Post-Polio Sequelae', polio survivors are 'competent, hard-driving and time-conscious overachievers who demand perfection in all aspects of their personal, professional, and social lives'. These traits fit the 'Type A' profile, which was originally intended to define those with a high risk of heart disease (the lower risk 'Type B' people were more laid back). Bruno and Frick conclude that polios 'demonstrate significantly more "Type A" behavior than do nondisabled controls' and that the 'chronic stress' associated with this may initiate or exacerbate the late effects of polio.

> It is possible that adults and even children who exhibited 'Type A' behavior and were experiencing stress were more susceptible to infection by polio viruses because of stress-induced immunosuppression. It is also possible that to survive the acute polio infection and then thrive despite paralysis in a totally inaccessible world, the special drive of the 'Type A' personality was required. It might also be the case that persons with disabilities must learn 'Type A' behavior in order to succeed in a 'barrier-full' society . . . Social prejudice might require persons with disabilities to become 'hard-driving overachievers' – personally, professionally, and especially physically – to be accepted by peers and employers.[96]

Behind the attitudes *of* the disabled are the attitudes *to* the disabled. In the abstract, at least, these are generally negative. As Nancy Frick writes in a paper on the late effects of polio and the psychology of a 'second disability', 'Literature portrays disabled persons as unhappy, hopelessly enmeshed in the difficulties of their disability, bitter and manipulative, or more "in touch with God" because of their suffering.'[97] Mary Johnstone, a woman in her sixties in a short story by the Canadian writer Alice Munro, is ostensibly a good example of the second category. She has been leading 'the annual overnight hike of the . . . Canadian Girls in Training' since before the war.

> Mary Johnstone was a woman you were hardly supposed to mention in Carstairs without attaching the word 'wonderful'. She had had polio and nearly died of it, at the age of thirteen or fourteen. She was left with short legs, a short, thick body, crooked shoulders, and a slightly twisted neck, which kept her big head a little tilted to one side. She had studied book-keeping, she had got herself a job in the office at Douds' factory, and she had devoted her spare time to girls, often saying that she had never met a bad one, just some who were confused.[98]

When one of the girls on the hike goes missing overnight, Mary Johnstone insists on treating her disappearance as a childish prank. After breakfast another girl suggests they go and search for her friend, but Mary says, 'Dishes first, my lady,' and complacently carries on with her time-honoured routine, obliging the girls to sit in a semi-circle and listen to her 'Sunday-morning-of-the-hike sermon'. 'What Mary Johnstone told the girls in her talk was always more or less the same thing, and most of them knew what to expect . . . She told them how Jesus had come and talked to her when she was in the iron lung . . .' Mary Johnstone is not exactly a caricature, she is too subtly portrayed for that. She may be a figure of fun to the girls she patronises, just about tolerated but 'crazy' all the same; but to the narrator Maureen, she is more sinister: someone to be avoided because, despite her inanity, she appears to be able to penetrate

Maureen's façade and discover her hidden faults and flaws. This contradiction neatly encapsulates the social attitude of 'normals' to disabled people, who are simultaneously objects of fear and derision.

> Maureen felt . . . that she was being put on the spot and could do nothing about it, that a challenge was being issued, and that it had something to do with her lucky marriage and her tall healthy body . . . her rosy skin and auburn hair, and the clothes she spent a lot of time and money on. As if she must owe Mary Johnstone something, a never specified compensation. Or as if Mary Johnstone could see more lacking than Maureen herself would face.

In a complex story about perverted sexuality and murder, Mary, the innocent and yet knowing Mary Johnstone, lectures the girls (who snigger behind her back) about the facts of life – facts that have played no active part in her own life. Hers is a prepubescent world, which is why the girls are so important to her. A precocious child among adults, she can play at being a bossy adult among children. When she calls Maureen *Mrs Stephens* 'as if it were a make-believe title and she were thinking all the time, It's only Maureen Coulter', that is because to her it *is* a make-believe title and Maureen is forever young, frozen at thirteen or fourteen, the age at which Mary's life was derailed by polio.[99]

One thinks of Janie, the girl Bentz Plagemann met at Warm Springs in 1944, who tried to stop the clock by going to the candy store every year and making new friends among the next crop of high-school girls.[100]

The sexuality of the disabled was a taboo subject until recently. (In this, as in so many other ways, *Rehabilitation Gazette* broke new ground, carrying articles about it, which others have followed.) Polio affects sexual performance only insofar as a serious lack of mobility curtails the range of possibilities. One polio paraplegic says, 'I don't know what it's like to make love with your legs, you know, and it drives me crazy!'[101] An 'upside-down' polio (one who can walk, but whose arms are useless) remembers being summoned into the office of the director of the respiratory centre he was in for 'the famous sex education lecture' when he reached puberty.

> It was really incredibly funny. When he was asking me about whether I ever had erections and I said, 'Yes, of course,' he said, 'Do you ever masturbate?' I said, 'I can't use my hands.' I mean, I knew I was really getting at him; he hadn't thought of that! And he said, sort of mumbling, 'You know, if you just think about it long enough, sometimes . . .' I said, 'Well, I've never been successful at that.' I wasn't having any of it.[102]

Both these men got married, as so many have done, after having had polio in childhood or adolescence.

Some women, like Mary Johnstone in Munro's story, have not been so fortunate; while their sexuality may be no different from anyone else's, their deformity has effectively precluded marriage. Unrequited love is a painful experience for anyone; for a polio quadriplegic – male or female – confined to an iron lung it can be close to unendurable.

When a group of students was asked to 'rate disabled persons on twenty-four character and personality traits', they did not dismiss the project as so generalised as to be absurd, but solemnly concluded that disabled persons were 'more conscientious, intelligent, kind and altruistic, as well as less aggressive, less able to adjust to new situations and more likely to engage in reverie than the average person'.[103] In a word – though this was not the word they used – *neutered.*

The strongest influence on the attitude of children with polio to their disability was that of their family. During the Fifties, the years of the 'polio *zeitgeist*', the sociologist Fred Davis undertook a study (largely funded by the National Foundation) of 'polio victims and their families' in Baltimore. 'In our society', he writes, 'crippling not only signifies a relative loss of physical mobility but also suggests social abnormality, isolation, and, in the eyes of some, visible manifestation of inherent malevolence . . .'[104]

The focus of Davis's study is on what happened when the polio child returned home after a long stay in hospital. 'Like honeymoons', he writes,

> reunions are of notoriously short duration. The extent to which the parents' optimism was sustained during the months following the child's home-coming depended on a number of factors, chief among them the amount of physical and functional progress the child was able to demonstrate.

He distinguishes between two groups he calls the 'Progress Group' and the 'Stationary Group'. The improvement shown by children in the former 'became for the parents a miniature re-enactment of the classic American success story'; while the lack of it in those who remained 'functionally stationary' typically caused their families to vacillate 'between optimism and pessimism, hope and resignation, conviction and uncertainty'.

Parents in both groups were uncertain how to treat the returning child, 'because of his having had polio, his having been away from home, and, most important, his being handicapped'.

> In general . . . they tended to be more protective of the child and tried to help him much more than they would ordinarily have done. At the same time, however, the parents were concerned lest this 'spoil the child', making him too dependent on them . . .

There was the danger of a 'double-bind' situation, in which parents would 'verbally support the child's attempts to do as other children did while at the same time emotionally discouraging these attempts by holding out affectional rewards for non-performance'.

In the families he studied, Davis detected 'two broad stratagems of adjustment, normalization and disassociation'.

> Families that tended toward normalization denied the manifest social significance of the handicap rather than the handicap itself . . . Families that leaned toward disassociation sought to insulate themselves from contacts, situations, and involvements which might force them to recognize that others, and they themselves, regarded the crippled child as somehow 'different'. Both stratagems, of course, had inherent limitations . . .

Davis points out that these stratagems are not mutually exclusive, and the following anecdote about the eleven-year-old Marvin Harris (a pseudonym), whose 'keen sense of stigma and resultant self-hatred led him – often with the tacit support of his parents – to reject "normals" as well as other handicapped children', reveals the strong urge to 'normalization' even in a family 'that leaned toward disassociation':

> Marvin broke his impaired leg in a fall, and the leg was placed temporarily in a long plaster cast. Marvin was not altogether unhappy about this, because when he was wearing the cast strangers often did not know that he was crippled. When Marvin was riding in a cab one day with his mother, the cab driver noticed the cast and asked, 'How'd you break your leg, son? Playing football?' Mrs Harris replied that this was indeed what had happened. When they got out of the cab Marvin said to his mother, 'Am I glad you didn't tell him I had polio! They just think I have a broken leg.'[105]

A later generation of psychological commentators would label this behaviour 'passing' – 'hiding their disability from other people'.[106]

Dr Frederick Maynard and Sunny Roller of Ann Arbor, Michigan, identify three 'characteristic styles of living with a chronic disability' among 100 polio survivors who took part in an attitudes survey they conducted. They are categorised as 'Passers', 'Minimizers', or 'Identifiers'.

> Passers had a disability that was so mild that it could easily be hidden in casual social interactions. They could pass for nondisabled. Minimizers had a moderate disability that was readily recognized by other people. They often used visible adaptive equipment or had to do physical tasks differently in order to optimally function. They typically minimized the importance of their physical differences. Identifiers were severely disabled by acute polio. They generally needed wheelchairs for independent mobility. Some also needed respiratory equipment. They needed to fully identify with their disability in order to make major lifestyle adaptations and successfully cope.

Paradoxically, those with the least physical disability were the most distressed by the changes in their condition wrought by the late effects of polio; their whole coping strategy was undermined: 'Passers . . . often have their self-image threatened because they cannot pass any longer.'[107] Nancy Frick, writing on the psychology of a 'second disability', tells the story of a mildly disabled polio who was so upset by his doctor's advice that he should use a cane for walking that he told a friend, 'I would not want to live like *that*!'[108]

Sandy Hughes Grinnell, a medical researcher in Pasadena, California, graphically recounts 'A Post-Polio "Normal's" Reconciliation with the Ghost of Polio Past'. To begin with, as is so often the case, Grinnell did not associate the fatigue and upper and lower back pain she was suffering with the polio she had had as a child thirty years earlier; and even after the connection was suggested to her she could not see why she should be experiencing post-polio symptoms when she had apparently made a complete recovery. The very experienced Dr Jacquelin Perry of the Rancho Los Amigos medical centre 'put her hand on my shoulder and said, "My dear, you must understand that polio comes like a thief in the night, and when it leaves it takes part of you with it."'[109] Post-polio normals, Grinnell writes, may look 'solid and healthy on the outside', but on the inside, they are 'like slices of Swiss cheese . . . with big and little holes'. She feels that 'perhaps the more severely disabled members of the polio community have unintentionally contributed to the dilemma that many post-polio normals find themselves in' by being as sceptical about their claim to polio-related complications as many doctors are.

> Possibly, just as we 'normals' fear them as reminders of the things that could have happened, they may look upon the rest of us with resentment because we symbolize the many people who fared so much better than they did at the outset.[110]

Certainly there are resentments of this kind, as one respiratory polio quadriplegic reveals when he talks about some of the other people who came to the support group he attended.

> Several people were there who had polio-impaired walking. I don't want to sound cynical, but you would've thought that was the worst thing in the world that could happen to anybody. All they could focus on was, if they go shopping they get too tired, and somebody needs to do something about it! One woman gave piano lessons, and you would've thought it was the end of her life because post-polio syndrome is making her fingers weaker and she can't play as long, and she is going to have to stop playing in church . . .[111]

Polio lesions, like war wounds, may be graded according to their severity, creating a kind of hierarchy. This is a harmless pastime, and may even be a necessary psychological strategy for survival. Maynard and Roller write of 'Identifiers' that they 'have needed to more fully integrate their disability into their self-image in order to create successful and meaningful lives'.[112] Identifiers, they found, 'most strongly endorsed the statement, "high achievement is a requirement for survival as a disabled person"'. They also have most to lose through polio's late effects, since 'the smallest functional forfeiture can be extremely distressing to a person who has been chronically severely disabled'. Indeed, it may make the difference between partial independence and total dependence. So a dismissive attitude towards those with lesser disabilities is only to be expected.

The 'Minimizers', those with a moderate disability, most strongly empathised with the survey statement, 'I feel uncomfortable around other disabled people.' Maynard and Roller interpret this as being a 'phobic-like reaction' to the dangers now facing them: 'they may feel that to personally begin using a wheelchair signals joining a social group that they have previously devalued, and/or implies defeat, helplessness and not fighting vigorously enough against polio's disabling effects'.[113] Denial was an essential component of Passers' and, to a lesser extent, Minimizers' stratagem for coping with mild or moderate disability, and it served its purpose. As Dr Richard Owen says, 'I don't think denial is all bad. I feel sometimes that post-polio fatigue is like the fatigue in depression, and I think that probably denial protects you from depression – Freud said that neuroses protect you from reality, didn't he?'[114]

Owen is 'not convinced' that there is a 'polio personality' – 'except in the sense that many of us are driven to want to be successful almost to prove people wrong'.

> There are some socio-economic factors, like the fact that polio was much more commonly a disease of the middle and upper classes. So that was sort of a selection process: it chose people who were likely to head towards success anyway . . . And there was the factor of a lot of tension from grown-ups as you were coming out of acute polio. I mean, I think there were teachers who gave me advantages because they didn't want to hurt my feelings, or crush my spirit. In some ways, a lot of us are like only children, whether we have siblings or not.[115]

The anthropologists Jessica Scheer and Mark Luborsky point to the ambiguity in polio survivors' self-image: 'By labelling themselves as "overachievers", instead of high achievers, they may at the same time conclude that, along with much of society-at-large, they did not expect themselves to achieve much at all.'[116] Scheer and Luborsky define 'the concept of a polio personality' as 'a value-laden symbol used by some survivors to explain the source of their current polio-related health problems'. It takes hindsight for these survivors to 'perceive themselves as "living too hard"', since no one 'anticipated the long-term physical consequences of their behaviour'.[117] At the time they were just trying to lead normal lives.

Gini Laurie, who outlived her husband Joe, died of cancer in June 1989, shortly after the Fifth International Polio Conference – which she insisted on attending despite the seriousness of her illness. Her successor as director of GINI, Joan Headley, wisely does not attempt to emulate her powerful predecessor.[118] Gini *was* GINI; she was the '*grande dame* of the independent living movement',[119] inspirational but also autocratic. Joan wants the organisation to be 'the link between the consumer and the health professional'. Though no greater a lover of bureaucracy than Gini herself, she sees the need to professionalise GINI now that it is no longer a 'one-person organisation'.[120]

Like the support groups for which she provided a network, Gini performed an essential consciousness-raising role. Such an energetic, committed and well-loved activist would never have become redundant, but by the time she died this fundamental task had been achieved. More than anyone else, perhaps, she had been instrumental in putting polio back on the map. She could die secure in the knowledge that others would take up the torch whose flame had guttered and almost been extinguished following the National Foundation's change of direction at the end of the Fifties. Post-polio conferences and the unresolved medical questions surrounding the Late Effects have ensured a continuing interest in this disease – at least as long as the old polios of the Forties and Fifties are still around.

In the West polio is largely a disease of the past, even if its effects linger on; but in many parts of the world, particularly in India and China, it is still rife; and repeated claims by the World Health Organisation that it will be eliminated by the year 2000 seem optimistic. In the United States, there are an estimated 641,416 survivors of paralytic polio; in India alone there could be as many as twelve million, and the subcontinent may account for forty per cent of new cases worldwide.[121] Dr Geoffrey Spencer, who visited China with an orthopaedic surgeon colleague to learn about a knee operation being performed there (see Introduction), writes:

> It has been estimated that there are still over a million cases of polio in the world per annum. This figure cannot be more than an informed guess because notifications from those countries where polio is still rife are admitted to be grossly incomplete. For example, China reports around 18,000 cases a year and this is certainly a serious under-assessment. Vast numbers of people in the densely populated rural areas undoubtedly die without the condition being even recognised. Second children are often not notified at birth, nor their diseases diagnosed. Admitting to a second child leads to

> severe parental penalties, so the death of a second child could be a merciful release to state and parents alike. There is no national immunisation programme, and none is at present feasible. Elimination of death, if not disablement, from polio must be a relatively low priority in countries already faced with severe overpopulation. India has already recognised that the introduction of smallpox vaccination substantially increased this problem and it has been suggested that there may be as many as 300,000 to 400,000 new cases of polio per annum there.[122]

In England and Wales, by contrast, there have been less than 100 new cases in the past two decades; and in the six years between 1985 and 1991, there were just twenty-one, of which thirteen were vaccine-associated and five were 'imported' (two from Pakistan, two from Bangladesh and one from Morocco).[123]

Albert Sabin was guest of honour at GINI's Third International Polio Conference in St Louis in May 1985. He told his listeners that throughout the world there was 'more paralytic polio now than we had in the United States or in the temperate climate countries of Europe before the introduction of vaccine'.[124] The WHO then estimated that there was an average of 400,000 new cases of paralytic polio a year in the tropical and subtropical countries. (By 1990 the official figure had dropped to 116,000 cases worldwide,[125] but none of these figures can be relied upon, given that in developing countries they are often based on 'lameness surveys' of children and the 'reported numbers of poliomyelitis cases [may] represent only 5% to 10% of all cases'.[126]) An article in the *Journal of the American Medical Association* suggests that 'the high rates of polio in developing countries have required a reassessment of the traditional theory that paralytic poliomyelitis is a "disease of civilization", resulting from improved sanitation and hygiene in western countries'. As Sabin pointed out, 'it is quite evident you can have tremendous amounts of paralytic polio . . . [even where] sanitary hygiene is very bad'.[127]

Yet Sabin's observation that most of the epidemics in tropical countries were among infants of two to three years, suggesting to him that older children had already developed immunity, seems to contradict his refutation of the 'traditional theory' – at least insofar as the older children are concerned. What is not clear is why there should suddenly be *epidemics* among very young children. A possible explanation might be that there always were, but they went unreported. A tempting interpretation would be that sanitary hygiene has improved in these tropical and subtropical countries to the extent that they are now roughly where the temperate-climate countries were a hundred years ago. Yet in Toluca, in Mexico, Sabin and his colleagues found both 'an extremely high rate of poliovirus dissemination early in life' *and* an annual average incidence of paralytic polio higher than in the United States during the epidemic years 1951–5.[128]

According to traditional theory, one was supposed to preclude the other. Polio, it seems, has by no means yielded all its secrets.

Sabin's oral polio vaccine (OPV) is still the vaccine of choice throughout most of the world, even though its performance in tropical countries leaves something to be desired. There have been problems keeping it at the right temperature, and interference from other viruses prevalent in tropical climes has reduced its effectiveness. Salk's inactivated vaccine (IPV) has now come back into contention, especially since its antigen content has been increased and it has been successfully combined with DPT – the diphtheria, pertussis and tetanus vaccine.[129] Some experts favour a combination of IPV and OPV for maximum safety and effectiveness, provided the cost is not prohibitive.

In 1985 the Pan American Health Organisation set itself the goal of 'eradicating the indigenous transmission of wild-type poliovirus in the Americas by 1990'.[130] The three key elements of this campaign, which succeeded in reducing the number of *confirmed* cases of polio throughout the region from 800 in 1985 to nine in 1991 (though one wonders just how many cases went unreported in those countries where children run wild and might just as easily die of polio as be shot in the streets), were:

> Firstly, the use of 'immunisation days' or mass campaigns, which aim to immunise the whole child population with one dose over a very short time, ideally one day, and then to repeat this two months later . . . Secondly, 'mopping up' operations, which consist of identifying parts of the country or continent with a continuing high incidence of disease and targeting them for intensive [house-to-house] vaccination. For example, in the Americas 95% of cases were found in only 5% of districts.
>
> The third element is high quality surveillance . . .[131]

Brazil embarked on its mass vaccination programme in 1980. With almost twenty million children under five years of age scattered over a vast territory, it had huge logistical problems. Yet, according to Sabin, it 'organized an army of about 320,000 volunteers at 90,000 vaccination posts that were within walking distance' of every child; and this had to be repeated, not just after two months, but twice a year thereafter.[132] This intensive coverage has been repeated throughout the Americas and is now the model for the global eradication of polio.

The involvement of Rotary International, 'a private voluntary organization with membership in 160 countries', in what is called the 'PolioPlus' effort to eliminate the disease has increased its chances of success, if not by the year 2000, by 2005 – which is Rotary's centennial.[133] But even if one can believe that – as the *1993 State of the World's Children* report of the United Nations Children's Fund asserts – the number of children in the world paralysed by polio has been reduced from 'more than half-a-million' to 'more than 100,000' in a decade, and that the number immunised against polio has 'leaped up' from twenty to eighty-five per cent in the same period, the fight is still far from over.[134]

Even in the West there is no room for complacency, as outbreaks of paralytic polio among the unvaccinated demonstrate. The communities most at risk are the very poor, whose inadequate access to medical care means that their children often slip through the immunisation net, and those who reject vaccination on religious grounds. In 1979, there was 'an outbreak of polio from wild-type virus' in the Amish community of Pennsylvania.

> This outbreak was linked, through an Amish family that had moved from Ontario to Pennsylvania, to the same virus strain responsible for 1978 epidemics in the Netherlands and Canada. While still in Canada, the family had come into contact with healthy members of the Dutch Reform Congregation who were visiting from the Netherlands, and who had presumably introduced the type 1 wild virus into the Canadian province. There were 110 cases of acute polio in the Netherlands epidemic (80 paralytic), and 15 acute cases in the United States and Canada (all paralytic). With the exception of one boy, who received vaccine five days before the onset of illness, none of the victims had been vaccinated.[135]

In 1992, there was another polio epidemic among unvaccinated members of religious groups in the Netherlands. Since the Netherlands is one of the few remaining countries to use the Salk inactivated vaccine for routine immunisation, these outbreaks provide Sabin supporters with an opportunity to argue for the superior utility of the oral vaccine in providing 'herd immunity' by spreading a harmless form of the disease even to unvaccinated children. But whether the future belongs to Salk or Sabin, or a combination of the two, or some 'new vaccine strains derived from the Sabin vaccines by recombination, further mutation, or gene insertion',[136] remains to be seen.

Part II

Lives of the Polios

The idea of a crippled child of any kind gets people all stirred up. And, by and large, polio survivors seem to be a courageous, determined people who evoke pity or admiration. They don't hide themselves away – so it seems. They make heroic efforts to be self-sufficient. They go on to college and become president of the United States – or, at any rate, parents themselves. With any kind of help, many make a winning fight.

Greer Williams, *Virus Hunters* (1960)

The Other Boy was always with me. He was my shadow-self, weak and full of complaints, afraid and apprehensive, always pleading with me to consider him, always seeking to restrain me for his own selfish interests. I despised him, yet he was my responsibility. In all moments of decision I had to free myself of his influence. I argued with him then, when he would not be convinced; I spurned him in fury and went my way. He wore my body and walked on crutches. I strode apart from him on legs as strong as trees.

Alan Marshall, *I Can Jump Puddles* (1955)

In England

This section is made up of seven life-stories. It could have been three times as many, since I talked at length to more than twenty people in different parts of the country, every one of whom had an interesting story to tell. The majority of them contacted me in response to a note I put in the Bulletin *of the British Polio Fellowship. Others were already friends, like David Widgery, or public figures, like Ian Dury. What these seven had in common was that they had all contracted polio during the epidemic years of the late Forties and Fifties, before the arrival of the Salk vaccine.*

The lives complement or contrast with one another in a variety of ways, and – to highlight both their similarities and their differences – I have interwoven them and divided each into three parts, beginning with the onset of the disease, following this up with the stages of recuperation and rehabilitation, and concluding with the long-term effects on the life. As far as possible, each story is told in the individual's own words.

1

David Widgery never forgave his family doctor for misdiagnosing him. 'You'll think it's polio,' she told his mother. 'But it isn't. What you've got to do is make him do exercises and snap out of it.' His father dutifully had him doing press-ups and leg-strengthening exercises – the worst possible treatment during the active phase of the disease.

At one moment the six-year-old David was lying on a rolled-up carpet while his mother hoovered around him, unable to move when she asked him because he felt so tired. The next, he was in one of a row of glass cubicles in hospital, saying goodbye to his father.

> They were little goldfish bowls; my mother was allowed to come to the door and read me stories. She read me stories of the sickleback sailor and I

> hung on her words. Mothers were only allowed in for half-an-hour a day or something. You cried yourself to sleep, and you'd wake up and then you'd see another little boy crying himself to sleep in the next cubicle. It was awful.

A cleaner who did not know how ill he was made him get out of bed; he fell to the floor and could not get up again. He dimly perceived that something was wrong: 'Like when the car lights won't come on when you press the button, something had gone and I didn't realise at all . . .'

After a period of quarantine, he was transferred to a children's ward. He was the only one there with polio and was picked on by the other children: 'We used to get issued with sweets and they used to throw the sweets at me – quite a sadistic sublife on a children's ward.' The staff seemed to have little knowledge of polio, and a *laissez-faire* attitude: 'He'll get over it, sooner or later.'

> Then there comes the glorious era of physiotherapy, where there are all those beautiful – even at that age I realised it – buxom ladies with belts of different colours like karate belts which linked in complicated ways in the middle, and sort of pushing and pulling . . . The only other thing I remember about being in hospital for so long was that I had to teach myself to read, and I still can't spell because I taught myself wrong.

A few years earlier – in 1949, during a hot August – seven-year-old Ian Dury had been going to the local swimming pool (just as David Widgery had in 1953). He went to Cornwall to stay with his granny, who was a doctor's widow. One afternoon he felt giddy and lay down on the settee while his granny made him a cup of tea. The last thing he remembered was being unable to get up from the settee; then the ambulance came.

He was in isolation in Truro Hospital for six weeks:

> My mum got the milk train down to Truro because they told her I wasn't going to make it. It was in nissen huts, barrack huts, an RAF-type hospital, and they let her come to the outside, and they turned the light on over my bed – you know, she saw my little white face on the pillow: a four-o'clock-in-the-morning job. But I was still there next morning.

Both arms were affected, as well as his legs, but not his right hand. So when he was put in a body-plaster – 'six weeks, like, in the bottom half of it with bandages to hold you straight' – with his arms in the air, he was able to hold a dinky toy in it: 'And if you dropped it you were fucked if the nurse didn't like you . . . Oh yes, prisoners to hot milk in the morning, this bloody little jug thing coming towards you; it looks like a teapot but it's for pouring into people's mouths – "Wah, wah, don't want this!"'

After six weeks Ian was transferred to Black Notley Hospital in Essex, near where his mother lived. He was in a big ward there for a year and a half, with the children all at one end of it. If they were naughty, they would be wheeled up to the other end and placed under the surveillance of a man in an iron lung, who watched them in the mirror above his head: 'Shocking, they did things like that.' Worse still:

> Every Tuesday and Thursday they put you into this manipulatory situation. Obviously, they thought they were doing you good, stop you ossifying or whatever the word is. But I mean this geezer used to get hold of my left ankle and my left thigh and, *kkkkkrrr*, put my heel up to my arse – keep it loose, fucking hell. It was called the screaming ward and you could hear people screaming on the way there, and it was you when you was there, and you could hear the others on the way back – I imagine, I can't really remember, I've blocked it out.

Ian has no recollection of learning to walk again, though he must have been able to get about before he went to Chailey Heritage Craft School for Boys and Girls: 'I don't think I would have been able to go to Chailey if I hadn't been mobile. The main ingredient at Chailey was that you had to be able to look after yourself.'

Maureen O'Sullivan had polio when she was just three months old. Her earliest memories are of 'shuffling about on the floor' (she could not walk at all) and 'being lifted and carried and cossetted' by her mother. Her father worked on the railways and she was the second child in what was to become a large family, living in Kent. Her troubles began when the next child, her younger brother, was born: 'I needed everything doing for me, and was very jealous of this baby boy that had come along and also needed lifting; I was very nasty to him and would push him off the bed at any moment.' Her harassed parents, in their mid-twenties, sought advice on what to do and were told of a place for disabled children called Chailey Heritage in East Sussex. Maureen was taken there when she was three:

> I remember the day. It was poppy day. My mother put me down to play – as I thought – with other children, and I was thinking, 'Oh this is fun, it's going to be wonderful.' And then came the actual parting: it was absolutely horrendous, absolutely horrendous. There are times I remember, like you were punished if you cried: you were put in a cubicle because you upset the other children. For a child who'd known nothing but a mother and father, to be put away for crying and only allowed out if you stopped was, um, very difficult . . .

Jim Porteous was on a family holiday on the north-east coast when he was taken ill at the age of six:

> The symptoms were the usual headaches and that sort of thing. The doctor identified it as possibly polio and I went into the fever hospital in York, and spent a period in an iron lung before being transferred to the Adela Shaw Hospital in Kirkbymoorside.
>
> With polio there is a rapid deterioration, and a period of wondering when and where it is going to stop. With me it stopped at total paralysis, including interference with the breathing, but leaving some mobility in the neck and minor flutterings in the arms, hands and feet.

Jim spent several months in a head-to-toe plaster-cast and was in and out of an iron lung: 'The prognosis at that stage was really of the "vegetable brigade", that there was little chance that I would sit up, never mind doing anything useful.'

> I was fortunate that my parents were not overly happy with this scenario. My father had occasion to be in the States on business and came across details of the Sister Kenny method of treatment, which was being practised at only one hospital in the UK and that was Queen Mary's in Carshalton. When he came back from the States he went to see the local consultant, who advised against this method: 'No, this is a fanciful treatment. The best bet for your son is to be plaster-casted to avoid deformation as far as possible, and you may at some later date be able to nurse him at home . . .' But father was not prepared to accept that as an answer and got in touch with a chap by the name of Dickie Metcalfe [chief orthopaedic consultant at Queen Mary's]. Dickie Metcalfe said to father, 'Look, just tell your local consultant to take a running jump, and bring your son down here.' Which he duly did.

As a child of six, Ann Stevens had had TB meningitis. So the medical officer of the army unit in Benghazi where her father was stationed was not taking any chances when, at the age of nine, she fell ill with what he thought was probably flu. The family had been in North Africa for only a few months and Ann, who had been swimming a lot, started getting bad headaches and running a temperature. When her breathing became laboured, too, the doctor rushed her into hospital. But there was no iron lung in the military hospital in Benghazi, so the unit of Royal Electrical and Mechanical Engineers had to make one out of a coffin. Ann was a little frightened of this homemade respirator at first, and found that the black rubber collar which acted as an air-seal pulled her hair. But she soon became dependent on it and loathed being taken out, as they attempted to wean her off it.

> I can remember feeling very special. It was Christmas time and I was the only polio case in the hospital, and there were lots of young soldiers around who would come and read me stories; the wives' club made a collection for

me and gave me Christmas presents – it was just a lovely time for me as a child. I didn't understand what the repercussions were going to be.

As soon as she was out of the infectious stage and able to breathe on her own, she was sent home to the United Kingdom:

> I remember being very frightened flying back – my parents didn't come with me. I remember feeling very lonely and sad. It was a hospital plane and we were on stretchers; there were people above me and it felt very claustrophobic. We stopped overnight – I think it was Malta, at the hospital there. It was so frightening to be on my own and not know anybody, and they didn't seem to know how to look after me really. Then we flew back and I think I was another day in Tidworth Military Hospital . . . and then arrived at Alton, the orthopaedic hospital there. I just remember feeling on my own and very vulnerable.

Ann spent about a year in Alton. She had contractures in both arms, which were put in plaster splints, but only at night, During the day she had physiotherapy to straighten out the arms: 'I can remember lots of passive exercise.' She also had hydrotherapy, which she hated despite her earlier enjoyment of swimming:

> I was very frightened of the water – perhaps it was the breathing. I was okay as long as I had a really good buoyancy aid, like a lifejacket. I hated getting my face anywhere near the water, I still do. And the performance of getting undressed, getting in the water, and then getting dry again – all of that seemed such a hassle. Even at that age I didn't like being handled . . .

Simon Parritt's parents took him to the circus soon after Christmas 1955. Looking up at the high wire gave him a headache, so he had to look down at the ground for relief. By the time he got home, his headache was worse and he had a stiff neck and a very high temperature indeed. The family GP was sufficiently alarmed to call in a consultant, who got Simon out of bed and discovered that he could no longer move his arms. He was taken to a local hospital in south London but, because it had no iron lung, he was transferred to the Western Hospital in Fulham within twenty-four hours.

> My parents weren't allowed to visit me for a while; they were only allowed to look through a window in the ward. All I can remember is being in a darkened environment . . . and being fed through a tube – raw egg and milk, I think, was my favourite. I guess I was fed other things, but raw egg and milk was the only thing I could vaguely taste. How I could possibly taste it, I don't know. Maybe I picked up the smell or something. When I came out of hospital, I tried it and it was disgusting.

After a few weeks the process of weaning Simon out of the iron lung began: the lid would be lifted up and he would be left to breathe on his own, first for three minutes, then five, and so on. He found it scary, 'although I probably didn't tell anybody'. He was in an open ward and there were people at different stages of recovery, some in lungs, some on rocking beds.

> What was extremely unpleasant was the physiotherapy, it was so painful just to move – there was a lot of pain, it wasn't just stiffness. My physio used to walk in at nine o'clock, and I remember there was a central speaker with the Light Programme, or whatever it was in those days, playing *Housewives' Choice*, and the music for *Housewives' Choice* used to precede the physio by about three minutes: it was like the knell – or toll – of doom. Physiotherapy was the hardest thing. However, I progressed, spent more time out of the lung and eventually got on to the rocking bed. I'm not sure how long I was at the Western – probably three, four months. As far as I can remember, I was off all breathing aids and sleeping on an ordinary bed by the time I left.

Mary Gould was a young adult, nineteen years of age, and a resident in Halls at Nottingham University. In December 1951 she thought she was going down with flu: she felt unwell, had a headache but no temperature. She got little sympathy from the bursar, who told her that if she could not go to lectures she should work in her room.

> But I was conscious of feeling very hot and *oddly* ill, and later on I was sick and went to bed. I have a very vivid recollection of getting up in the middle of the night to go to the loo and having to lean my whole body against the swing door and thinking, 'How odd, I usually just push it.'

An intelligent doctor, who discovered that she could not drop her head on to her chest – a crude test of polio – sent her to hospital for a lumbar puncture. There she was diagnosed and promptly isolated and put into one of the new Nuffield iron lungs – named after their sponsor, Lord Nuffield.

> I was supposed to be dying, and my mother was summoned. I lost all breathing and I remember telling the doctor that I was finding it hard to get rid of a bit of phlegm in my throat. He looked terribly serious about that, and he told my mother that it looked as if it were beginning to attack brain cells. He said, 'I'm sorry we can't do anything more,' and she said, 'Well, we can pray,' and he said, 'You can do that, of course.' She duly prayed, and so did a lot of other people, and I didn't die that weekend and I didn't die for the next fortnight . . .

Mary's was an isolated case, there was no epidemic in the university (apart from any other consideration, it was the wrong time of year). No longer a child and 'robustly healthy', she pondered why she should have been vulnerable to the virus:

> I wouldn't have thought that I'd got any problems. Very socially assured, I was okay, I was confident I could handle it all. But I was aware – in a kind of terrible moment of despair that occasionally hit me – that I wasn't able to catch up with work, that I didn't know how to work properly. Okay, I'd scrabbled through the first year, and instead of coming in the top ten I was down at twenty-seventh or something. But I didn't know what to do about it, I was many essays behind . . . I thought, 'I won't be able to do this.' The idea of failing was just not on. You see, I was a person who passed exams. I got on and did things – I mean, I didn't have problems; I wasn't one of those people. When tutors said, 'Come and see us if you have any problems,' it wouldn't have occurred to me to go. It just needed a bit of willpower: 'I really must get down to it – next Monday.'

Both Mary's parents had been teachers. Her father was killed in the war, and both she and her brother were shunted off to boarding schools. At university, she had intense friendships – 'more with girls than boys really'. She would look at other girls, who successfully combined working hard for their degrees with busy social lives, and wonder how they did it, feeling that she had 'missed out on the key'.

> I can remember not knowing how to cope with the demands of my friends on me – a bit of jealousy and all the rest of it – of standing almost physically motionless in the corridor going, 'Damn, damn, damn' and, as I didn't swear, it was a big thing for me. I've sometimes thought that getting polio was a good escape. I can remember lying in the iron lung, knowing it was a bit serious, everyone being upset, but thinking, 'Oh, I'm so glad, nobody can find me here.' I didn't want to see anybody from college for ages. Not having visitors wasn't difficult for me, I didn't want them. I had the feeling that I was hiding, escaping, and I wouldn't have to do the essays.

2

There had been a mixture of adults and children in Simon Parritt's ward at the Western Hospital in Fulham; but Queen Mary's, Carshalton, was a hospital for children only. The treatment there was different, too: it was a version of the Kenny method, with hot packs, hot baths and physiotherapy. In contrast to his earlier experience, Simon enjoyed physiotherapy at Carshalton, mainly because it took him out of the ward, where his life was a misery. 'I looked forward to it. The physios were more affectionate towards you, they held you more. There was human contact, whereas on the ward it was very distant and unpleasant.' The ward was run on old-fashioned, military lines, with the sister very much in charge – 'I think of it almost like a concentration camp.' Visiting hours were few and far between, and rigidly adhered to. At the end of a session, the sister walked round the ward clapping her hands and saying, 'Come on, parents out.' Worse than that – though Simon was unaware of it at the time – parents had to wait outside the hospital until the gates were opened to let them in, 'just like prison visiting'. And during those precious moments when his parents were there, Simon was too frightened to tell them what went on for fear of reprisals.

> My mother was increasingly concerned because I was becoming more and more withdrawn. I didn't speak to them when they came in and then I used to cry when they left. She thought it was a complete character change. Later she said to me, 'I should have noticed, because when you were admitted they commented on what a bright, rosy little child you were, and' – she said – 'in five or six weeks you'd gone completely pale.'

So what was going on?

> The nursing care was just appalling. I was largely incontinent unless I had someone there to give me a bedpan or a bottle the moment I needed it, because I couldn't move at all. However I was strapped in the bed, and strapping in the bed was used as a punishment if you wet the bed or shit yourself or whatever. Looking back now, I can hardly believe that young kids, some of whom couldn't have been more than a year old, were smacked for wetting themselves.
>
> I had difficulty swallowing and breathing because I'd had bulbar polio, but when I didn't want my food they used to hold my nose and stuff greens down my throat – greens not being my favourite thing . . . I was put out in the sun at one stage and just left there, and I ended up with a sunburned face, quite severe sunburn. That was another way of punishing you – roll you out on the bed and leave you there in the sun. These are things that are difficult to pin down if you're trying to accuse somebody, but when you're experiencing them you know. Also things like leaving me unattended in the bath – I mean, I could've drowned because I couldn't hold myself up, I was still virtually completely paralysed.

When Simon's parents did manage to wheedle the truth out of him and then complained to the staff nurse, he suffered for it in all sorts of niggling ways: the hot packs were put on a little bit *too* hot; he was handled more roughly than usual; and, if he asked for a bottle, he was told to wait his turn. Eventually – and not without considerable difficulty – his parents succeeded in having him transferred to the Royal National Orthopaedic Hospital at Stanmore, in Middlesex, where the atmosphere was completely different: 'I remember wetting the bed and being terrified of what would happen, and they just couldn't work out what was wrong, you know.'

The nursing staff at Carshalton had warned Simon that if he left Queen Mary's he would never walk again; yet by the time he was discharged from Stanmore almost a year later he could just about walk: 'They had a huge back-end of a bus in the physio department, and once you could get up the steps of the bus you could go home.' But he still could not dress himself or go to the toilet unaided.

> I got better upside down. My legs got better than my top, so I was in the strange position that if I fell over I couldn't get up because I couldn't use my arms; yet I could get about. My problem was keeping out of the way of being knocked over. After I left hospital – apart from regular check-ups because my spine eventually twisted and they were keeping an eye on that – it was just a lot of denial that I was in any way disabled.
>
> I used to have strange dreams – my mother would be horrified if she knew. But when I came out of hospital, I used to dream that my parents were going to poison me, or kill me in some other way. I guess it was insecurity. For a while afterwards I had these dreams that the coalmen were chasing me and they were going to pour coal all over me. Maybe that was about going into the iron lung – or that's how I rationalise it now. I remember when I was eight or nine I used to have a dream about being one of the Three Musketeers, but whenever there was a fight my sword was always broken and I had to run away. It's fairly clear what that was about. But I didn't have many dreams of that sort after I was twelve or so.

As a result of the years he had spent in hospitals, Simon had fallen behind educationally. To begin with, he went to a girls' school (which took very young boys) and for a time he was the only boy in his class, but that could not continue.

> The problem was that the state system wouldn't take me unless I went into a special school [for the disabled], so it was either going to a special school or to a private school. I ended up in the private system, which I suppose – educationally speaking – was an advantage, but I don't know whether it was in other ways. I mean, prep schools are very into sport and I was excluded from that. I did join the choir; that was a major part of what I did. But I was always a bit behind.

Simon hated school, though public school was less of an ordeal than prep school: 'On the whole, the older you got and the more they treated you like an adult and you could choose what you wanted to do, the easier it was to deal with.' He left school at fifteen because his back was giving him pain and his breathing was getting worse. The local authority provided home tuition, and that enabled him to catch up academically, at least to the extent of acquiring one or two GCE O-levels.

The day after Boxing Day 1967, when he was just seventeen, Simon had to go back to Stanmore for a spinal operation.

> They put you in a plaster jacket, then for the first few weeks they stretch you bit by bit by having these bolts that push the bottom and the top half apart. Basically, when you pass out they can't stretch you any more. The pain is incredible – you think, 'Well, I've got to carry on, and it'll go.' Because initially it only lasts for about an hour. But the longer the stretching goes on, the more it becomes a problem. Then they put in these bone chips and fix it – or they did in those days; it's different now . . . Well, I suppose I was in for about six months altogether, and then had to rest in a Milwaukee brace for another three to four months. But it was very successful as far as my straightness went because I was stretched about three inches. The main thing was that the pain went, and without pain you have more energy. So I was a lot fitter and ready to face the world again, so to speak.

When he came out of hospital, he took more O-levels and went to art school in Croydon. This was 1968 – September '68:

> I was involved in quite a lot that was going on in art schools in the late Sixties, which was exciting. In our foundation year group we had an age range of seventeen to thirty-five, but it was probably the most cohesive group I've ever been in. So it was a good feeling, staying together for a year.

In 1969 Simon went to Goldsmiths' College in London for a three-year course in graphic design. His ambition was to do film and television work, and he was doing this in his spare time at Goldsmiths', driving all over London in his newly acquired adapted car, 'spending Goldsmiths' money'. 'The late Sixties and early Seventies was an interesting time to be at art school. I learned an awful lot about art in general, but not how to draw . . .'

When Jim Porteous made the train journey from Kirkbymoorside to London, *en route* to Carshalton, the window had to be removed from the railway carriage to admit his full-length body-plaster. But when he arrived at Queen Mary's, 'they just ripped the plaster cast off and threw it away – never to be seen again'. Then began the regime of hot packs and hot baths.

> To this day I can remember the smell of a hot pack. The other memory I have is of the scalding hot nature of them when they went on, the relatively short period of time when they were warm and comfortable, and the inordinately long period of time when they were like having a wet nappy on. They were changed by nurses who wore caps that would have been the pride of an origami master – individually hand-folded butterfly caps of the

> old variety. They were a fascination to young boys, who used to knock them off as often as possible. When these hot packs were taken off and replaced, you were treated with methylated spirits and powder to prevent bed-sores. And to this day, whenever we have a fondue and pour methylated spirits into the burner, I immediately go back to that.

Jim was almost totally paralysed. Over the next two or three years, constant physiotherapy paid dividends as far as his arms and shoulders were concerned, but not his legs. Yet 'the improvement made the difference between humanity and non-humanity, I mean, between self-respect and no respect'.

> The regime at Queen Mary's in Carshalton at that time left a lot to be desired. My mother tells horrific stories of coming all the way from York and not being allowed in. My parents travelled up and down to London as often as they could, at least once a week. On the other hand, I had a number of people I called Uncle This and Uncle That, and in later life I discovered they were just members of the Carshalton Round Table who took it upon themselves to come and visit patients in the hospital.

When he returned home to York at the age of ten and started going to school in a wheelchair, Jim found that his peers were about to sit the eleven-plus exam.

> I have both failed and passed my eleven-plus. I sat it fairly soon after I came out of hospital and failed it; and somebody on the North Yorkshire education committee decided – on application from my dear old father – that this was ridiculous. So I had another year at primary school and passed my eleven-plus at twelve-plus. Even now, in a relatively comfortable management position in a big firm, I still can't spell. I put it down to having missed that chunk of learning. The standard joke about me is that I spell *phonetically* with an F.

Once Jim had successfully negotiated the eleven-plus, he and his parents were faced with a similar dilemma to that of Simon Parritt's parents.

> A number of schools wouldn't take me, so the prospect was Welbourne Hall, a special school for disabled kids, which was like – you know – going back into the ghetto. The only other two options were Archbishop Holgate's Grammar School in York, which was totally unsuitable for a wheelchair – but the headmaster was prepared to have a go – and Gordonstoun. Gordonstoun said, 'Yes, okay, you pay the money and we'll cope with him.' The decision was left to me. Socially and financially, I might have been in a better position had I gone to Gordonstoun – I'm realist enough to know that the old boy network still works, and there have been some good old boys there, not least Prince Charles. But my reasoning, which was accepted by

> my family, was that I'd already missed out on four years of being at home and I didn't want to go away again. So I went to the local school and the education authority paid for a taxi backwards and forwards.

A physiotherapist came to his house twice a week for years, trying to get Jim on his feet and walking with full-length calipers joined by a waist-band, but it was a losing battle: 'I was moving myself by sheer brute force from my shoulders, and it was absolutely exhausting – okay, one night you might manage twenty yards, but it took you half an hour to get into all that gear, so what use was that?'

Neither Jim nor his parents allowed his disability to get in the way of a full social life. Being 'one of the lads' was more important to him than doing his homework. He joined the scouts, and has been involved with the scouting movement ever since.

> Between the ages of eleven and thirty-five I would do a week's camp out in the Dales with the scouts every year. In the early days I used to sleep on the floor and two lads would throw me into the chair in the morning; and I took a toilet chair along with me in the back of the lorry. I can remember going on my first camp with some lads on bicycles: I had one of those three-wheeler pump-action chairs and we just slung a rope around the back of these kids' bikes and we were off! I spent the night sleeping in this thing with the back down and the tent sort of built round me. One scouting project was to get me to the top of York Minster, which we did. Under normal circumstances my chances of getting to the top of York Minster are remote . . .

Because he was still having difficulty putting on his trousers and doing things of that kind for himself in his mid-teens, Jim was sent to the famous paraplegic centre at Stoke Mandeville for training in what, had it been Warm Springs, would have been called ADL – activities of daily living.

> I found Stoke Mandeville tough, I would liken it to a concentration camp in terms of its regime. You had a tight programme and you can bet your sweet bippy that the bloody table tennis is at the other end of the hospital from the club swinging you're supposed to be at in half-an-hour's time. That was not a happy time . . . It was mainly homesickness, missing the friends I'd spent four years getting to know – and moving in a circle of strangers who were also having trauma, but whose trauma was not a long-lived trauma like mine. It was unusual to find polio-disabled at Stoke Mandeville. It's mainly a spinal injuries place and I felt they didn't fully understand that I couldn't do some things, not because I didn't want to, but because I didn't have the balance to do them. If you break your back, okay, you don't have any feeling in your legs, but when you have polio you do have feeling in your legs; and therefore to throw your legs on or into something can actually hurt.

In his struggle for mobility and independence, Jim 'couldn't wait to get a set of wheels'. He could not drive a car until he was seventeen, but the three-wheeler invalid car came into the motorcycle category and he could learn to drive that at sixteen. So he got one on his sixteenth birthday and, as soon as he had passed his driving test, proudly drove to school in it, instead of a taxi: 'There was one wonderful occasion when somebody stopped a chemistry lesson I was in to say, "Sir, I'm terribly sorry, but Porteous's car has blown over."'

Jim's sixteenth birthday was in 1962, when pop music was young and the youth scene divided into Mods and Rockers.

> There were lots and lots of bands around. There was another guy in a chair in York at the time and he managed all the Rocker bands, and I managed most of the Mod bands. That was they way we got involved in the scene. Obviously you couldn't take part in the dancing, but the bands needed somebody who could speak on the phone and keep the books, who knew somebody with a van, or somebody who could raise £50 for an amplifier or whatever.

So Jim, too, had a role in the teenage rebellion.

All attempts to get Ann Stevens walking again failed. The physiotherapists strapped her legs into calipers, which they then attached to the back brace she had to wear, and stood her up: 'It was crazy, it really was; I was just like the steel man – all the stuff was heavier than I was.' In the end they settled for a wheelchair. After that, Ann concentrated on getting back as much muscle power in her upper body as she could.

> I remember the muscle charts, and the subjective assessments, you know, 'Is it a two or is it a three, or two-plus or three-minus?' I quite enjoyed all that. The two physios who were working on me became great friends; they were both lovely girls. I suspect there weren't many like me – I think I was fairly cheeky.

She also had to do occupational therapy – 'weaving things and stuff like that' – and schoolwork: 'But I couldn't write very well, I didn't have enough strength in my hands.' She had always been a voracious, if indiscriminate, reader. Now there was the question of where she should go to school and her parents had 'a battle royal' with the educational authorities.

> The director of education said I should either have a home tutor or go away to a special school. My parents insisted I go to an ordinary school. Fortunately, my parents won, but only because the head teacher at the local primary school where my brother went agreed to give it a try.

She went daily by taxi, but since the school was far from accessible she needed help to get into the classroom. She passed her eleven-plus and then had to find a secondary school which would take her. Once again, there was a head teacher at the local girls' school who was prepared to give her a chance.

> I've never been anywhere where I wasn't the only disabled person. I think a lot of the time I felt I was very different. Sometimes I felt uncomfortable about it; at other times I didn't. Girls all together can be extremely bitchy and they will single out any differences – and there were plenty of those. But I had very good friends and I don't remember ever being persecuted by the girls – or if I was, I'm certain I gave as good as I got.

It was harder with the teachers, particularly the games teacher, who insisted that Ann did exercises while the other girls played netball. Ann refused to co-operate and sat in the lobby next to a radiator: 'I got very adept at using my disability for my own ends; I would say things like I'd get chilblains and then spend my time sitting by the radiator making them worse.'

She could never have got away with it at home. Her father was determined she should do everything she could for herself.

> I can remember I had this particular dress which had tiny little buttons that did up all the way down, and Dad would make me sit on the edge of the bed and fasten every single button before I could get up. I really hated that, and I used to cry and get hysterical, saying, 'You can't possibly love me if you make me do things like this.' I realise now that it was precisely what I needed. His favourite phrase was, 'There's no such word as can't.' If ever I said – and I often did say – 'Oh, I can't do that,' he'd say, 'Yes you can, there's no such word as can't. Get on and try.' Which was brilliant, though I'm sure Mum went through absolute hell thinking it wasn't fair . . .

At the age of twelve, Ann had a tendon transplant from her third finger to her thumb; and when she finally left school, she had a spinal fusion (like Simon Parritt, she just missed out on the modern operation using steel rods, in the same way as they had both just missed out on the polio vaccine, having contracted the disease in 1956).

> They did it twice; they did one side and then the other. So I was on a plaster bed for six months. The actual operation wasn't very pleasant, but hospital's a great leveller, because everybody's in a similar situation, and I think I grew up tremendously when I was in there. I could actually be me, the same as everyone, instead of me different – I felt that very strongly. It sounds crazy, but I actually enjoyed it – once I got over all the horrible bits.

When she finally emerged from hospital, Ann decided that she did not want to go into further education, as she had intended; she wanted a job. She worked for the Ministry of Defence as a clerical officer for five years.

> It really wasn't demanding enough, though my social life was quite good. It was almost entirely men there! I had been in a youth club, through the church I used to go to, and that was good – though it would seem extremely tame to teenagers today.
>
> All teenagers go through the 'it's not fair' phase and as a young teenager I'm sure there were a lot of nights when I cried myself to sleep. But then I think probably I would've done that anyway. Teenagers do. If it hadn't been polio, it would've been something else, you know, 'Why have I got curly hair instead of straight hair?' I don't think I ever felt, 'It's not fair that I've got this.' There must have been times when I used that, or said that, but I don't remember feeling resentful about it.

Chailey Heritage was out in the country; the nearest town was Haywards Heath. Once dubbed 'the public school of crippledom' by an editor of *The Times* (with all that that implies of both privilege and hardship), it was taken over by the National Health Service after the Second World War. It consisted of several buildings spread over a large area, and Maureen O'Sullivan first went into the hospital wing, where she remained until she was nine, periodically undergoing surgery and regularly attending school. She found Chailey educationally competitive, with high standards:

> The bulk of us were physically disabled and I was one of the worst. It wasn't just my arms, or just my legs; it was both. I mean, I write with a pen in my mouth; and once you'd acquired those skills and could write like everybody else, the same was expected – quite rightly – of you. If anything you were expected to do better.

Maureen remains grateful for that: 'I have almost total independence through what was expected of me so early on.' Medically, too, Chailey could not be faulted. There was an orthopaedic surgeon by the name of Donald Brooks, himself a polio survivor, who came down from Great Ormond Street Hospital to perform operations.

> He was a magnificent surgeon, the things that he could do. There was so much dedication there: it didn't matter, in a way, what was the cost to Maureen – if it could be done, it was going to be done. My mum would get a little letter saying, 'Can we do so-and-so?' which she would just sign. Without being uncharitable to my family, I think there was an attitude of: 'We've made the awful decision, she's gone . . .' They kept in touch with me, but the staff would have to deal with the terrible things that could happen.

When Maureen first went to Chailey, the disabilities most of the children had were polio, TB, or spina bifida; the thalidomide children came later, when she was nine or ten. The emphasis was on physical achievement –

'there was no room for emotion or anything like that'. It was highly effective when it came to learning to walk.

> There was no way I could hold crutches and there were no electric wheelchairs in those days, so if I didn't walk I crawled everywhere. I was always falling over and cutting my face and my eyes . . . They once gave me a crash helmet. But I couldn't stop myself, you know, I was down on my face just like that. Black eyes . . . But it was just a question of being stitched up and off you went again. You can see why parents were not encouraged to join in; it would've been dreadful, dreadful.
>
> The thing that remains with me is how confident people were if they could use their arms, even if they got themselves around with crutches or a wheelchair; they seemed to have more strength and confidence if they could use their arms. Most of them could use at least one. The spina bifida were always below the waist, wore calipers. I can't remember anyone who was quite as severely affected as I was.

Maureen moved from the hospital wing into what was known as the Junior Dormitory, which was for nine- to twelve-year-olds. Apart from dormitories, there was a dining-room and a classroom on the site. From there she eventually graduated to the Georges, a huge monastic building a mile away from the school, to which she travelled daily by bus: 'There was a lot of discipline; everything was done by the bell.'

> When we got to be fifteen or sixteen, we had afternoons out on a Saturday. These were occasions when I became aware of how badly I walked, of how difficult it was to get up steps – at Chailey there were ramps – and of how vulnerable I was. I felt then that I didn't really belong in the outside world. I had huge insecurities about leaving; I didn't want to think about the time when I'd have to leave.

Maureen's teenage life was complicated by the fact that she fell in love with the school coach driver, who was twenty years her senior and had the reputation of being a womaniser, always after the nurses. It was an entirely innocent relationship; the man was kind to her and kept in touch for years after she left Chailey. But the authorities frowned on any suggestion of 'hanky-panky' and tried to keep them apart, even to the extent of ensuring that, if there were two buses and he was driving one of them, she got on the other: 'They were sure he was going to take advantage of this girl, you know.'

> Again, there was no space for the emotional things that happened to you. There was never anybody who did any type of counselling. Everybody was so busy teaching you how to go to the loo and pull your trousers up and dress yourself and feed yourself and all that, they were completely unaware that there was a young woman developing. 'Hey, I've got this

> disability, you know, how do I look to a man?' In later life it's very difficult, you don't know where you are in many respects.

Maureen learned to type at Chailey. She was pre-ordained by her disability for some form of clerical work and was provided with a Possum typewriter, which she could just about manage with her right hand: 'The movement of my fingers – a combination of movements similar to morse code – could operate the keys once I learnt the code.' She went on a school outing to Queen Elizabeth's Training College for the Disabled in Leatherhead and was frightened by the size of the place, the amount of walking she would have to do if she went there, and the intermingling of the sexes. 'Gosh,' she thought, 'it's kill or cure in here.'

At this stage, she wanted to become a nun, but bearing in mind her infatuation with the coach driver the staff were sceptical.

> I longed to be a nun. I thought, 'Please let me be a nun; I won't have to do anything, face anybody' – because nuns can still be cloistered away. They said, 'Well, if your vocation's strong enough, you can be a nun' – and I was given the part of Mary in the Christmas play to keep me quiet. I was open to it because my family are Catholics. But as I got older, I started to think, 'Well, if He's up there, why am I like this?' That's my argument even today, I'm afraid . . .

Ian Dury was at Chailey Heritage for nearly four years. He found it tough and he was often homesick: 'Every morning I woke up hating it.' But he was 'a fly boy'.

> Even though I was very young I didn't get no aggro from anybody older than me because I was quite able to divert that. I was good at the verbals, you know, and sussing things out. Also I always happened to be with somebody who could handle themselves, a mate.

Even before he had polio, Ian had been wild; he had refused to go to school and had been sent to psychiatrists. His mother had taught him to read before he ever went to school, so he had not known what he was doing there; he could not see the point of it. But at Chailey he was one of the literate élite: 'I remember writing letters every Saturday morning for all the ones who couldn't. You'd have a queue – the twelve of us who could write – ten each. "Did you get a parcel this week?" It was pretty harsh, you know.' He once saw a confused and disturbed child receive a beating for not getting into line properly: 'He was in another world anyway, but he got well done, he got thrashed by this cane-bearing second-in-command.' Nor was Ian himself immune, despite his survival skills:

> There was an orderly, his name was Hargreaves, he was a bastard, he had a large boot – a boot that big – clunking in, and he'd blow the whistle, ring the bell, whatever. One night I shit the bed. I woke up enough to shake the shit out of my pyjamas and fall asleep again. In the morning, you had the sliders went out first, who would slide on the lino on a blanket, who couldn't walk without appliances: the sliders went out – I was a slider then – and then the walkers . . . As the sliders were coming in from these low washbasins, the walkers would go out. And the walkers were going out when somebody spotted that I hadn't moved, right, and Hargreaves said, 'Why haven't you moved yet?' Ooo-ooh, he got everybody round the bed, all the sliders up on adjacent beds and all the walkers all round, everybody, the prefect, all gathered round the bed. He said, 'Pull the bedclothes down.' At that point I go . . . He said, 'Roll over,' and I rolled off. And everybody saw the shit and they went, 'Waa-aah.'
>
> There's a Dickensian little memory, but it was the same for everybody.

Ian had a tendon transplant operation when he was eleven. The operating theatre was in the girls' wing at Chailey, and he went over there with a boy of seventeen called Brian, whose leg was being amputated below the knee. There were just three beds in the ward and, in the middle of their second night there, the screens went up around the empty bed and a girl of eight or nine was brought in.

> She never stopped moaning and we had to read to her; and we were cursing her out really. When we got back to the main boys' hospital, this nurse was riding past on her bike and she said that poor little Wendy had died two days after we'd gone. The incredibly intense guilt we felt for not liking her when she was moaning – I remember that strongly; we both felt really bad.

But Chailey was not a closed society, like a borstal – 'it wasn't fearsome there at all; it was an incredibly healthy life because it was on a common, a bloody great big huge common, so we'd all be out in the fresh air'.

Ian passed the eleven-plus exam and went as a boarder to High Wycombe Royal Grammar School on a grant from Essex County Council.

> That was the first time I became aware there was something wrong with me, when I went there, not when I was at Chailey because there was plenty worse off than me there. I was aware of my position in the society of the Royal Grammar School, High Wycombe. Yeah, I entered the RGS as a chubby little mascot. That did feel like a weird environment, and it was – I still think so. It was the Round Table – you know, grocers' sons and stuff.

One of the day-boys there had cerebral palsy and was very severely disabled. But he was very bright, an Oxford scholar. Ian 'went the other way'. 'They were over-helpful at first, and then they realised I didn't need that much help – by which time I was a villain for taking advantage of it. It's quite a difficult thing knowing what you're supposed to do, anyway.' The RGS made him feel disabled in a way that he never did at home, where his holiday companions would adjust their pace to his, or, if they got too far ahead, simply wait for him to catch up.

> I had some kinds of extreme shyness, but also an associated who-gives-a-fuck, devil-may-care, what's-the-point-of-worrying-about-it attitude. I remember the first experience of sex I had was with a girl in a rain hut in Upminster Park. She was sitting on my right-hand side and my left hand couldn't undo her blouse. So I said, 'Would you mind coming and sitting on this other side, please?' She went, 'That's a weird one,' and moved over, though she didn't know why. She wouldn't see me again because I was too fresh. I was quite young, I was fourteen.

Ian went on to Walthamstow Art College. He wanted to be a painter; he grew his hair long and started going to places like the Partisan café in Soho.

> I met this geezer there called Larry, an amazing bloke, who had a pension from the Burma railroad; he was an habitué. And he said a true thing – though I think it's less true now – he said, 'If a disabled girl comes in here, all the geezers will look away from her. If a disabled guy comes in, all the girls will look at him.' Plus, if you're an art student and you've got long hair and you've got a bad limp, you're a romantic figure in a way.

In hospital, David Widgery was made to do occupational therapy and discovered he had a talent for basket-making: 'I set up a small basketwork business and farmed out orders; it had absolutely no relevance to my leg, but I did a lot of basketwork.' He did exercises to strengthen his leg and he also had electro-galvanic treatment. Large copper plates were placed on either side of his knee and the muscles artificially stimulated in the hope of maintaining muscle bulk until the nervous impulse reasserted itself naturally. That was the theory, at least.

> It's now largely discredited, but it was damn painful. It was a pulsating electric shock which convulsed the muscle, and they would turn it up and it got more and more painful. I remember watching them turning up the knob: it was real Nazi-time, you'd get up to the limit of pain you could bear and the muscles of the calf and the thigh would go into spasm.

Wax baths were another peculiarity, but at least they were painless. Then there were calipers, which rubbed in all the wrong places and were hard to put on.

> I used to go every day after school to physio and have this hour of treatment and then strap myself back into the caliper. I had to count up to 100 and the idea was, could I strap it on before I got to 100? I did 98, 92, 87 . . . while the physio was doing something else.

David's parents could see that he was not making much progress, so when they heard of an orthopaedic surgeon who was a pioneer of reconstructive surgery they took David to him.

> He usually had someone from Czechoslovakia or the Soviet Union to whom he was demonstrating, and he would always work very closely with a physiotherapist and a person who measured your feet size – he was very interested in feet size and the way your foot stuck on the floor. So he'd always trace your sole and he would turn to the Czechoslovakian and say, 'Absolutely clear, left inversion . . .' And he did these clever operations which released tendons that had tightened up. My foot used to point downwards, so he elongated that tendon. The old-fashioned thing was to graduate from a caliper to a spring which stopped foot-drop and he had this theory that if you released the Achilles and transplanted some of the existing tendons across the toe you could improve your eversion.

More complicated – not to say controversial – was the surgeon's idea of stapling the knee of the unaffected leg in order to slow down its growth and prevent lopsidedness developing. David's parents worried over the prospect of stunting his growth. But for David, the ultimate loss of three or four inches in height was more than compensated by the immediate relief of no longer having to wear 'those awful clodhopping orthopaedic boots with very large heels'. It was, however, a very painful operation.

> And I felt very fed up that I was back in an orthopaedic ward with about six Hell's Angels who taught me about masturbating and all sorts of interesting things like that. The orthopaedic wards weren't for polio boys of eleven and twelve – you were meant to have a fractured hip or something. I suppose these operations were interesting to the surgeons because obviously it's an interesting technical problem. And I became quite interested in orthopaedic surgery as a career – 'Well, that's clever, isn't it?' – you know, you've got a lopsided person and you stop the growing side and it levels out.

But it was *all* technical, *all* a 'hammer-and-nails job'; no-one talked about the emotions involved: 'I was never given any help on that side at all.'

> I don't think it occurred to doctors of that period. It occurred to my mother, but not my father; my father just thought one had to grit one's teeth and make the best of a bad job. I remember walking with him in the High Street

> in Barnet when I came out of hospital and he would squeeze my hand harder when he felt I was limping more than I needed to.
>
> I think my parents felt guilty that I'd got this thing, and felt there was a hygienic precaution they'd probably forgotten. A lady down the road got it, which was quite a good thing; she got it in her arm. I think my dad was rather cross that his eldest son wasn't going to be a sportsman, wasn't going to be a ballroom dancer, wasn't going to be a skiier . . . I think any father would be a bit fed up that his eldest, his only son gets a disease which is going to make him imperfect for life really. A lot of things mattered in those days; having a limp wasn't really good news.

Because his mother bore the brunt of his disability, David 'became more interested in traditionally feminine type of things'. He became very bookish and compensated for not being good at games by a 'retreat into sarcasm and literary pretension' – though, at his father's insistence, he also swam and did weight-lifting. Adolescence was problematic enough without the additional burden of a withered leg.

> It's a particularly bizarre imperfection because you are all there everywhere else. It's not a progressive disease like multiple sclerosis; hopefully it's not going to start getting worse. It's as if you've suddenly been branded and a quarter of your body's gone AWOL and you're never going to get it back; and you have to work out a way of dealing with it.

David had great difficulty in coming to terms with it. He would provoke fights with other boys and then extricate himself because of his limp, which only compounded the offence: 'Then you'd be teacher's pet because you were a cripple.'

> It was a real exercise of living by your tongue. You couldn't compete in any beauty stakes, and you couldn't compete in any physical stakes on the football field or the cricket field, and you couldn't even ride your bicycle up the hill. All that was left really was sarcasm and being best at answering back at the teacher.

Such was his sensitivity to his bodily imperfection that he could not even walk down the high street without feeling that people were looking at his limp or his caliper and laughing at him, though he knew that it was not so.

> I'm amazed when people who've known me for ages say, 'Oh gosh, what's wrong with your leg?' I sometimes say, 'I had a football accident.' People are terribly unobservant, aren't they? And polio people are very good at disguising if they want to.

The attitude of most women – that it did not matter that much – was 'an extraordinary revelation' to David:

> I can remember going out with women early on and expending a lot of effort on concealing the leg and so on – even, when I was a teenager, of having all sorts of systems inside my trousers where I would put a single trouser-leg to bulk out the thinness. Really pathetic things . . . The highest praise was when a girl would come up and say, 'You dance quite well, don't you?' when you're wiggling round on the spot sort of thing. That's why I found Ian Dury so emancipated. Suddenly there was this sex symbol who had a limp . . .

For Mary Gould, progress was infinitessimal:

> At the very beginning I had slight movement in a right little toe but lost it; slight movement came back in the extensor and abductor of my left thumb, tiny bit of flection in my right thumb, tiny bit of extension in two other fingers and an odd muscle in my abdomen; and weak neck muscles began to strengthen up. But the old muscle chart, which I disliked intensely, was full of noughts.

After Mary had been in hospital for two years, her mother was given to understand that there was little more that could be done for her.

> She'd meanwhile heard of a Mr Lancelot Walton, an orthopaedic surgeon in Woking who'd gone out on a limb about polio and developed a theory that if he could stimulate the blood flow on the surface, he could perhaps revive nerve cells which he was convinced were not rendered useless for life but merely knocked out, as it were. He'd let you sweat and stew under hot packs for ten minutes or so, then take them off and do a particular kind of massage he called 'skin stretching'. He would pick up the skin and roll it in his fingers and stretch it, convinced that if he got the circulation on the surface of the body going, then the deep blood vessels must of necessity be opening up, loosening and freeing muscle fibres. He'd sometimes use hot ultraviolet for the same purpose – quite a heavy dose that would really redden the skin. And he improved my breathing enormously by working to free the rib cage.

Mary was grateful to Sister Kenny's old ally, Lancelot Walton, for much more than sympathetic and imaginative treatment. She came under his care at precisely the moment when he – supported by the Rotarians and the British Polio Fellowship – was opening the residential home at Silverwood in Cobham, Surrey, that 'staging post' on the road to recovery for those who were lucky enough to get into it, and one of the earliest experiments in independent living.

> It was only for polios, but mixed sex and mixed disability. There were about eighteen of us, a very mixed bunch – there was an ex-teacher and an ex-

labourer. We did generate a good feeling. We were all given a lot of respect as people and a lot of consideration for our different wants. I used to have a nice time, doing the crossword with a friend and occasionally being taken out. Quite a lot of people there could walk, just limping badly. And the people who were not so disabled could do things like getting hot drinks for the others – there were a couple like myself. It was marvellous because they would get you a book, or move this, or get that, or show you the other. One or two people had old-fashioned wheelchairs you worked with your arms or very modern electric ones, and they would go off to the shops with a shopping-list from the rest of us. It worked very well, but gradually it became obvious that it wasn't going to be so easy to leave . . .

Even before she had left Woking Hospital, Mary had been urged to resume her university degree course, but she postponed making a decision. At Silverwood, she came under pressure again, this time from a couple who befriended her there, a disabled man who was doing a PhD and his able-bodied wife, who paid for their room by working on the staff. They shamed her into it by saying, 'Mary, you're going to get your degree, of course you are, don't be ridiculous.' So she wrote to the Vice-Chancellor of Nottingham University and to her old professor there, who encouraged her and said that she could start again by correspondence if she could manage to come up once a term to see her tutors. She had a friend in Nottingham, a nurse, who would help her at that end.

I used to ring up the Red Cross and say, 'Can I have somebody to come with me on the train in the luggage van from London to Nottingham?' The Friends of Silverwood by then had a very good scheme. If you wanted transport, you could ring for this super taxi with a lovely driver who would lift you in and out and take you to the local train to London; and you had to have somebody up there to help lift you and your chair into the van. It worked better in those days, possibly because not so many people wanted to do it.

The university worked out a complicated system for Mary's exams. She had, of course, done a number of essays, which demonstrated her capacity to sustain an argument. Now she had to perform under exam conditions. The papers were cut up and pasted on to a blackboard where she could see them, and a departmental secretary sat smiling in readiness to take down the thoughts she dictated. That was the first stage. For the second stage, her notes were pinned to the board and she was confronted by the examiners, who said, 'Right, Miss Gould, when you're ready . . .', and she had to answer the exam questions verbally.

The thing about speaking, especially when you're nervous, is that you forget if you've made a point. I had the feeling that I'd either missed out vital things or constantly gone back over the same point, like a gramophone when the needle's stuck. Every now and again they would scribble something, and I never knew what it was they were scribbling. The awful thing was that it was up to me to say when I'd finished.

But Mary passed her exams and, in due course, attended the graduation ceremony, where all the new graduates trooped up to the platform to receive their degrees from the assembled dignitaries.

They worked their way through to the G's and then my name was called. Suddenly the dean was on his feet, yelling down at me, 'Jolly good show, Miss Gould' – you know, half a mile away. And I thought, 'Do I say anything? If I do, I'm going to have to yell.' I decided the politest thing was just to call out, 'Thank you very much.' It reminds me of the clergyman who once yelled down the church, 'Do you come to me, or do I come to you?' I think I said, 'I come to you.' It's such a farce, isn't it?

Once she had her degree, Mary wrote to the county council, saying: 'You helped me get it, so what can I do with it?' The council responded by sending round a careers adviser, who arranged an interview for her at the local technical college.

I learned later that the Principal, though he pontificated that he didn't see any reason why there shouldn't be something, had absolutely no intention of doing anything. But a lecturer I'd chatted with in the passage had said to another, you know, 'We've got to give that girl a job.' That was how I got invited to do two lots of two hours or something, assisting with English A-level teaching. And it gradually built up over the years.

3

Ian Dury wanted to be a painter; that was why he went to art school. One of the first things he learned there, however, was just how good you had to be to have any chance of success. After a while, he found himself more drawn to song-writing than painting: 'Obviously, it's easier to be a good writer in pop music than it is to be a good painter – there aren't many good lyric writers in pop music, hardly any at all.' He gave up painting when he no longer felt like doing it.

His first band was called Kilburn and the High Roads.

I was doing a gig in the Nashville once – it's called the Three Kings now – and after the gig I'd just come down to have a drink (this was long before we got any notoriety). This guy goes to me, 'You're putting that on.' I went, 'Whaaat?' He goes, 'You're putting that on, that limp.'

I don't wave it about. Roger Daltry said to me, when the Kilburns supported the Who: 'Cor, if I had a bad leg I'd fucking wave it about.' I said, 'No you wouldn't, Roger, you'd leave it alone.'

Far from parading his disability and using it to get ahead, Ian reckons that it held him back:

> If you've got any brains at all, you don't want to look like you're doing it because you got polio. If someone wants to say it's compensatory activity, well, okay, it's compensatory activity. But I was always the noisiest boy in the classroom, I was always a show-off.

In the late Seventies Ian Dury and the Blockheads – the successor to the Kilburns – had three Top-Ten popular hits, including the No. 1 single, *Hit Me With Your Rhythm Stick*. People started recognising him in the street; they might confuse him with Bob Geldof – pre-Band Aid – but they knew they should know him. He became frightened to go out on his own: 'I read somewhere that Paul Macartney – I think it was – if he's hassled in the street walks briskly away; well, I'd fall flat on my arse.' He never goes out alone in London because he no longer feels free. Whereas before he was known he felt as though he owned the street, now he feels vulnerable. Performing in public is different; when he is on stage he is doing his job: 'It doesn't show as much on the stage anyway, you can kind of glide through it more; you know, I do my little moves, work with my microphone stand . . .'

During the International Year of Disabled People, Ian received letters and tapes from people in sheltered homes, saying how lonely it was when the staff went home for weekends. One sent a tape with a song that went, 'This is the international year of the disabled,/Now we're very grateful, we've got sports to do and activities . . .' What could Ian do with such stuff? He could hardly show it to EMI.

> The Year of Our Disabled Lord 1981 I was getting lots of requests. I turned them all down. We had this thing we called the 'polio folio', and we used to put them in there – we'd say, 'What is it – lesbians in wheelchairs? Put it in the polio folio.' Instead I wrote this tune called *Spasticus Autisticus*. I said, I'm going to put a band down the road for the year of the disabled: I'll be Spastic and they can be the Autistics. I have Blockheads and that means they're autistic anyway. And my mate goes, 'No – Spasticus Autisticus, the freed slave.' Great, I'm Spartacus. So I wrote this tune.
>
> I put in the second verse, 'So place your hard-earned peanuts in my tin/ and thank the creator you're not in the state I'm in./So long have I been languishing on the shelf/I must give all proceedings to myself.' When they said, 'Are you going to give it away to charity?' I said, 'No, I'm not, the second verse explains that.' I thought it would be a war-cry type of item. But it wasn't allowed to be played anywhere and people got offended by it – everybody except the spastics. All the spastics went, 'Yeah man, what a tune man, yeah, right.'

Some years ago, Ian was invited as a disabled adult and celebrity to talk to teenagers at a special school. He found that they thought they were going to get better, that this was part of being an adult.

> They were very disgruntled, some of them; they were day kids, and there was one called Lee and he said, 'How old are you?' I said, 'Forty. How old are you?' And he said, 'Fourteen.' He said, 'Is it as bad when you're forty as it is now?' I said, 'Yeah. In fact it gets more difficult because, as you get older, your youth and your strength are going. But you do get your mind prepared to deal with it in another way.' I said, 'But it ain't no different at all. It don't get no better. All you get is better able to handle it if you can.' I mean, some nights I take my caliper off and throw it across the room, swearing at it. But I never cry, not ever, never once. I cry if I get a spot on my nose . . .

Ian Dury has led a full life. He lives in a flat overlooking the Thames in west London. Separated from his wife, he has a grown-up son and daughter. In addition to being a pop singer and lyric writer, he has had a secondary career as an actor.

> Comparatively, I've had it easy. I don't believe in nothing, but if I did I'd say thanks to somebody that I'm not worse off than I am. I remember driving along the highway with one of our roadies, and he saw someone and he said, 'Aw hell, I'd sooner be dead than lose a leg.' He was looking out the window at somebody, completely forgetting that I was . . . What he didn't know was that he wouldn't sooner be dead.
>
> I think all prejudice is based on a kind of weird fear, a sort of paranoia. But any prejudice that goes towards disabled people is purely the inability of people to visualise themselves in that situation – apart from Hitler, who just wanted to keep the streets clean, let alone the race.

David Widgery resented it when people said, 'I suppose you've got a bit of a chip on your shoulder.' So he would say, 'It's not my shoulder, it's my leg.' From an early age he got involved in politics. In the late Sixties he wrote for and, later, edited the underground magazine *Oz*; he was a committed and active socialist and an initiator of the Rock Against Racism movement; but first and foremost, he was a doctor in the East End of London.

> It's often conventionally said that people on the left have a chip on their shoulder, that they're misfits, or they've got a physical problem, or they're Jewish, or something like that. I think the reason I became left-wing was that I read a lot, and I read a lot because I had a lot of time on my hands. In fact, I was converted to left-wing ideas by reading Bertrand Russell and Bernard Shaw and people like that, and I think probably by my mother's influence – she was a left-wing Methodist verging on Quakerism, which was about as far left as you could go in the Thames Valley in those days.
>
> I don't think my leg had anything to do with my political views at all. It

> certainly had a lot, however, to do with my views about the National Health Service and about being a doctor. There was no one medical in my family: my mother was a primary school teacher, my father had no qualifications at all. I wouldn't have ever dreamt of getting into medicine, nor would I have managed to get in, if I hadn't had this very close experience of the health service as a patient. It certainly has made me very pro-NHS, because I know for an absolute certainty that the treatment I got would never have been available on a commercially based system; I would have had to have been a millionaire to get the surgery I had. The fact that, by hook or by crook – by my mother's determination, or whatever – I got a series of very successful reconstructive operations for nothing on the NHS made me feel there was a debt of honour and I had to pay my dues. Becoming a doctor was partly to do with that, and my political defence of the NHS – which I spend a lot of time doing – is partly to do with that.

The NHS was a socialistic set-up – to each according to his need – an area of British life not dominated by the cash nexus, unique in all the world; hence its hold on the affections of the British people, who might also have some kind of collective memory of the postwar polio epidemics and 'the feeling of powerlessness against invisibly spreading disease'.

David did not, as he had once expected, end up as an orthopaedic surgeon, but as a GP – one of whose functions was to vaccinate children against polio. If he encountered resistance among his Bangladeshi or Vietnamese patients, who did not understand why he wanted to squirt pink liquid on to their tongues, he would – like some old soldier – roll up his trousers, show them his withered leg and say, 'Now look, I got polio, and if I'd had what you're going to get now, I wouldn't have.' He became a fanatic about vaccination because he felt he should have been vaccinated, that the introduction of the Salk vaccine was delayed too long in this country: 'So the bit of medicine I've ended up in is very much preventive-oriented.'

His political stance may not have derived from his experience of polio, but that experience enabled him to identify emotionally with other people whose self-image was damaged or damaging: 'A lot of people on the left seemed to find that difficult to understand – and that people are crippled inside, as John Lennon says. Yet to me it was so bloody obvious.' He was, as a result, an enthusiastic supporter of liberation movements, whether of blacks or women.

Although he realised that being 'normal' was no guarantee of psychological health, and he came across people who made him think, 'Well, I may not be able-bodied, but I'm not so fucked up as you are,' he still found it impossible to accept his disability, comparatively mild as it was:

> No, I still haven't come to terms with it. I don't think I will for a while. I get long periods when I forget about it, but as I get older I do get more

> fatigued . . . Last night I was humping three bags of medical equipment round an estate in Bow, getting lost and having to climb lots of stairs. I got halfway up the second lot, I was terribly tired and my back was hurting – I've got a bit of a back problem because of the asymmetry – and I suddenly thought, 'I'm in my mid-forties and I don't like the look of this. I won't be able to do this in ten years' time.' You suddenly realise that the sort of mental energy that a lot of people who've had polio, or other kinds of illness, manage to acquire is going to be drained by the physical weight you're having to carry, that you probably won't live as long as everybody else, that you'll get more exhausted more quickly, that you'll probably get home-bound, or chair-bound, or bed-bound sooner . . .

This did not happen in David's case. Less than a year after he spoke these words he met an untimely death – as the result of a domestic accident which had nothing to do with his polio.

After the cloistered life of Chailey, Maureen O'Sullivan initially found the freedom of her existence at Queen Elizabeth's Training College for the Disabled terrifying. She was there in 1973 to 1974, the winter of the miners' strike in Britain, when there were frequent power failures and power cuts. The flickering lights triggered fits among the unfortunate epileptics, as Maureen discovered one night when she went to the college cinema:

> I just happened to be sitting beside this chap. He was a great big man. The lights went off and he was thrown on the floor and he was all arms and legs, this strong man, and I couldn't do anything. I just sat there thinking, 'What is going on here?' From then on I avoided like the plague anyone who wasn't obviously disabled.

Because she still had difficulty dressing herself, Maureen had to get up an hour earlier than everyone else. She would arrive at her eight o'clock class exhausted from the effort of getting into her clothes.

> Luckily, there was a girl who'd come from Chailey along with me who had spina bifida. She was quite timid and she felt safe with me because I wasn't. So I said, 'Well, you dress me and I'll look after you.' It worked quite well. But had she not been there, God knows what would've happened. You were very much on your own.

Practically, Maureen made good progress. She learned speedwriting, holding the pencil in her teeth, and typed well enough to pass her secretarial examinations. But social and emotional problems were ignored, just as they had been at Chailey.

> There was the famous day when I went into Leatherhead and had to sign something in the Post Office. I picked up the pen in my teeth and signed it, and this woman behind me started making remarks: 'Oh, how marvellous.' Of course, the whole of the Post Office was looking at this eighteen-year-old girl with a pencil in her mouth. I didn't go out after that.

After she had completed her course, Maureen returned home to Kent. She found a job in the typing pool of a perfumery shop and drove to work in a battery-driven three-wheeler invalid car, which she steered with her foot: 'It was like a milk float: it went 15 m.p.h., and above 13 you couldn't hear a thing, it made such a din.' Having spent her entire childhood in institutions, she found it impossible to slot into home life:

> My mother would dress me; I went to work; I came home; I had my tea; I did nothing around the house; then I'd sit in the lounge and take care of my younger sisters while my mother went to work. It was always quite a close family, still is: they all live in a nucleus with each other. But I was absolutely suffocated mentally. My mother probably had feelings like: 'What can I do for her when her mind is racing around and she can't?' Very soon I thought, 'I can't live here.'

With the help of an occupational therapist who had befriended her at Chailey, Maureen eventually got a job as secretary to the matron of an old people's home in Brighton. Despite its inaccessibility – there were steps everywhere, and her room was upstairs – Maureen did not hesitate. Freedom was paramount, and she desperately needed to get away from her family, who were smothering her with kindness without realising the extent to which this was merely exacerbating her sense of uselessness and dependence. Now she could go out in the evenings. One of the helpers at the old people's home would take her to a nightclub in Brighton, where she enjoyed the atmosphere but stayed in her wheelchair to avoid wearisome explanations when she was asked to dance.

> Then I got the cranks instead, the ones who'd say, 'You'll be better soon, dear.' But one night we went there and this chap came to talk to us. He seemed a nice man. We had a few drinks together. He could see that I was in my chair and I needed help with my drink. He was quite an emotionally insecure man. I went out with him for a while, then I moved in with him.

There were repercussions. Both the home where she was working and her parents were religious and did not approve of her 'living in sin'. By the time she married Peter, she had moved on to another job, working as a typist for a finance company. There, for the first time, she encountered some prejudice among the audio typists she worked with: they complained about having to get her coffee and would not help her when she struggled into the office on a windy day in her inva-car (she now had a petrol-driven one, which went 50 m.p.h. and sounded like a motorbike). So she resigned forthwith. By this time she was expecting a baby.

> Of course, everyone raised their hands to heaven when I made the announcement that I was pregnant. My mother said – I'll never forget, it was the first thing she said – 'How are you going to pick it up?' I had no idea how I was going to pick it up. I said, 'Oh, I'll find a way.'

In the fourth month, when she started to bleed, there was a general feeling of relief which only made her the more determined not to lose the baby. She stayed in bed for a week and the bleeding stopped. Five months later, Emma was born after a sixteen-hour labour which was made the more painful by Maureen's inability to push.

> Once I'd had her, it was quite instinctive for me to pick her up in my teeth. I didn't even stand there and think, 'How shall I get hold of you?' I just bent down and picked her up in my teeth and held on to the front of her clothes and carried her like that and nursed her and put her head in the crook of my knee to feed her.

Such improvisation carried through into all aspects of nursing. Being unable to hold the baby and rub her back to get her wind up, Maureen would place her between her knees and rock her gently backwards and forwards until she burped that way. Changing a nappy would take her twenty minutes, 'then she'd wet it by the time I'd done it up, or she'd kick me in the face'. But Maureen was fiercely independent and resisted offers of help from social services because she thought, 'Yeah, well, helping means doing.'

Since it was illegal to take a passenger in an invalid carriage and her husband did not drive, a trip to the local supermarket became a major expedition with Peter having to take Emma on the bus and then meet Maureen in her trike and take out her wheelchair and push her round: 'It put an enormous strain on our relationship.' In addition, Peter was having a hard time adjusting to the arrival of a child.

> He had quite a short fuse if she cried. I'd say to him, 'Walk away from it.' His whole reaction was nothing like I would have guessed it would be. Whether is was jealousy or what it was, I don't know, but he didn't have it with the baby. This wasn't common knowledge at the time. I thought to myself, 'If you start shouting about that, there's me, a very disabled woman with my baby, and a man who's on a short fuse . . . You keep quiet, Maureen, and work twice as hard.'

As Emma grew bigger, Maureen could no longer pick her up in her teeth so easily; nor could she bathe her on her own. Peter and Emma's mutual fear or antipathy ruled out his doing it for her, so Maureen – through her doctor – got in a young girl to help her. Housing and transport problems were eased with the intervention of her MP and help from charities, including the British Polio Fellowship. Maureen got a Mini and moved out of her ground-floor flat in an old people's block – 'You always get that, disabled people lumped together with the elderly, because of the access' – into a house with a stairlift and a garden in which Emma could play. And Maureen was pregnant again.

As before, she started bleeding at four months, and this time she really bled. Peter was good, as he always was to her, and once again she got over the setback – until one day in her sixth month when she was doing the washing-up.

> I went to the loo and sat down and there was this avalanche . . . It was a Saturday, thank God, and Peter was there. I expected him to say there was a little head, a tiny head because I was only six months. I'll never forget, he said to me, 'There is this long thing wiggling about in a bag.' I thought, 'Aaahh.' It was the baby's leg. So the ambulance men came and I was tilted upside down. They looked at me, and I could see they thought, 'What's she doing pregnant? She is so disabled.' They go to get you under the arms to help you on to a chair, and of course you can't get me under the arms. But they got me on to this thing in the end, and I went into hospital.

They kept her in hospital on a drip to stop the contractions. Because her arms were so wasted, they put it into her right hand, which removed her last vestige of independence; she could not even read. So she just lay there and worried about Emma, who had just started going to nursery school in the mornings, and Peter, who had just been made unemployed.

> On the Tuesday evening, I started to have what I knew were contractions, and I was told I had constipation. So I was panting away with this constipation – 'You can't possibly have contractions as you're having this stuff to stop them.' At about one o'clock in the morning I finally just yelled, 'I'm going to have this baby.' I could feel myself pushing down and they said, 'No, no, you're constipated.' 'Well', I said, 'get a bloody bedpan then, but I'm going to have a baby.' They got the bedpan and of course the waters burst and everything shot across the room. There was one mad dash then to take me to the theatre, me crossing my legs, saying, 'I can't have it yet, it's a bit little, not big enough yet.' I had been told very candidly that if the child was born, there was very little likelihood of it being alive. I'd have to have a caesarian or the strain would kill it.
>
> It had been the wrong way round initially. Lucy took about ten minutes to be born, and she'd turned herself round the right way. They were ready to knock me out for a caesarian and I said, 'There's no time, I've been shouting for hours, there's no time to do anything like that.' And there she was. She weighed one pound thirteen. She had skin like cigar paper and her

> eyes were still fused and her ears were folded over, she had no substance to her ears. She was like a little frog.
>
> I'd never seen anything like this baby. It was a baby, I could just about see that it was a baby. Her salvation was that she came out and gasped, let out this tiny little cry that was like something miles away. She was whipped away from me quickly and wrapped in foil and taken upstairs where they had this baby unit. They were all rushing round, tidying me up and that. But you can't believe what's happening to you really. They took me up to this floor where she was and suggested that she was christened because there wasn't a lot of hope for her.

Maureen blamed Lucy's predicament on herself, on her disability. At one point she said to the doctor, 'This has happened because I'm disabled, hasn't it?' He told her, 'There's more than one baby in this unit, you know, and you're the only mother who happens to have a disability.' This was some comfort to her. Her other worry was that Lucy herself might be disabled.

> I thought, 'If there's going to be anything wrong with her I don't want her to live.' That was quite strong in my mind. If she'd been handicapped and had to be sent to Chailey – I really couldn't face that.

Yet every time someone came to the door, Maureen was terrified she was going to be told that Lucy had died. She had her christened and each day Lucy survived she grew more optimistic – too optimistic, in the opinion of some. Lucy was 'all wired up to everything' in an incubator, and Maureen would not let her family see her because she did not look like a baby. Babies were in the news: Prince William had just been born and all the newspapers carried pictures of Di with her chubby little son. Maureen felt the irony of it as she stared at her own little sprog and thought, 'Oh God.'

> Day by day she got stronger and she came off the ventilator. Normally you put your hand through the porthole, but I was allowed to have the side of it down and her slid out on a little Jay-cloth, so I could touch her. The nurses were very nice and they showed me how to clean her mouth with little cotton buds that actually filled it up. Her arms were like your finger. I sat with her for hours and willed her to live. She had one setback, when she got an infection, and apparently if they get an infection it's pretty much touch and go – they'd say, 'She's only got to catch a cold and that's it.' Peter and I sat with her . . . She was in there for three months and came out on the day she was due to be born, weighing four pounds by then and looking enormous to me. People said, 'Isn't she tiny?' They had no idea of what I'd given birth to.

Although Maureen had a family aid to help her, it was a difficult time for her and Peter with two children under three, and one of them requiring constant attention: 'It was just all children; our lives were just all children.' Maureen was afraid that Peter would snap under the strain; she encouraged him to go out and look for work, which he did. But there was an incident in which he had made Emma's head bleed and Maureen had to take her to hospital.

> I said, 'Come on, Emma, we're going to play a game' – because I was in one hell of a position – 'you're not to tell the doctors what happened. We're playing a game, and you say that you fell over.' 'Oh no, my daddy did it, my daddy did it.' God, kids are so honest, aren't they? I said, 'Now don't tell them, otherwise they're going to take you away from mummy, that's what's going to happen: we won't be able to keep you.' That frightened her. She had butterfly clips put in her head.

Maureen did not really blame Peter; she could understand the pressure he was under, and his remorse was genuine. But her children were so precious to her – she had gone through so much to bring them into this world – that this was the last straw. She had become involved with a man with spina bifida whom she had known at Chailey. He had been married to an able-bodied woman who had left him some years earlier. He had no experience of children but he wanted Maureen to live with him. Now she had to confront Peter, who took it badly, though the only violence he threatened was to himself.

The situation was unbelievably complex. In addition to the usual messiness involved in changing partners, there were the dealings with the council, the question of housing – who went where – and the entry of another disabled person, who could not be expected even to change a nappy.

> We moved from Brighton to Newhaven, to a ground-floor flat with a garden we'd exchanged. I lived with him for about a year, then I married him – this was really to make him feel more a part of things because I could see there was not a great deal of bonding going on with the kids. In retrospect it was a mistake. He felt useless because he couldn't be a proper husband to me as he saw it; he wouldn't be a dad and he couldn't have kids, so he got quite demoralised. By the time we parted, when we were talking about parting, he was ill in bed a great deal. He'd also had this – what I thought was a blackout. Then his mother told me, 'Hasn't David told you he has epilepsy?' I thought, 'Aaaahhh.' A couple of weeks later I'd just put the children to bed and he had a fit. I just thought, 'Oh my God.' I mean, we'd all been in his car. It put a completely different complexion on everything.

The nightmare of a failed relationship played itself out around a dog – or, in the end, two dogs. David got himself a collie: this was his baby;

Maureen already had her babies. But because he was often ill and the dog needed exercising, he suggested they get another – as a playmate.

> I had by this time got myself an electric wheelchair and, never-say-die me, I said, 'Well, if we get another one, then I could walk it while you walk Kim – it's something we can do together.' Dear God. So we went and bought Squeaker. The kids loved Squeaker, and Squeaker decided that she loved me, and she wasn't very fond of David. So we still had two sides. But we did go out together, I did take them out. I thought, 'Whatever must we look like, him sitting in his chair and me going along in my wheelchair with a bull terrier attached to the side of it!' Oh dear.

Separation was inevitable, and no less problematic than getting together had been. Once again, it revolved around moving house. David had not wanted to go with Maureen when she moved to Peacehaven, and she had agreed that he should not. 'But the doctor knew better; he said, "If you just give it a few months in the bungalow, then I'll write to the council."' They were still living together – apart – almost a year later. Eventually, the council found him a flat and he moved out. So Maureen became a single parent with two children – and a dog.

> Having polio plays a big part in not having a steady partner and not being to men as other women are. In the first instance, it takes a different kind of man to look at you in that light. They might feel, yes, she's an attractive lady, or she's not, but there's more to it than that if you have such a disability: how could she *do* anything, you know? It's terribly difficult. I suppose the way it happens is, you meet people through something you're doing, and once they know you, if things progress, it would be that way rather than walking into a room and a man seeing you and thinking, 'She looks nice.' They don't do that when you have polio, or when you have any disability. Sexuality in actual fact is very little to do with what you're looking like in the first meeting, but that's what men cotton on to. It doesn't leave you feeling too good.

For someone as disabled as Maureen, even the spontaneous expression of affection was well nigh impossible:

> People never think about not being able to put your arms round someone. It's very, very difficult to live with that. If you want to welcome somebody, for instance, if you can't put your arms out, you sort of stand there – and if they grab hold of you, you feel a bit like an ironing board, you know. Or you have to say, 'Well, give me a cuddle then,' or something like that, make a joke of it. You have to vocalise something when you shouldn't really have to. There are lots of things that people wouldn't think about that come from not being able to use your arms.

Maureen's father died of cancer a year or so later. At the end of June 1992, she wrote: 'I sat there staring at him; he was such a lovely man and he told me just before he lost consciousness that he was so proud of me, something he had never said to me before.' The shock of her father's death and her mother's new-found need of her revolutionised Maureen's feelings about her family: 'I feel that I "belong" to my family now, a feeling I have not experienced since parting from them when I was three.'

After graduating from Goldsmiths', Simon Parritt went to work for a company in Tooting that bought and sold second-hand paper machines. During the Seventies the company expanded; it even had branches in America. So there was plenty of design work – catalogues, brochures, advertisements. But Simon still hankered after a job in film or television; he applied for anything and everything – 'production assistant jobs, and assistant to the assistant to production assistants'. When Channel Four started up and was advertising for commissioning editors, he tried that. He 'had a go at the disability thing'. Nothing came of his efforts, except for one interview with a company producing educational programmes, and 'they just took my show reel and I never saw that again'.

Having failed to break into the world of film and television, Simon got into sex therapy and relationship counselling almost by chance. When a friend of a friend for whom he had done some designs offered him a free weekend workshop in lieu of cash, he was not enthusiastic. But he went along out of curiosity and, when the company he was working for failed in the Eighties, he started training in counselling skills and underwent both individual and – later – group therapy.

As far as girlfriends and personal relationships were concerned, Simon did not think he had a problem:

> Although I say it myself, I was a very attractive young child and, growing up, I always had a lot of attention from staff. The most attractive patient gets the most attention; and I had lots of care and attention – which maybe was due to my looks, my compliant personality, or whatever! I'd been around girls all the time I was in hospital, I'd even joined the brownies – they only had brownies on the girls' ward I was in. I'm not sure I was an official member, but I was allowed to wear the toggle! The real difficulty came when I was thirteen or fourteen. Up to then I'd had a fairly normal sexual development, you know, playing around with girls – the sort of thing kids get up to their parents would die if they knew about. Then my back began to twist and I became less well and my peers started doing things that were more physical, like going away on trips and playing games together and so on – I always claim it was because I couldn't use my arms, and making relationships is about grabbing people and putting your arms around them and being a bit more physical than I was able to be. I could talk a lot, chat people up, and do very

well in that area. But when it came to making a move from being very friendly to being intimate, it broke down and I wasn't able to deal with that.

With the prospect of an extended stay in hospital for the spinal operation, Simon postponed confronting the issue. He thought, 'Oh, when my back's better I can sort things out,' and, 'When I get to art school it'll all be different.' He had several girlfriends, but no proper sexual experience until he was in his late twenties.

> Because I was so confident in every other field I found it difficult to admit that there was an area where I needed help. My problem really was that I was scared – it was fear of rejection. Okay, so everybody gets rejected, but if you're disabled you've got to put up with a lot more rejection. Sex itself was not a problem – there was no physical problem – only the fear that girls would reject the look of my body.

His mother's over-protectiveness had not helped. Her advice, when he finally got round to asking for it, was that he should not get upset about it, that he just had not met the right person yet, that all would be well if only he were patient – advice that now, working as a sex counsellor, he would regard as wrong. He ignored it anyway. He wrote to Dateline, though he did not go so far as to join up, and then he wrote to SPOD, for whom he now works – 'It was originally Sexual Problems of the Disabled, but now it's all got to be politically correct and they've changed the name to Sexual and Personal Relationships of People with a Disability.' Then, when he was twenty-six or twenty-seven, a woman for whom he was doing some graphic design work took the initiative and they ended up in bed together.

> It was a confidence-giving experience. I must have packed a lot into the seven months or so the relationship lasted. I had a lot of frustration to get rid of, and a lot of stuff to work through. After that I just went crazy for five or six years. In the early Eighties I did things that everybody else was doing in the Sixties. I reckon I got it in just before AIDS broke . . .

Among the affairs he had were two with disabled women: one was a Ugandan – 'I've always fancied black women' – who had had polio; the other had a facial disfigurement and had lost her arms. He is still friends with the latter.

> It was surprising how difficult I found the Ugandan relationship, being with somebody who was disabled. Lots of people have said to me, 'With somebody who's disabled they understand what's going on, and it's easier.' But it isn't, because you both have the problem. I couldn't use my arms and she couldn't use her legs, and rather than us helping each other – like going to a restaurant was a nightmare, because where I could gain access she

couldn't, and people seemed to . . . I remember going to an Indian restaurant and they virtually fed us!

This relationship ended badly, largely because Simon could not handle it: 'It was a bit like being confronted by your own disability.' Because he could walk his disability was not immediately apparent, as hers was; and his awareness of her disability made him pussyfoot around issues where normally he would have been direct. 'I was actually called a bastard, which was probably a breakthrough for me because I was considered a nice guy!'

In the end, feeling that he was ten or fifteen years behind everyone else, he put a couple of adverts in *Time Out*. He tried different approaches: in one he said, 'Disabled, but no nurses please'; in the other, 'Anybody who thinks disabled people are boring, sexless and in wheelchairs need not apply.'

I might as well tell the story. Having met people who advertised in *Time Out*, I knew – because you compare notes – that women get a lot of replies and men don't get that many. I got about twelve from each. Then you get the odd trickle of loonies – the copied letters and the strange ones. One letter I got was a page torn from a notebook with this little lecture about how being disabled shouldn't be . . . – you can imagine the sort of thing. It said, 'You must think of yourself as special, like the royal family.' It was signed 'Sylvia' and it came from Gloucester or Stratford, somewhere like that – I don't know how *Time Out* got up there. I thought, 'This is a really irritating letter,' and put it to one side, while I went out with a few of the others. But I decided to write back to this one and say, 'Now look here. One or two points. 1, I'm not feeling sorry for myself. 2, Being disabled isn't like the royal family . . .' etcetera, etcetera. Back came this letter saying, 'Terribly sorry, let's start again.' So we started to write to each other and after a few letters it turned out she was only nineteen, which was rather a surprise. I said, 'How did you manage to write like an old woman? I thought you were some forty-year-old busybody.' By then, of course, I was into the relationship by letter.

Then she came to London and phoned me. Her mother was living in London, and I went to tea. I'd had a car crash, so I couldn't drive there and my sister took me. She said, 'I can't believe this: you're in your thirties and you're going to tea with this nineteen-year-old girl and her mother . . .' Well, we met, had tea – which was a rather strange experience – and carried on writing. Eventually she offered to come down and do some decorating for me, because she'd got some time off. She moved in, the decorating never got done and she never moved out again! She said, 'I couldn't do the decorating because I thought I'd have to go once the decorating was done.' That was in '86, and we married in '88.

Despite the 'no nurses' clause in one of Simon's ads, Sylvia gave up a good job to train as a nurse. A previous long-term girlfriend of Simon's and his sister had both become nurses, too. So he feels obliged to accept some responsibility for these career choices.

> We'll probably have kids when Sylvia finishes her training – that's if I haven't had a vasectomy. I always wanted kids, but when I got to my late twenties and early thirties I'd sort of given up on the idea – not completely, but I wasn't in a relationship, I'd only recently become sexually active and I was still being a bit wild in those days, I suppose. It sort of bypassed me, and now I'm in the situation where I'll be forty-two this year and it's a bit of a shock to the system. A bit of me really wants kids and a bit of me thinks, 'I'm disabled and I'm not getting any younger – bones creak a bit more and you age a bit quicker – so how will I cope with that?' So that's an issue . . .

One year on:
Sylvia finished her training, but there were no jobs when she qualified, so she stopped taking the pill and became pregnant. Simon saw it as 'another phase that I'm going through late in life, if you like', and he worried about being able to cope physically.

> I do handle kids. Very small babies are quite easy to handle. It's when they're in-between, not very small but not big enough to give you some assistance to lift them up. When my nieces want to be picked up by me and I can't do it, that's emotionally difficult.

Being both quite severely disabled – 'My breathing has been borderline okay for years; I always say that if I really relaxed I'd probably stop breathing' – and ten or fifteen years behind his peers has meant that he is starting a family and wanting to do more psychological research at precisely the moment when he ought to be slowing down. But to do so would not only involve relinquishing a good salary just when he needed it most; it would also be flying in the face of the polio myth.

> I recognise that the way to look after myself is to key down, to take rests, not to drive myself and work six or seven days a week, which has been my pattern, part of the way I've been brought up, part of what polio rehabilitation was about – pushing yourself and not quite listening to what your muscles are saying. I've sold the image of myself as someone who is able to compete on an equal – or nearly equal – level with other people, and as someone who doesn't accept help unless I really need it. So to key down is very difficult . . .

On the day on which Mary Gould had been delivered into Lancelot Walton's care at Woking Hospital after a six-hour journey from the Midlands, her mother (who had accompanied her in the ambulance) had staggered into the nearest café and sunk into a chair, desperate for a cup of tea. The waitress came and told her, 'You can't sit there, madam, it's reserved.' She was tempted to walk out, but she was so exhausted that she allowed herself to be shown to another table and sat next to a woman who herself had come in tired from shopping. It seemed like a providential meeting. The woman had a Bible study prayer group and she assured Mary's mother that when she left Mary would not want for visitors.

Every year from then on, this woman organised a birthday party for Mary at her home, which was up a few steps from the pavement. One year, when Mary was in the residential home for polios at Silverwood, she got a rather garbled message from a member of staff: 'Somebody rang up and said they were very sorry but the gentleman who helps you up the steps where you go for your birthday party or something can't do it, but there's somebody else who's got a ground-floor flat and says would you mind going there instead, dear.' This was the prelude to the most important friendship in Mary's life.

> The day came and I got myself in a taxi and followed the instructions and arrived at the front door of this friend who wouldn't mind having my birthday party there. Dormy and I liked each other on sight. I left my birthday cake behind and she came over and returned it, and then said, 'Would you like to come out for a drive?' It really developed from there. I went to stay for a weekend, and then after due thought and prayer and everything, we announced that I was going to move in with her. Everyone else had kittens and said, 'You don't know each other,' and, you know, 'Burning your boats . . .' and Dormy getting lumbered with somebody, and all the rest of it. But it worked very well.

Dormy had come out of the Wrens a few years before to look after her 'rather cantankerous, elderly and rheumaticky mother' – which was why she had a ground-floor flat. She needed something to do, and since her mother did not like her going out, some friends had suggested, 'Why don't you start your own typing, secretarial sort of business?' She started in a small way and then, a couple of years before Mary moved in, her mother died. She got new machinery and Mary helped her develop the business, while in return she ferried Mary to and from the technical college, initially lifting her in and out of the car but, after they had bought a – newly invented – rooftop hoist, using that.

Mary had always been a Christian; she had been brought up 'to take Jesus for natural'. But when her mother, ten years into retirement, went into a depression, she felt powerless to help her and became interested in the idea of counselling.

> There was – and is – an organisation called the Clinical Theology Association, which was started by a psychiatrist called Frank Lake. He'd been a medical missionary and he came back to this country and trained as a

> psychiatrist. He was horrified at the fact that he could go along to church and after church people would cluster round and when he said, 'How are you?' they'd say, 'Oh, I'm so depressed,' you know. He thought, 'This is ridiculous, we've got Good News, Christ for everyone, and yet I shall never see these people in my consulting rooms.' So he started to look both at modern psychological understanding of the dynamics of behaviour and at what the Christian gospel had to offer, and discovered – not entirely to his surprise, but to his satisfaction – that the two melded together superbly. And he constructed a two-year course, with twelve seminars in each year, building a conceptual model of human development on the model of Christ.

There was a tutor living nearby, and a counselling course was held in Dormy and Mary's front room. Useful as this might be in giving her life a sense of direction, it did not give Mary immunity from a crisis of her own; and in her mid-forties, she went through a 'black patch'.

> I'd been helping someone coming out of a nervous breakdown and I'd got myself far more emotionally involved than I realised. We got home from a holiday and there was this letter saying, 'I won't be coming down any more because I think I need to be independent.' I thought, 'Oh jolly good, that's super.' I wasn't prepared to find that I suddenly hit a land where everything else looked grey and colourless and I was acting being me behind a sheet of plateglass. I had to have a good old cry every day before I went to college, otherwise I couldn't trust myself to keep control for the rest of the day. I thought, 'What on earth is all this about?' Then gradually I began to see it was all about loss: losing somebody I was emotionally involved with was losing Daddy, and then that modulated into losing my physical mobility. I hadn't mourned my Dad and I hadn't mourned my disability either – because we were all wonderful, weren't we? We had this wonderful willpower. So the idea that we ought to be sorry about it was probably too frightening even to contemplate.
>
> Having spent a lot of time saying I wasn't the marrying type, I came to realise in my mid-forties that I would have liked to have got married and I would have liked to have had children. Okay, so I can't see myself washing socks with great enthusiasm, and I'm not very good at routine, and I get easily bored . . . But maybe we would have managed that somehow. I rather distrust myself, I very easily get emotionally involved. Although I don't like to admit it, the external discipline of polio has actually been quite a safeguard for me – I'm not quite sure how safe I would be let loose. But maybe I'd have managed that too; it's one of the unknowables . . .

When Dormy reached retirement age in 1982, they moved to rural Hampshire and Mary, who had resigned from her job at the technical college and was reluctant to take on tutoring work, started practising professionally as a counsellor. She found that her disability gave her certain advantages: it made her less frightening than an able-bodied person might be; it also enabled her to establish a rapport, particularly with older people who could no longer do what they had done before. They would say, 'I expect you understand,' or, 'You must have had something like that,' and she could say yes without the slightest scruple.

For Mary, trying to keep up with people has been 'rather as if I were swimming through thick water while they're all strolling along on dry land – you know, you're doing it, you're keeping up, but it takes rather more out of you'. She feels her limitations most when she is unable to give strength and support in ordinary, everyday ways.

> Okay, Dormy's fit and well and that's jolly good. But she's human, she gets splitting headaches sometimes, and I can't even say, 'Sit down, I'll get you a cup of tea.' Not only can I not say that, but I actually have to say, 'I'm terribly sorry but I want to spend a penny' – or I need moving, or going to bed, or getting up, or whatever. So not only are you not able to give, but you have to make it worse by having demands that you can't do anything about.

Yet crucially, in Mary's view, there is also a positive side to dependence:

> I think we are fundamentally dependent, you know – we didn't make the world, we didn't make our life, etcetera. I spent years thinking that even though people said I'd given them something they were probably just making it up. But that's ridiculous, too. I know and acknowledge that I do give, so it's not all one-way. I think there's a pool of humanity: we put into it what we put in, and we take out of it what we need. Of course, at times you feel a bit fed up and narky and 'Oh, flipping heck'. But it isn't to my mind an ignoble place to be.
>
> If I may talk in religious terms again, Jesus demonstrated that it's okay when he said to the woman of Samaria at the well, 'Would you draw me a cup of water?' The difficulty is to admit one's need. For a large number of years I didn't really want to acknowledge it. I can remember sitting at a place in Silverwood and saying to myself, 'Come on, you've got to admit that what you actually are is a cripple in a bathchair. Now will you shut up and stop pretending you're not?'
>
> There's a basic realism that you've got to come down to, and realise that's okay, it's all right to be there.

Ann Stevens is not religious. There was no blinding flash of insight, just a gradually growing disbelief, encouraged perhaps by the study of Religious Education – which was then basically Christianity – for GCE

A-level: 'Because you then think about it much more academically, it seems less likely to be true!'

Nevertheless, when she applied to go to teacher training college, it was suggested to her that as a disabled student she might consider choosing RE or modern languages as her subject rather than biology, which she was determined to do. The labs were on the second floor and there were no lifts, so how would she cope? She said, 'There are lots of male students and if I sit at the bottom of the stairs and look helpless, I'm sure they'll help.' The authorities agreed to give her a trial, and she was there for four years.

> By then I was twenty-five, I was a mature student. I was expecting the great academic experience. Having felt bored and that I hadn't been challenged, I thought that this was going to be a real challenge and that I was going to enjoy it; and in fact I didn't. The only part I enjoyed was teaching practice. I was unfortunate, I think, because the group I was in was mostly married women who had completed their families and were now coming into teaching. They were about forty, which doesn't seem old now, but their attitude was, 'Where can you get the cheapest tin of baked beans?' – that kind of thing. I didn't find that at all stimulating.

There was a tutor with a particular interest in teaching disabled children, who tried to help Ann in various ways, by making transport arrangements and placing her in schools. But he also used her as a guinea-pig in his lectures and asked her questions like, 'How did *you* find it?' which she hated – in consequence, she avoided his courses. 'He really didn't understand and he was so condescending – to the children in school as well as to me. I found him totally objectionable.'

Ann got a job in an accessible primary school. The age group appealed to her; she preferred teaching the child rather than the subject.

> It went fine, there was never a problem – apart from the problems all teachers have. People were always saying, 'You won't be able to use the blackboard, you won't be able to do this or that.' But there are so many ways round it. I just used an overhead projector, which is not a problem – in fact it's an advantage because I'm facing the class, I haven't got my back to them. Most of the things that were put up as obstacles could easily be turned to advantage – for instance, I was always at the children's level, I was always sitting down, I didn't suffer from backache, and I also had eye contact all the time, which was good. Discipline didn't worry me; it's very easy to identify the children who might be a problem – they're usually boys. I would appeal to their better nature, get them to do things for me. It's usually when they're standing in lines, queuing, that someone starts to kick or punch, and if they were pushing me in my chair, I knew they couldn't be kicking or punching anyone else. I used that sort of strategy.

Ann was a teacher for fifteen years. She gave it up when primary education ceased to be child-centred and became more bureaucratic and prescriptive. She would not have minded working long hours had she felt that the time was well spent; but filling in forms and reading superfluous documents was not her idea of teaching. So she took an office job with British Telecom, dealing with customers' complaints: 'The main advantage was that I wouldn't be bringing my work or worries home with me.'

She had married when she was still teaching, at the age of thirty. She did not want children, and her husband already had five from his former marriage. They lived in a flat at first, then moved into a bungalow; but they had to sell it when the marriage failed:

> We were married for ten, eleven years, then we got divorced – and we're still divorced! Brian has his house and he comes here weekends and quite a lot of evenings, and we get on much better now. It's unconventional, but it works. We had big problems. He had problems with the business which caused drinking problems and made it very difficult for me. I felt very insecure and just had to get away. I think some time on his own made him realise how his life had deteriorated and so made him stop drinking. From that point on he started to go forward. He realised what mistakes he was making; I realised what mistakes I was making. We're still very much together – if he's not coming over we talk every day on the phone. We do shopping for each other; I still do his washing. He cooks for me and I cook for him. We go away on holiday together.

Travel and holidays abroad, which Ann loves, have opened her eyes to the kinds of mobility and access attainable where there is the will and the means to provide them, unlike in this country.

> We went to Canada last year and it was a different ballgame altogether; the access is just phenomenal. There are so many people in wheelchairs – well, in Vancouver there were. Everywhere is accessible. They consider it everybody's right to have access to public transport, every public building, all shopping malls . . . The buses have tail-lifts and special places. It's just treated as, you know, 'Why shouldn't they?' It's not like dial-a-ride, where you have to make the effort, it's public transport. Nobody thinks anything of it. Everything's just so much easier: you don't have to think, 'Is there going to be a parking place?' There will be – and it's going to be near the door, and all the doors are automatic; you know there aren't going to be any steps. Here you have to be really determined to do something. Okay, but there must be so many people who're not. It's too easy for them to say, 'Well, I can't do that.' It's a shame. I'm a great letter-writer, I'm always protesting because I feel so strongly about it.
>
> One of my bugbears at the moment is the things that are so-called 'specially adapted for the disabled' and are done by people who don't know.

> They spend a lot of money and it's all wrong. At work, when I first arrived there they said, 'We've got a disabled loo.' So I said, 'Oh lovely, that's really good. I wonder if I'll be able to use it.' They said, 'Of course you will.' I said, 'Would you think me awfully rude if we could just pop down there to check before I actually come to work? I suspect there may be one or two things that might be difficult.' They were very sceptical about this, so I said, 'Well, I suspect that I won't be able to reach the light switch, I suspect that I won't be able to open the door because it'll be on too tight a spring . . .' and so on. Of course, when we got there I was right and they couldn't believe it. At the moment they're spending a lot of money on adapting a second one. It's unbelievable, the architects who design these things . . .

Ann learned to drive when she was twenty, but it was a traumatic experience. The weakness in her arms made it almost impossible for her to steer and brake or accelerate with hand controls at the same time. She took her driving test five times, and it was not until the law was changed and she could be tested on an automatic car that she passed.

> I had to be lifted in and out of my car, and I still do to this day. But it meant that I could do favours for other people. Mum didn't drive then and if she wanted to go somewhere I could give her a lift, or I could give people going to college a lift. It was very satisfying to be able to do something like that for other people.

She is so accustomed to not being able to walk that Ann regards it almost as an irrelevance: 'It's irritating, the access problem to buildings, but that's not really a factor.' If she had a wish, she says, 'it wouldn't be that I'd never had polio; it would be that my arms and back were a bit stronger. I can't even sit up without a back support. It would be wonderful not to have that . . .'

One year later:
Ann had recently started waking at four every morning with a pounding headache. She put it down to stress, and tried to relax and breathe deeply. This worked to the extent that the headache would go, but she still suffered panic attacks, which she ascribed to her 'time of life'. Finally, when she could no longer explain it away, she went to the doctor; he agreed with her self-diagnosis of asthma and prescribed accordingly. But in the back of her mind Ann recalled that an ex-physiotherapist friend had told her of a respiratory unit somewhere in London. So she contacted this friend, who gave her details of the Lane-Fox Unit at St Thomas's Hospital, to which her GP was happy to refer her.

The Lane-Fox was a revelation to her. It took just one night to cure her 'asthma'. After testing her blood gases, which were 'absolutely horrendous', Dr Geoffrey Spencer suggested she use a positive-pressure ventilator at night.

He said, 'Try to get used to it tonight. If you can tolerate it for two or three hours I'll be pleased.' It took a while to give in to it, but once I'd gone to sleep I slept right through to half-past eight the next morning, missed breakfast, missed everything, didn't hear anything. I had the best night's sleep I'd had for well over a year. I stayed in for nearly a week to get used to it. I hadn't realised how badly it had affected me – incredible. After one night I felt ten years younger.

The Lane-Fox was 'like a home from home'. The staff understood what to do without having to be told and the equipment was first rate.

While I was in there, there was a total change in my attitude. It was like Saul on the road to Damascus inasmuch as all my life, my disabled life, I've liked to think of myself as not being disabled and I very rarely went to anything or did anything where disabled people were. I just felt different there – what I mean is that I *didn't* feel different there, I felt like I was one of the group and I didn't resent being disabled so much. It made me look at life differently . . . you know, it's the corny line: every day is precious. It really made me appreciate what I've got and enjoy every day, because there is going to be an end. Suddenly realising that, okay, my muscles were becoming weak – I hadn't realised they were – made me ask lots of question there of the staff, and particularly the physio: 'What's the prognosis? What's likely to happen? How can I best hang on to what I've got for as long as possible?' I hadn't given that a thought before.

It changed my attitude to my relationships as well . . . Brian and I have decided to get married; we're getting married in September. *Re*married after – what? – four and a half years of divorce! It changed my attitude to a lot of things and made me examine what was important.

Jim Porteous worked for two years in a firm of stockbrokers – 'the snag in stockbroking is that you need a lot of money to be a partner' – before he joined the sales force at Rowntree's in York: 'After that it becomes just a normal sort of business career, except that I was using a wheelchair.' In the Sixties you could get a job in Rowntree's marketing department if you gave a good account of yourself in an interview; nowadays you would need first class honours from Oxbridge: 'I'm now the national sales promotion manager of the biggest confectionary company in England and I wouldn't get my own job if I applied for it.'

Throughout his career, Jim has consciously used his disability and the wheelchair as 'branding'. In that way he has turned it to his advantage. In the early Seventies he applied to move from UK sales into export sales.

Later, much later, I learned that some people were not too happy with the idea. Bear in mind that this was 1972 and it was still fairly unusual to have a disabled person in this sort of management position. Apparently there were two areas of concern: one was, 'Will he be safe, and comfortable?' and the other was, 'What sort of image does this present of the company?' I'm delighted to say, they grasped the nettle on both counts.

For ten years, Jim travelled all over the world in his wheelchair – to Africa, the Caribbean, Japan . . . He made quite an impact – 'When we were negotiating on some of the licensing operations abroad, the people with whom I was talking would remember me intimately but wouldn't have recognised the chairman if he'd walked past them' – and he enjoyed himself: 'I think it was Michael Flanders, of the Flanders and Swann duo, who once said, "Being in a wheelchair is no problem, provided you live First Class," and I totally agree with that.' When he was travelling on business, he wanted for nothing. If he needed a car, with a driver to lift the chair into the boot, he had only to ask for it. If a hotel room was unsuitable, he could change not just the room but the hotel too, if necessary.

Travelling eighteen hours in a jumbo jet to Japan is no problem, provided you're going First Class and there are three stewards to every person. I do have a fear of being separated from the chair. I can only describe it as – how would you feel if the bottoms of your legs were removed and you had to travel with them somewhere else? Now try explaining that to an airline on a tourist flight to Majorca. They'll just bung it in the hold and you won't know whether the arm will be still on it when you get there. But if you're paying to travel First Class with British Airways, you can say, 'Don't take my chair away, put it in that cupboard' – and they will.

A First Class ticket in life does not solve all problems, however. Like Ann Stevens, Jim castigates architects 'who don't have the problems but think they know the answers'. He has been in any number of 'individually designed' disabled persons' hotel bedrooms 'which are so completely over the top they would do credit to a chronically sick and disabled ward – when all you need are a few minor modifications'.

But, credit where credit's due, the London Tower Hotel is a masterpiece of organisation. It's got nothing to do with the facilities in the room, though these are immaculate – the ability to open the door from the bed is wonderful, when you think of the palaver you have to go through to get out of bed when somebody knocks at the door. The main thing is that the staff have been trained to talk to you. As you go into the dining-room with your able-bodied wife, the waiter talks to you, not to your able-bodied wife. They haven't got a lowered reception desk but the receptionist stands up so that you have eye contact, and they don't have a pen on a chain that you can't

> use. It's a minor point, but somebody has obviously looked into it. The doorman knows how a chair goes in a taxi . . . I could go on for hours about this.

Jim enjoys public speaking, and one of his speeches, which he calls 'Standing on your own two feet in a wheelchair', is very popular in Women's Institutes.

> There are a number of things you can start off with. One of them is: 'My old headmaster said, when I left school, that with my cheek I was bound to go down in history – and how right he was. I went down in History and English and Maths!' Once you've got a smile on somebody's face you can go off on a different tack, talk about access and put the needs of the disabled across . . .

Jim lives with his wife and two children in a bungalow in a village outside York. He has been married twice; his first marriage ended after a dozen years when he and his wife parted by mutual consent.

> There were no children. Part of the problem was that Chris, my first wife, did not feel that was what she wanted and I jolly well knew I wanted children. But we parted friends and Chris and Tony, the chap she married, are amongst our best friends. They come round regularly, and we go round to them. I've just been gobsmackingly lucky with the two major ladies in my life. And when I was courting, some nights I was lucky and some nights I wasn't – as weren't we all? I suppose I've had the odd one-night stand that was out of sympathy, but what the hell . . .

He readily admits to compensating for his disability in various ways: he likes to talk about his job; he wants to be asked for his advice; he yearns to be respected. He enjoys driving around in a big car like a Jaguar and having people look at it twice – 'It's nonsense, it's silly, but it's something that's very important to me.'

> The one occasion when I am totally equal is when I'm driving. When I drive in London, the first people to cut me up are taxi-drivers; when I'm in a wheelchair, the first people to stop and let me cross the road are taxi-drivers. Now if a taxi-driver sees you as a 'cut-uppable' person, you're ordinary.

For all the compensatory swagger, perhaps this is the ultimate luxury – sheer ordinariness. As Jim Porteous says, 'When I'm in my car I'm ordinary.' If he carves up another car on his way to work in the morning, and gets a mouthful of abuse from the outraged driver, he will more than likely smile to himself. 'I must be one of the few people who would actually take being given the V-sign at a set of traffic lights as a compliment . . .'

In America

I went to the United States for three months in the spring of 1992 to research the history of polio in libraries and other institutions, and to talk to individual polio survivors about their experiences. But how was I to contact these polios? I knew of no American equivalent of the British Polio Fellowship; the March of Dimes was no longer concerned with polio; and I didn't know what, if any, other organisation had taken its place. While pursuing my initial researches in the library of the Wellcome Institute for the History of Medicine in London, I had come across a mini-obituary, scarcely more than a mention, of a woman I'd never heard of called Gini Laurie, whose claim to fame (of a sort) was that she had set up an international network of polio survivors.

I found an address in a reference book and wrote off to St Louis rather as a castaway might send a letter in a bottle – more in hope than expectation. So I was pleasantly surprised to get a helpful reply from Joan Headley, who was Gini's successor (see below), enclosing a copy of the thirtieth anniversary issue of Rehabilitation Gazette, *which consisted of potted life histories of forty severely disabled individuals, most of them polios. I selected a number of these, who – from the internal evidence of their stories – would be more or less on the route I intended to travel, and whose lives sounded interesting, and wrote to them via GINI. All who replied – about half of those I'd written to – were ready to talk to me, and account for four of the stories in this section.*

Of the remainder, three came through personal contacts in America – one of whom, Leonard Kriegel, was a friend from a previous visit to the United States.

Jean Johnson was eleven years old when she contracted the disease after going to a girl scout camp in the summer of 1949. She lived in Butler, Pennsylvania, just north of Pittsburgh. There was a severe epidemic in the Pittsburgh region, in which forty-nine people died in the space of five weeks. Jean was rushed to the Children's Municipal Hospital, where – she found out later – she was put into the last available iron lung: 'They didn't have an iron lung for the child that came in the door after me, and she died.'

When she came out of quarantine, she was taken to a place which was soon to become famous for its connection with Dr Jonas Salk – the D.T. Watson Home for Crippled Children. It was the place where 'everyone wanted to go'. It was up on a hill; it had its own school; and it was the teaching wing of the University of Pittsburgh for physical therapy.

> I was put in a bed on the second floor. A nurse came in and gave me a plate with spaghetti on it, and I was totally paralysed, except I could move my right hand and my left foot. And I could breathe all right. She said, 'We're not going to coddle you here. You'll have to eat your dinner.' And I couldn't reach it. There was no way I could even get one strand of that spaghetti. I'll always remember that. And then she came in and took it away.

Such cruelty was untypical, however. Most of the nurses and physical therapists were kind and caring. And even crippled children found ways of getting back at the bad nurses.

> They would be going past our quarter and we would slide the forks from the trays and hit them in the ankle when they weren't expecting it and it was dark. Then we'd go back in the ward and they didn't know who had done it. We'd go in and leave the showers on and put the plugs in the tubs. We'd spill our food on our beds and then say, 'Oh, it was a mistake,' you know. It's a strange thing, I can only remember a couple of names from that time, but we all knew each other from our abilities. When we pulled one of these capers, you'd get somebody with hands to help somebody that could use their feet. I mean, we had to take a combination of bodies to be able to pull these things off, so that you knew people by what they could help you with, or what they were able to do, not by what they *couldn't* do. Which, of course, is entirely opposite to the view of the outside world.

Not all the children in the D.T. Watson Home had polio; there were some with displaced hips, some who had cancer, some born with arthritis, some with muscular dystrophy. One after another Jean's classmates in Sunny View (which was the name of the school) suddenly ceased to appear and, when she asked where they were, she would be told they had died. Comparatively, polio was a good thing to have: 'With polio you always grew up if you got past the messy part.' But there were polio suicides. Two boys in their late teens who had been football heroes in their high schools killed themselves; and the father of one of them had a heart attack when he found his son dead and himself died on the spot. The staff tried to keep the children in ignorance of this, but Jean heard it from the other boys in the boys' ward.

Gradually and painfully, she learned to walk again. By the time she went to college, she was walking with a cane – and fibre-glass braces to keep her legs straight. She went to university in Miami, of all places: 'southern Florida, the home of the beauty babes, of the revealing bathing suits and sun – and of course I look *terrible* in a bathing suit'. She would go to fraternity parties and sit by the pool, with her emaciated arm and emaciated leg, and no one would talk to her, no one would even acknowledge her existence. She had been a pretty child and, being the only girl for many years, she had held court to her admiring boy cousins. So she had grown up with a high opinion of herself, until disease knocked it out of her.

> In this kind of culture – in most cultures – men like perfect women; they like their arms and legs to match. And that was quite a shock to me, because I still thought of myself as quite a catch! And yet I came out into a culture which gave me much more resistance. Right away, you knew where you stood with men and it was just devastating. I mean, it was terrible. And nobody ever talked to me about it, nobody ever said, 'This is what you should expect . . .'

Salvation came in the shape of her husband-to-be, a fellow student who had lost his teeth and the front part of his jaw in an accident: 'We were two walking wounded when we met each other – he'd had his car crash; I was in braces.' They were married in 1960, but Jean was in no hurry to have children. She studied painting and printmaking at the Art Institute in Chicago while her husband completed his PhD at Northwestern University.

They moved to Rhinebeck in New York State (where they now live) when he got an academic job at Vassar; she also worked full-time for two years as an art teacher. Their two daughters were born in 1967 and 1976.

> Now I didn't want to have two children right off because I can't run and I don't have much stamina. I thought, 'Well, I'll wait till this one's in school and then I'll have another one. And that's what I did. I didn't really have what I would call labour either time. I think one of the reasons is that when I was being rehabilitated I was taught how to let go of my muscles: 'Forget your muscles, go some place else.' And that's what I did during childbirth. When I felt a contraction coming, I would just go numb and let the contraction take care of itself.
>
> So even though I can't run, coming through childbirth both times gave me a wonderful sense of accomplishment. The second child was wonderful, of course, but the first child was, I think, the height of emotional involvement that I've ever experienced in my life. It was scary, in that I didn't know what was going to happen, and I didn't know if I could control it. But I could, and I was able to produce this child.

Jean is conscious of wear and tear and the effects of 'what they call Post-Polio Syndrome' on her: 'Sometimes when I get up in the morning it's as though somebody is lightly pressing their hand on the top of my spine.' But in some ways her disability is less noticeable now, more masked than it was. She will never be entirely reconciled to the loss of 'the body that I had'. Yet there have been some gains to counteract this loss:

> Everyone I knew when I was at college is now divorced. Some of them are divorced twice over. Sometimes my marriage isn't happy, you know, sometimes it is happy. But the point is, I have faith in myself and the person that I'm with, I don't believe that person is a dud. If I hadn't been touched by the disease and had grown up in the restrictive, Victorian, anti-Jew, anti-black, anti-anything-different-than-we-are kind of atmosphere, and not stumbled into learning something different, then I would have married someone that would have fitted into that, and I believe I would have had a very dull life. I haven't had a dull life; I've had a very interesting life. I think that in large measure it's because I was taken *out* of my life. And I'll always be grateful for that, I really will.

Leonard Kriegel is a professor of literature at City College in New York and the author of several books, starting with *The Long Walk Home*, his polio memoir, which was published in 1964. Almost everything he has written is autobiographical, from his memoir and his very Sixtiesish essay, 'Uncle Tom and Tiny Tim: Some Reflections on the Cripple as Negro', to his latest collection of essays, *Falling into Life*. Polio has been the critical experience of his life and Hemingway the formative literary influence – his 'nurse', as he puts it in his first book.

He acknowledges that, from what he now knows, Hemingway 'was an awful, goddamned human being'. Nevertheless, 'that whole rhetoric that Hemingway had, that the world will break you but you have to be good in the face of it, was very important to me when I was seventeen, eighteen, nineteen . . .'

Nineteen forty-four was an epidemic year in New York. Lennie, eleven years old, went to a summer camp for two weeks. Four boys were sharing a bunk, two got polio. Jerry died; Lennie survived, but he lost the use of his legs . . . He was moved from what was then a small country hospital in Cold Spring, New York, to the Willard Parker.

From there he was transferred to the New York State Reconstruction Home, in West Haverstraw, NY, where he enjoyed the 'hot pool therapy' and endured the salt tablets he had to take.

> Even today, more than four decades later [he writes in *Falling into Life*], I still shiver at the mere thought of those salt tablets. Sometimes the hospital orderly would literally have to pry my mouth open to force me to swallow them. I dreaded the nausea the taste of salt inspired in me.

His first wheelchair – his parents bought him one when they were still scarce because of the war – brought him 'immense freedom'. Then he was fitted for braces – double long-leg braces, with a pelvic band. They were 'a pain in the neck to put on,' and though – as he recalls in *The Long Walk Home* – standing should have been another triumph on the scale of the wheelchair, it was not:

> At first, all I could feel, as I stood hanging like dead meat on the shoulders of John the Barber and Charley the Bracemaker, was a sudden surge of new consciousness, frozen, just for a moment, in a dark corner of my terror. The legs are dead, it whimpered, boring into some quiet alley of my twelve-year-old mind. Nothing more. Just a momentary thrust of recognition. And I quickly killed even that. Because consciousness of this kind, the death of myth, never helps a child – and I was, after all, a child. And being a child was my salvation then, for something in me instinctively knew that I had no choice but to play their game, to lose myself, or my recognition of the death of my legs, at least, in the courage that was expected of me at that moment. I was standing – on alien shoulders, to be exact, but standing nonetheless.

As long as he was in the rehabilitation centre, Lennie used his wheelchair, as it was 'a much easier way to get around, particularly in an environment where everybody else was in a wheelchair'. But as soon as he got home, he stuck the chair in a corner of his room and did not use it again until he broke his wrist in his thirties.

In what was then (much more than now) a city of neighbourhoods, Lennie lived in a working-class, white neighbourhood – Jewish, Italian, Irish: 'It was a good place to come back to.' Family and friends gathered round and there was a strong sense of community. But despite this family and neighbourhood solidarity, Lennie's was still a lonely struggle.

> I created, to a large extent, my own rehabilitation – without meaning to pat myself on the back. I mean, one has the idea with a disease like polio that one can will oneself into living a successful life. And it's almost as if your will becomes your kind of thrust against the world. At least, that's how I remember it, though I don't think I could have verbalised it in this way at that age. But what seems to me to have happened when I was sixteen and seventeen was an almost instinctive recognition that I had to will myself into being, that I had to kind of will a self, to create a self. And I think that that's the *one* gift, if that's the word, that one can take from polio, or from a disease like it.
>
> What I wanted, above all else, was to be able to define who I was. I mean, I lived a peculiarly vivid fantasy life for a long time. I'd been a fairly good athlete as a kid, and it was very difficult to give that up. So I would fantasise, I would listen to ball games and imagine myself, you know, playing ball. I think that reading helped me also. But what I finally realised

> was that I wanted to live on my own terms – which sounds rather melodramatic, I know. What I meant was not just the usual thing, that I would not be a burden on anyone, but that I'd also do the things I'd been told I couldn't do, like I would *travel*, I would marry, I would have children, I would – you know – live.

Lennie would, indeed, do all these things. But first he had to learn a hard lesson: to accept that he would never walk again without braces and crutches.

> I started to scream and cry and bang my fist on the window, I remember. There was nobody in the house, thank God. But right after that I very methodically sat down and thought, 'What do I have to do?' It was a month before my seventeenth birthday and I decided that what I had to do was to build up my arms. I started lifting weights and working out on my arms, and building up these massive shoulders because I realised I had to walk on my shoulders. So I would walk five, ten miles a day sometimes. When I think about it now . . .

There was a children's playground a few blocks from where Lennie lived. He would go up there at night, when the children had gone home, and practise pull-ups on the monkey bars. It was at this time that he started to become aware of women, and he would show off at parties in the East Village, leaning out from the landing, up several flights of stairs, hanging on by the strength of his arms: 'I remember this tremendous sense of joy I had, the feeling I was the strongest, toughest man in the world.' But there was anger as well as joy in his drunken, wild behaviour. Lennie has never ceased to be angry and he regards rage as a healthy emotion: 'I mean, you're always measuring the life you have against the life you might've had and – wacky as this sounds – at some level I still resent the fact that I can't play ball and box any more.' If there is something adolescent in all this, that is understandable:

> I never had an adolescence. I went from being eleven to being an adult. For one thing, I was on home instruction, so I did my studies by myself in the house. Then I went to college. There was no in-between, no secondary school. I married when I was twenty-four. And all my social life came together suddenly. I mean, there I was, feeling that I could do anything – everything seemed to be heightened.

Disease, Lennie wrote in his first book, 'is a sharing'. By which he means, it does not only affect the sufferer, but also mothers, fathers, wives, husbands, lovers – anyone whose life is intimately bound up with the person who has it.

> Before she died, my mother would say things about 'what happened to me when you got sick'. At first I'd get angry, and then finally one day I just

> said, 'Yeah, well, a few things happened to me also.' What she was talking about was something I don't think I really quite understood at the time, and that was the sense that she'd been held in judgment, that there are no such things as accidents in life, that all things are punishments from this vengeful kind of puritanical or Jewish God, and that somehow one gets one's merits. It didn't make my mother a superstitious, Jewish woman, but she came to it as a superstitious, Jewish woman. Deep down my mother believed that my having polio was God's judgment on her – which is a very common attitude, not only among orthodox Jews but among orthodox Christians as well.

Paradoxically, Lennie reckons that polio affected his younger brother's life more adversely than his own:

> My brother didn't have polio, but my mother's overly protective instincts were incredibly magnified by my having had polio. And he had a terrible struggle, which resulted in years of analysis and years of therapy. I had the polio to fight against, he didn't.

Lennie's wife Harriet has been affected by his polio, too: 'It made her a very tough human being, it brought out the best in her.' Their marriage has not been without problems – 'I don't think any marriage is' – but, like Jean Johnson's, it has endured. 'It's quite possible that polio made it a better marriage than it might otherwise have been.' Their two sons are now grown up and out in the world, but theirs – like his parents' – is a close family.

Lennie is not dissatisfied with what he has achieved: 'I've written somewhere that the dirty little secret of anyone who's lived with polio – or any severe physical disability – on intimate terms is that after a while you come to feel a certain contempt for people who haven't.' Yet he cannot face the future with equanimity:

> My arms are still very, very powerful. But the day's going to come when they're not going to be, and that scares the hell out of me. I'm terribly frightened of becoming dependent. I'm not afraid of things physically, but I'm terrified of the idea of helplessness.

This fear finds powerful expression in a recent essay, 'In Kafka's House', in which he tells the story of a crippled Dutch friend – not a polio – who is imprisoned in a room by his wife, with the connivance of their two growing sons, for several days. Michael, as he calls this friend, is plonked down on a mattress in an otherwise almost bare room. There are no toilet or washing facilities and the door is unlocked just once a day, when food is slid across the floor to him on a tray. One afternoon, his wife and sons go out, deliberately leaving the room and the house unlocked, and his car keys in the ignition, so that he can make his escape, crawling out of captivity in all the stink of his soiled clothes like a disgraced infant. He goes to the woman with whom his wife suspects him of conducting a love affair; she takes him in, bathes him and, much later, marries him – thus confirming, if only after the event, his wife's suspicions.

In that same essay (which demonstrates how, as a writer, he has outgrown the baleful influence of Hemingway), Lennie offers what might ultimately serve as his own epitaph:

> The man who is 'successful' at creating a life out of the after-effects of disease – as, however immodest it is to write, I have been – discovers that he must, sooner or later, fight against an inflated notion of what it is he has achieved. When a mutual friend praised Franz Rozenzweig's courage in living with the pain and suffering inflicted on him by a long struggle with cancer, Freud, doomed to undergo the same long struggle, is reported to have said, 'What else can he do?'

Josephine Walker had contracted the disease in Philadelphia in the same epidemic year that Lennie Kriegel had got it in New York. She was then six years old.

> It was *the* most profound thing that happened in my young life. I remember the night I got sick. I remember my father returning from a business trip and coming up to say goodbye to me. I remember the ambulance coming and taking me off alone to the hospital. We were all put in quarantine for about two weeks, when nobody was allowed to see us.

When she came out of quarantine, Jo was transferred from the Municipal Hospital to the Children's Hospital and was allowed to see her parents for the first time: 'I remember being *absolutely* hysterical, which I'm sure upset them enormously.' But neither doctors nor parents talked to six-year-olds, so there was a complete silence on the subject of the illness: 'Things seemed to happen, but nobody discussed it.'

> My parents did everything for me that was needed *physically* – I was held and carried around by my mother for many years. They were in total denial about the fact that there was an emotional component to this. And so they pretended, after a while, like it didn't happen, other than the fact that I needed – you know – a little bit of medical help. People didn't talk about it; they didn't talk about the implications of it for my life. They just kind of let me go.

After a disastrous spell in a school for crippled children – 'the other children had disabilities that impaired them in so many academic ways and I was not impaired in that way at all' – Jo went to a small, independent, Quaker school, which provided her with just the kind of protected environment she needed. It was the best thing that happened to her: 'Education was my lifeline.' She went into higher education, choosing to go to Wellesley College, in Massachusetts, 'which was a large institution, with a huge campus and, being in New England, tons of ice and snow'. Her mother, on the one occasion she tried to interfere, was unsuccessful in preventing her going there.

> And in fact, it was probably *not* the best place for me. I was miserable for three years of my life. I finally made an adjustment in my fourth year. But I was under *incredible* physical stress the entire time. Had to cut myself out of activities and things, because I just could not keep up. That was totally unrealistic on my part. But I had to prove something – especially when challenged.

Like Jean Johnson, Jo discovered that 'women had to be perfect'. She tried to participate in activities like dancing and dating, but she did not seem to derive the same pleasure from them that others did. She was far too anxious, always striving to please others: 'I mean, whatever somebody else wanted was far more important than what was right for me.' Her professional life – in counselling, and then in educational administration – went ahead in a reasonably satisfactory manner. But her emotional life was going nowhere. In her mid-thirties, she sought help but was unlucky in the person she found, who was abusive: 'So that was a disaster, three years of extraordinary expenses for nothing.' At forty, she got married, 'and found another abusive person – this has been part of my pattern, that I find people that are not going to be helpful to me.' The marriage lasted ten years – 'ten years of enormous physical burdens put on me, that I could not handle' – before it broke up.

> The famous word today is *dysfunctional*. To me, as I look back now, I see the dysfunctionality in my life, and I do think that it grew out of not having the emotional support and help that I needed at an early age to cope with it. I can't blame anybody. There might have been some families or parents that were stronger themselves. My parents had some problems, and I think the fact that they were caught up in their own problems made it more difficult for them to see beyond those things to what I needed as a child with this issue. They addressed it as far as they could, but they really wanted it to go away. And they thought that if they gave me a private school education, you know, and dressed me nicely and kept me healthy, from that point on it was up to me.

Not surprisingly, Jo did not want to know other polios, or anybody who reminded her of her condition: 'If I walked down the street and saw somebody else who was handicapped, I died of mortification.' The very word *handicapped* was unacceptable: 'I put handicapped licence plates on my car three years ago, and this is bizarre, because they've been available for twenty years, but I wasn't going to go round with that label.'

At the age of fifty, Jo had reached an *impasse*. Then she hooked into the post-polio network; she read a couple of articles by psychologists on the 'polio personality' and felt like 'jumping for joy – because I felt somebody really understands'.

> I've finally gotten myself involved in some good, substantial therapy, addressing the fact that I really had this disease, and it really had that effect on my life, and I have been able to say, 'Yes, it did this and it did that.' All along, obviously, it was having a profound effect on shaping my attitudes – mostly in terms of driving me to be as normal as I possibly could be. I mean, all of us put such effort into living and trying to do everything the way everybody else did. Hence the burn-out syndrome. I can understand it completely now.

The sense that she was not alone, that many others were fighting the same battle, revolutionised Jo's life. She might still bitterly regret the losses – 'like every woman's dream of having a family' – but:

> What I have accomplished in the last five years has just been a miracle, because I have *really* thrown out one life and brought in another one. I have made enormous transitions in terms of getting established on my own, buying a house, changing my life so that I'm doing what I need to do for me, and finding out what *I* want.

Dr Lauro Halstead – whose major contribution to the post-polio debate is documented in Part I (see above, pages 209–14) – works at the National Rehabilitation Hospital in Washington, DC. Unlike many polios, whose growth has been stunted either naturally or artificially through surgery, Lauro is a tall man, well over six foot. But then he did not contract polio until he was full-grown.

Here, in his own words, is the story of how it happened:

> I had finished my first year at college and was travelling in Europe. I'd been going to live with some people in southern France, whose names I had, because I was interested in studying French. I was hitchhiking and I got a ride with some Portuguese and got to be quite friendly with them; and they ended up inviting me and the companions I was travelling with to Portugal. It was there, in Oporto, that I began having the prodromal symptoms, which in my case were a little diarrhoea and a feeling like I had the flu. But I didn't pay a lot of attention to it because for a *turista* it's such a common problem. I was eighteen, and I was very healthy and very strong. In fact, the summer before I had been selected to be a model for an exercise book because I'd worked out and had very well-developed muscles.

Any event, we had tickets – myself and the three companions I'd picked up along the way – to catch an early morning train from Oporto to Madrid. After a few days lay-over in Madrid, I planned to go back up to France. From Oporto to Madrid was a twenty-four-hour, narrow-gauge, train trip, which turned out to be one of the worst nightmares imaginable. Portugal's quite a mountainous country and there are lots of tunnels and ups and downs. And it was a coal-driven locomotive, so there were all these cinders and black smoke. No air conditioning. Third-class carriages. The windows were open, so it just filled with this black soot. And peasants would get on and off with all their animals and the usual things. This is back in 1954.

When I got on the train in Oporto, I was able to lift my suitcase and put it up on the luggage rack. But during the course of that twenty-four hours I could feel myself getting weaker and sicker. And there were precious few amenities on that train. It did make a few stops along the way and I can recall wanting to urinate and having great difficulty in getting off the train and finding any place whatsoever to urinate and then getting back on. By the time we reached Madrid on the Sunday morning my right arm was quite weak and I could not raise it above my head to reach for my luggage, so I grabbed it with my left arm.

Well, being a college student, and with three companions who were also hitchhiking and trying to do Europe on five dollars a day, we decided to look for the cheapest *pensión* we could find and, rather than hail a cab, we just started out on foot. Sunday morning in Madrid things were very quiet. After three or four false starts – because we would try a *pensión* and then decide it was too expensive, or too this, or too that – we ended up going to a place on the fifth floor, as I recall it, all the while carrying this damn luggage. No elevator. And I was just about collapsing. I said, 'This is it, we're going to stay here regardless.' Which we did.

So I got to bed and my colleagues, who now realised I was sick, started to talk to the *pensión*-keeper, the guy at the desk, the manager. Eventually, through their broken Spanish, they were able to explain the position. I'm a little hazy on the details, but I think they had trouble getting a Spanish physician. Now one of the guys I was with had been in the US military, and there was an air force base outside of Madrid. He thought he would call over there to see if he could speak to a medical person. So he did call over there and spoke to a Colonel Patterson – I think his name was – whose first question was whether I was connected to the military. Of course I was not, so he said, 'Well, I would love to help but I cannot.'

Time dragged on and, for whatever reason, they weren't able to get a Spanish physician. But in the meantime this Dr Patterson finished his chores for the day, went home, and found he could not sleep that night, because he sort of put the situation together and realised that here was an eighteen-year-old American kid, on his own, in the middle of Madrid, and he could guess what the problem was. So he came over in middle of the night – maybe it was like 10 or 11 p.m. – because he'd kept the number, and found me and made the diagnosis. Then he got a Spanish doctor – a friend – and together they got me into a Spanish hospital.

Now, in those days a general hospital didn't admit people with infectious diseases. So they had to invent some reason why my right arm should be paralysed. It so happened that earlier in the summer I'd been in England, riding a motor cycle, and I'd had an accident in which I hadn't been hurt but could've been. So they said, 'Well, he had this bad accident on a motor cycle a few weeks ago and has this weakness in his arm.' They admitted me without diagnosis. Of course they started taking X-rays . . . I mean, I was now very weak, and they were asking me to stand up so they could take chest X-rays, and I kept collapsing on the floor, and they couldn't understand my Spanish – it was just wild. Anyway, they finally put two and two together and realised that I wasn't post-accident, but something else. In fact, I think they did a lumbar puncture, guessing it might be a polio problem, and diagnosed it.

Someone – probably this Dr Patterson – had alerted the American embassy, who in turn called my folks back home near New York City, saying that I had this diagnosis and that my prognosis was poor. In those days, getting a passport was a lengthy process, but my parents had friends who pulled strings and my mother borrowed some money from an aunt, and she was on a transatlantic flight within forty-eight hours.

In the meantime, they hussled me out of this general hospital, and the only place that would take me was a German nursing home. So I ended up in a German nursing home which – I don't ever remember seeing what it looked like from the outside, but it must have been just like a regular house – had two or three stories and a narrow staircase that turned at right angles. And they had no way of getting me up these stairs other than putting me in a straight-backed wooden chair. By now my neck was also – not paralysed, but very weak – so I couldn't hold it up. My right arm was paralysed. My legs were weak. And they put me in this chair and I was sort of tipped back, with my head falling, and some strong people carried me upstairs and put me in the second-floor front room of the nursing home, where I was in the care of a Sister Arnie, who turned out to be a superb nurse.

I started to get better and my mother arrived within a couple of days. And then, as polio often does, it had a second phase. Just as I started feeling better, the paralysis got worse, affecting my chest and involving my breathing. Again it was a Sunday – Saturday night, I think – and I found myself starting to hallucinate. There was no oxygen or anything. It took quite a while for the sister running this place to track down my mother, who came. She was very alarmed, and she tried to track down the Spanish doctor. Of course, everyone eats out on Saturday night, so tracking him down was quite a struggle. Then there was the question of finding oxygen, as I was now hallucinating and not with it . . . It seemed like weeks, but it was probably more like six or eight hours before they finally did get a little canister of oxygen.

The next thing they did was try and find a place to put me. There was one hospital that had so-called iron lungs: it was called the Hospital for Children of Jesus, a 400-year-old hospital. There were iron lungs, but they were made of plywood. And they were made for children, not for adults. Well, I was six foot four at the time. So they took me over and sat me – doubled up – in this little crate. It did the trick, though, it did assist.

I was in what was probably a basement room. As I recall it, and as my mother has described it, there was nothing else in the room. No other patients. And there was a ceiling about forty feet high, with frescos. It was a classy place. The only problem was, it was run by Catholic sisters, who were much more interested in saving my soul than my body, so they were always out in the courtyard at the back, singing and praying.

The first night I was there – I had no call-bell, no way of reaching out to anybody – the lung stopped working. It turned out that in those days Franco was trying to save electricity, and all the electricity in the city went off for four hours every night. Fortunately I was just able to breathe enough. But I started to hallucinate again and I thought I saw this lovely comfy bed over in the corner. The janitor, or somebody, came by, and I had just enough Spanish to convince him that I would be better off out of this plywood lung and in that beautiful featherbed – as it looked to me – in the corner.

I mean, I was a 190-lb, strapping eighteen-year-old, and he pulled out the cookie tray – you know – that I was on, and somehow got my arms around his neck and dragged me, literally dragged me, across the cold stone slab of a floor. And it turned out to be a cot! My muscles were so painful and aching that lying on that tight little piece of canvas was like lying on a bed of nails. I just couldn't conceive what the hell was going on. So then I had to convince him that I would prefer to go back to where I was. And the poor man dragged me back.

The sisters couldn't bathe me, as they were not allowed to touch a male body. My mother tried as best she could, but she couldn't give me an enema, and I needed an enema. So they called Sister Arnie, back at the German nursing home, who came over and gave me an enema, if you can believe that. My mother did a lot of the other essential things and got me food.

Eventually, I started to breathe a bit better, and then Dr Patterson reappeared. He decided I really needed to get out of there. But there was a ninety-day quarantine for anybody who had an infectious disease. So through his contacts in the military he arranged to have me flown sort of incognito to a big hospital in Germany.

Basically what they did was, they arranged a dawn raid, and they kind of crashed the gates of the hospital, these six young and husky orderlies. The sisters were just standing there with their mouths agape as, with enormous efficiency, they came in with this stretcher, wholly without asking anyone's permission – like they were kidnapping me – and tucked me on the stretcher, and then double-timed it to the ambulance.

Just at that point, the medical director of the hospital showed up. He started to bitch. But I was out the door. By now, sirens were wailing, and I was in the back of the ambulance. Then I went to the US air force airport, where they had a plane waiting, with the propellers going.

They were concerned there might be repercussions, because they were breaking Spanish law. And it was very important that we stayed on the good side of Franco. So they'd concocted some story that I needed special treatment at this hospital in Germany.

I stayed there for a week, until I could breathe twenty-four hours on my own. Which was the criterion for being flown across the Atlantic. I no longer needed a lung. I was in a proper hospital bed, but I was also in quarantine from my mother, who had been feeding me and bathing me – now she couldn't even visit me!

Back in the States, I went to a rehab hospital near my home, outside of New York City, where I stayed for . . . I basically got sick August the first and got home sometime at the end of August. But my recovery was fairly quick. I was in a wheelchair for six months, but I was walking during some of that time. The strength in my legs starting to come back, the left arm had got quite strong. The right arm never came back. But essentially I was ambulatory by Christmas. By the end of January I was down to just a short leg brace on my right. And I was well enough to go back to college on a part-time basis.

I was right-handed, so I had to switch to being left-handed. My legs grew very strong, so I could climb stairs without difficulty. In some ways my legs were even stronger than before, I spent so much time exercising them. I used to leg-wrestle my room-mates, some of whom were soccer players with very strong legs, and I could out-wrestle them. Like many people I felt I had, if not *beaten* polio, at least overcome it. I felt like I was able to do anything I wanted to, except, of course, use my right arm. I was one-armed. I finished college, and I went to Italy and studied Italian literature for a year and travelled round Europe again. Came back and did four years of medical school . . .

Lawrence Becker is professor of philosophy at William and Mary College in the old colonial town of Williamsburg, Virginia. He is also a respiratory 'upside down' polio. He contracted the disease in an epidemic in Nebraska in 1952, when he was thirteen; he was completely paralysed and he could not breathe.

> Not being able to breathe was frightening, but what was more frightening was the iron lung, and that was because of a radio story that I'd heard – illegally. I was not allowed to listen to a programme called *Inner Sanctum*. There had been a story a year or two earlier on that programme, a little mystery story, about a man getting in an iron lung to demonstrate it for a March of Dimes rally, and being murdered by his wife and her lover – they locked him in it and turned up the pressure! Of course, I didn't know that you can't kill someone that way. It took my father some talking to get me in it, even though I was in a great deal of distress.

Larry's father was a Congregational minister, and his parents were both immensely supportive; 'Essentially they reorganised themselves around me – at great cost to themselves, which I was not allowed to see.' When, after eleven months in hospital, he was moved to the nearest respiratory centre, which was 160 miles away in Omaha, they also moved, his father taking another church only thirty miles away. They would visit at weekends and, though Larry came to discourage these visits – 'I became acclimated to the environment in the rehab hospital and they seemed intrusive' – he never doubted the value of familiar support.

> I was sort of in the last era of major epidemics, and the polios I knew all seemed to be quite forward-looking, reasonably cheerful, unless they were entrapped in terrible family situations – and I had some friends like that, and it was just *awful* . . .

As Lennie Kriegel has written, 'Disease . . . is a sharing.' It was not only Larry's parents who were affected by his polio; his sister and younger brother were, too. His sister became a nurse, his brother a carpenter.

> He's suffered the most for it, I think. But he never blames me or the family. It's just that as a child he suffered a great deal from being the hale and hearty one. He's very large – six foot two – a football player and that sort of stuff. Also, I suspect that, in the case of both my siblings, there was something of the intellectual life being reserved for me. I was to be that one, and they were to do other things.

But for a while, Larry could do nothing: it took a year and a half to get him on to his feet for the first time; and before that, he had to be weaned off the iron lung.

> The weaning process was quite brutal. It was with a stopwatch, literally. It was perfect behavioural conditioning. They would give little prizes for each advance: 'Three minutes . . . All right, you can have a football!' By that time I didn't want a football, I wanted a chess set . . .

The first time the doctor tested his breathing, he did not tell Larry what he was doing; he merely went to the back of the iron lung, turned it off, checked his stopwatch, then came round and watched Larry turn blue: 'It was as if somebody had pushed my head under water.' After that experience, Larry would not allow anyone to turn off the lung – even when he could breathe for hours unaided and was not actually in it, but lying on a hospital bed nearby.

At the respiratory centre in Omaha his vital capacity – that is, the strength of his deepest breath – was measured, and it was discovered that he was operating on only 250 cc's for eighteen hours a day (3,500 cc's would be normal for a man his size). Various things were tried, including putting him on a rocking bed, to boost his breathing during the day. When he was finally weaned off the lung, he went home.

> I was clear of the iron lung for about three months – maybe four or five – and then I started having tachycardia. I faked it: I told them I was sleeping, and I didn't report the one incident of tachycardia I had in the hospital. I thought I was going to die and then I didn't, so I just forgot about it. But it was agonising because I would get exhausted and I would sort of fall asleep and stop breathing, and then the CO^2 level would go up and I'd wake, start to breathe a little bit, then drift back, in and out of a kind of twilight. And that would still happen today – without the respirator.
>
> Anyway, they sent me home without a respirator, without anything. I was often drowsy and I'd sort of drop off in the daytime, so they gave me Dexedrine to perk me up! Eventually, when it got really bad and I had this tachycardia that I finally acknowledged, my local GP just said, 'You're going back to Omaha.' So I went back and they did more studies, and then the doctor wheeled me into his office and said to me with great trepidation, 'I have bad news – you're going to have to sleep in a respirator for the rest of your life.' He thought that, given the way I'd fought everything, I would be beside myself. But I was relieved. I was so exhausted.

Warm Springs did not welcome respiratory cases. The regime of hydrotherapy and outdoor activity was too robust for those who were struggling just to breathe. But Larry went there for functional rehabilitation – all paid for, as in the respiratory centre, by the March of Dimes. He found it an amazing place; it had the most wonderful brace shop, where things would be made to your own design; it had a lovely dining-room, where meals were served on white tablecloths . . . Less impressive was the inherent racism of the system, where all the professional staff were white and all the menials, the attendants, black. There were other kinds of prejudice, too.

> By the time I went there, maybe only a third of the patients were old polios. Lots of spinal cord people and lots of arthritics. And there was a real pecking-order – the polios got the best of everything. I felt sorry for the other guys. They were doing various kinds of studies on, for instance, the

> distribution of intelligence or IQ scores. They were convinced, for some reason, that polio survivors were at the top. They really reinforced this notion of an élite, and they would say to others – in our presence – 'The polio guys never complain, why aren't you like them?' When you think about it, people with arthritis, with any sort of degenerative disease, who never know when they wake up in the morning whether they're going to be able to move – it's very different from polio which, when we had it, we thought was just going to get better, I mean, 'It can't get any worse, it can only get better.' Little did we know . . .

His father was astounded that Larry never seemed curious about the 'reason' he got polio. But for Larry there was never any question of blame; he did not think he was being punished: 'I didn't think it was God's fault.' He regarded it simply as very bad luck.

> In terms of coping with it, I think being two and a half years with other people who are relatively cheerful and work hard at their rehabilitation, and having a family that simply assumes that you are going to go on, sort of helps. The worst things were after I came home, being with other adolescents my age. That was bitterly difficult at times, outside of a school setting, which I didn't have much of anyway – for instance, in church group parties, or picnics, that sort of thing, where kids would club together . . . I remember weeping about that one night, and my parents were instantly very understanding. I think they just sort of manoeuvred me away from those situations – I didn't take much manoeuvring, because I didn't want to put myself into them. Which meant, of course, that I had no social life – and no adolescence.

If he had been asked about his life goals before he had polio, Larry would probably have said that he would become either a physician or a clergyman. His father, who was his idol, had started out as a pre-med student before taking holy orders. After he had polio Larry came under pressure – though never directly from his father – to consider the church as a profession. He used his disability as an excuse, saying that if he could not take communion services or give baptisms he did not want to be a clergyman. There was no answer to that. When he chose philosophy, his father gave him a beer label advertising 'Hemlock', which he still has: 'I've got it in my office, and a student of mine this year asked me whether it was from the Hemlock Society, and that's given me pause.'

He had learned to write – and type – with his feet, which he found useful when he went to college in his home town. From Midland College, he went on to the University of Chicago as a graduate student. He had two good fellowships, and he lived in a Congregational dormitory in the Chicago Theological Seminary. He hired two seminarians to look after him in return for paying their board and lodging.

> I was pretty spry by that time. I mean, I walked to and from classes, which horrified the doctors – and now horrifies me – across ice and snow and so on. Then I went up and down long flights of stone steps, which is very dangerous for me since, even though I was fairly stable, without the use of your arms you can't catch yourself or anything. But luckily I never fell.

In the library, he would put the books on the floor to read them: 'I have very good eyesight – used to, anyway.' But the advantage of studying philosophy was that so much of the work could be done in the head.

When he was at the rehabilitation centre Larry had been told by psychologists that he should not try to be an academic – certainly not an academic or clinical psychologist, which was what he then wanted to be.

> The general idea was that patients in various kinds of long-term care facilities often want to become care-givers, and they thought this was a manifestation of that. But the way they argued it was really quite damaging. Essentially, what they argued was that clinical psychologists need to look as normal as possible, right, and unless you want to confine yourself to rehabilitation medicine you shouldn't try to do that. Well, after they beat me down on that – although they didn't persuade me entirely, I was fairly ornery – the second argument was: I certainly shouldn't try to be an academic because there was a good deal of discrimination, they said. They didn't tell me about the long tradition of blind law professors, for instance. Maybe they didn't know. But what they tried to do was persuade me to become a lawyer. They said that all my tests and so on showed that I was good at that. I think what they meant was that I was very argumentative!

When it came to getting an academic job, Larry was lucky. In 1965 he applied for a one-year replacement post at Hollins College, a small liberal arts college in Virginia, and was summoned to an interview.

> It turned out that the president's wife had had polio as a child and had lost the use of an arm. All the president said was, 'It looks like it's going to be very difficult for you,' and I said, 'Well, I suppose it will be.' He said, 'Do you think you can do it?' I said, 'Yes.' He said, 'Fine.' I did a year, and it was extended to another year. Then it turned into what we would now call a tenured track position – that is, a real position, not a replacement one. They just expanded the department. And after a few more years I was given tenure.

His wife-to-be, Charlotte, was a senior at Hollins during Larry's first year on the staff there. At first they did not much like one another and it would anyway have been improper to date. It was not until she graduated and went to work in Chicago that they began to write. After a year they decided to get married: 'We never had a proper date, just correspondence.' They have now been married for more than quarter of a century. But they have no children:

> We could have had children, and we thought about it, right, but for reasons having rather little to do with polio, I suspect – although I think the fact that I couldn't help much, and hold children and so on . . . I think if I hadn't been an upside down polio I would have felt very different about it . . .

Larry and Charlotte travelled to England, where Larry spent a year at Oxford, studying under the legal philosopher, Herbert Hart, in the early Seventies. Since then, however, his mobility – though, fortunately, not his breathing – has declined, and when he was at Stanford in 1984 he finally 'broke down' and started using a wheelchair: 'that was a bitter pill to take'. For some time he had been having problems; he dreaded leaving his building in college as that meant negotiating the three or four steps outside. Now he has a three-wheel cart that goes in and out of buildings.

> I had a real bad patch, psychologically, when I didn't know how far I was going to slide. Now it seems to have stabilised, reached a plateau; and what I've got left I can live with because it's not that much different. I think I'm at peace about that now. But it was a bad couple of years.

Work has been both the mainspring and the mainstay of Larry's life. In addition to his teaching, he has written a book on *Reciprocity*; with Charlotte, he has edited a two-volume *Encyclopedia of Ethics*; and he plans a further book on 'the old and now neglected philosophical topic of "the good life"'. He reckons he has had a good life – 'not the best life one could imagine for a human being, but a good one'. But two things still trouble him: one is 'a certain flattening of affect' that comes from developing the patience required to cope with severe disability.

> The other thing is the inhibition about children that comes from not having the use of my arms specifically. Now that's not polio, that's a particular arrangement, or distribution, of disability for me. Of course that's idiosyncratic, too, because I have friends who are quadriplegic and had children. But I just wasn't quite able to face that one . . .

Joan Headley, executive director of GINI, describes herself as 'a sort of left-over hippy, not a real organisation person'. She has no memory of having polio, which is hardly surprising since she was fifteen months old at the time and had little in the way of after-effects: 'I never thought of myself as being disabled – ever.'

Her early memories reflect a certain ambiguity over her position on the able-bodied-to-disabled spectrum. For instance, she recalls queuing up for the polio vaccine:

> The little boy standing next to me said, 'Why are you taking this?' I looked at him and I thought, 'Why shouldn't I?' Then all of a sudden it occurred to me: all this fuss was for something that I had had – and it didn't even hit me.

If that recollection highlights her unawareness, another memory from her kindergarten days tells a different story. On Halloween all the infants in her school had to put on costumes and walk through the other classes.

> That was the big deal, that you'd go through and nobody would know you. I can remember the day that we were supposed to do that, I wouldn't put my costume on. I refused to do it. I can remember thinking, 'This is ridiculous, everybody's going to know who I am – 'cos I'm going to be the one that limps.' I cried and all that stuff, and they got my sister and she finally persuaded me to put it on. I can still remember, though, the teacher kept asking me, 'What's wrong?' And I never told her. Even my sister asked, 'What's wrong?' She sort of shamed me into it by saying, you know, I was embarrassing the family. It's funny, because I can remember thinking, 'Gosh, they don't even notice it.' So I naïvely went through my grade school years thinking, 'It doesn't matter, nobody notices.'

It came as a shock when she arrived at college and went to pay her tuition fees, only to be told by the woman at the desk, 'Oh, I don't think you need to pay all this – you can get money from Vocational Rehabilitation because you're handicapped.' Needless to say, she took the money.

After college, she became a teacher and taught for nearly twenty years in rural Indiana, forgetting all about polio – until one day a friend told her she had heard that people were getting polio a second time, which Joan took to mean that the virus was being re-activated.

> Well, to me it was like being hit in the stomach, I had a physical reaction to her telling me that. She, being a librarian, found Gazette International as the resource, and I wrote and got the blue post-polio handbook and then got the *Polio Network News*. It came out quarterly, and it was interesting because I didn't really know anybody who'd had polio, except two people distantly – but then I didn't talk to them, I'd just seen them on the streets. I enjoyed reading that every quarter; it was my connection with people I didn't know.

She still thinks of that now, when she is putting together an issue of *Polio Network News*: 'I try to put stuff in there for whoever's up there in Wolcottville, Indiana, who doesn't know any other polio survivor . . .'

Joan was about to be forty. She had been in love – or thought she had – but that had not worked out. She was unmarried and had no children.

> The hardest part of my life has been getting past the child bearing years. I don't know if that had anything to do with the polio with me, or not. I

wanted children, I really, really did. That's why it's so important for me to be at my niece's graduation, because my sister's children are sort of like my kids. Their father left when they were very young, so I always tease them about being their dad, or whatever. Anyway, I had them because my sister so nicely shared them. But growing up in the Fifties you were supposed to get married and have kids, so it was really tough not having them. But I survived through that and I thought, 'Well then, if I'm forty and I'm still alive and I don't have kids, then there's no reason to stay teaching in rural Indiana.' So I decided to do something else.

She told her principal that she had had enough of being 'Miss Headley' and teaching in Lakeland School Corporation. She took the computer home for the Spring break and wrote off to schools in other areas; she also considered going into further education and learning to train teachers. But during that week she got her copy of *Polio Network News* and saw an advertisement for director of the International Polio Network. She thought, 'I could do that,' sent in her standard letter, with an additional paragraph saying that she had had polio. By the time Gini Laurie phoned her, six weeks later, she had forgotten about it.

Gini always talked in superlatives – she had this 'lovely lady' who worked for her and she'd met this 'wonderful physician' and they were getting married and clogging off to California, 'madly in love'. And 'we do need somebody to replace her', etcetera. And 'we're having this conference and we'd just love you to come and meet all these people, so can you get to St Louis?' I kept thinking, the whole time she was talking, 'Why does she keep saying St Louis?' Because my *Polio Network News* came from Maryland Avenue, and I was sure this organisation was in Maryland! That was how out of tune I was with it.

Joan could not get to the conference until the very last day, but – 'Gini never took no for an answer' – she found herself looking out of the window of an aeroplane flying to St Louis and wondering what she was letting herself in for.

I walked into the Sheraton – that was the largest conference – and there were 600 people there, wheelchairs everywhere. I don't think I'd ever known anybody in a wheelchair before. I dreamt about wheelchairs and stuff that whole night . . .

After this initial encounter, Joan was invited back for a formal interview with the board – though, as she was to discover, the board was little more than a rubber stamp for Gini, who laid down the law and made all decisions herself.

Gini had sent me newspaper clippings, so I had read quite a bit about her. It sounds sort of naïve to say, 'Well, I liked her,' but I did. She had so much

> energy. She used to embarrass me when I first got here because she would do the same thing as she did with my predecessor: 'Oh, we have this new lady working for us, she is *so* brilliant! I can't wait for you to meet her.' She finally stopped doing that because she said she saw my face one time when she said it – I must have looked horrified. It did two things: it sort of gave you confidence, but it was also a lot of pressure. She was fun to work with, but I had mixed feelings.

When Gini was dying, she told Joan that she did not expect her to make GINI her life in the way that it had been hers. If Joan wanted to, that was fine; but if she did not, that was okay too: 'I think she knew in her own mind that she had started something that was going to go on, no matter what happened right here.'

Joan has picked up the mantle and kept the show on the road, but in her own way, which she recognises is very different from Gini's: 'I don't want it to be a Joan Headley organisation, I really don't.' She has tried to make it more inclusive than it was in Gini's day, when there was a strong sense of a cabal, or coterie, from which anyone who did not meet with Gini's approval was rigorously excluded.

I have this urge to do good things for people, but I also believe in the concept of self-help. I'll help people up to a point, as long as they help themselves. I don't have a strong urge to reach every polio survivor in the world and make sure everybody has braces, I really don't!

Thomas Rogers described himself as 'a breathing-impaired quadriplegic'. On paper headed 'Thomas Rogers Company/Investments', he had written in all modesty: 'I don't think I have done anything noteworthy other than the basic drive to be self-employed.' He has several part-time attendants and a specially adapted bus in which he travels between his home in Moline, Illinois, the family holiday cottage by Lake Michigan and a place in Florida where he has business interests and spends a month every spring.

Tom had had polio in 1953, when he was nineteen and a strapping youth of nearly six foot three, weighing 190 lb. He had been at Cornell for a year, majoring in engineering. He had gone to a summer camp in Michigan for two weeks – 'It was a wonderful summer, it was two weeks of sun . . .' His mother had wanted to see a cottage there, which she thought was vacant.

> We'd gone there. The cottage was closed, but we'd climbed in. I got in there and I realised there were four boys staying there. They'd been basketball players, and one of their best friends had died of polio. So they all went up there to get away from it – polio was the disaster that had struck

> their friend. Of those four boys, I think three of them came down with polio . . . So if there was a time I could trace the infection to, that was it.

When he learned that he had polio, it did not mean much to Tom: 'It was one of those things that mothers scared kids with.' He had an uncle who had had polio as a child; he had a withered leg and a bad limp – 'but he was such a neat guy that it didn't seem to bother him'. With Tom, however, it was turning out to be much more serious; his breathing started to go and he was relieved to get into an iron lung.

> I was left pretty well totally paralysed and not breathing. When they took me out of the iron lung my breath really failed me, so I started gulping. Nobody taught me this, but I became very proficient at it. I had no respiratory power, but within two or three months I was out of the iron lung, transferred on to a rocking bed.

The rehabilitation process lasted about two years, at the end of which he was able to frog-breathe the entire day – and he kept that up for twenty-five years: 'I think I was probably the champion frog-breather, because I have absolutely no respiratory power, none, yet I can frog and sing and yell all day long!' He had nursing care when he first came home, but mostly he was looked after by his parents.

> It helped a lot that my father was an engineer. My mother had incredible drive. She was a woman that never did what *they* did – in fact, if *they* did it, she wouldn't do it. She was kind of a nonconformist. And my father was steady as a rock, and just the sweetest – why, he was the greatest gentleman I ever knew. They had a great sense of fun, which was important. I realise I took a lot of their life . . .
>
> I was never thrown out. My father had a good enough job that he could support me. My needs were fairly simple – just somebody to come in and get me up in the morning, and occasionally somebody to throw me in the car and take me around. But mostly I was home, a walk in the neighbourhood, or just sit out in the driveway. That was my life, which was just slowly driving me nuts. So that's when I started my business.

He started his securities business in 1958, but for some years he made very little money, no more than $1,200 a year. His parents helped him set up a mail-order business, largely to keep him occupied; but that never took off. After 1964 his securities business did, however, and he had never looked back.

> My business has grown, but it comes in squirts. I can get big slugs, and then I don't get anything. So I'm often on borrowed money. But I have the assets to borrow. I live aggressively. I invest aggressively . . . I should give you my *modus operandi*. I operate in two areas – oil and gas – with partners, because you need somebody else to be your arms and your legs. My strength is in

selling the idea and raising the money – and that's the key element in any venture, because without it nothing goes. I've had a lot of very loyal friends that have stuck with me through thick and thin. We work on an unusual concept: we don't get paid anything until the investor gets all his money back. So I can look anybody in the eye and say, 'I tried. I've worked hard. I made nothing. I know you didn't make anything, but we all suffered together.'

When Tom started, his neighbour Mildred Larson would come over in the evenings to help. She still had a job then; but since she retired a decade ago she has worked the same long hours that he works: 'Without my full-time gal, Millie, I couldn't have done any of it.'

Yet, workaholic though he may be, Tom knows how to relax as well.

I'm an avid card player, I play cards all the time. I learned how to play bridge with the help of the parents. A friend of the family was a bridge teacher and she came . . . I have a simple wood rack with two slits in it. Whoever else is playing just sets them up and I call them out. I've been in a hearts game that we play every week for twenty years – every Monday night I'm in town.

The key to Tom's enjoyment of life, not to say survival, has been the system of care he has evolved for himself. Because he has needed *everything* doing for him, he has given the organisation of help top priority. Even when his parents were still living – and his father lived to be almost ninety – he did not rely entirely on them; he recognised their need for respite as well as his own craving for variety. His conviviality and generosity has meant that he has never wanted for helpers, and over the years he has built up a team.

I have my most stable force right now. I have Tim [Garrity], who is my bus driver. He has a Master's in Pharmaceutical Therapy. He works nine months of the year. But he does get his summers off, so he's fresh when we do these extensive and exhausting trips. We're doing more and more big business projects, and he is my legs and my arms and my eyes. Then I have a young man who has been with me much longer – about twelve years part-time. He's off now as he went to school; he works the weekends getting me up. And then I've always had a student who lives in the house, that would get me up for emergencies and often work on a Saturday and take me to parties – we call those, 'Pick up and drop'. It's like taking a large bag of potatoes in a wheelchair and dumping them at a party, then coming back and picking up the potatoes later on. The more people I get that we keep, the less training I have to do, the less talking through the job. It's so much easier to have a known quantity than an unknown quantity.

He had one woman for thirty years, and when she finally retired he lost more than a helper; she had become a close friend and confidante. 'She was wonderful; she knew how to handle me beautifully and she was a great deal of fun.'

Fun, you get to feel, is a crucial ingredient in Tom's life – fun, spiced with an element of risk, and exuberance.

> I love to cross-examine. I love to know where you've been, what you've done . . . When I meet someone, I usually cross-examine them. Most of the period when I was a youth, if there was a good-looking girl you could hardly wait to get your hands on – and I couldn't do that – I almost did it verbally. I've always gotten away with being quite audacious. I can say things to people that they wouldn't say to each other. I always have fun with young friends because I'll cross-examine them and ask them embarrassing questions – questions their mothers and fathers would not ask. I don't mean that you trade on being a polio, or handicapped. If you're reasonably bright, and can pick up on people, they grant you a lot of leeway. I think I've used every damn inch of it . . .

Tom's life still hangs by a thread, and he knows it: 'Any serious illness is going to take me fast.' In 1977, when he was in Florida, he developed a cough which rapidly turned into pneumonia. 'I was never so scared in my life – I was in intensive care for about five weeks and that was the lowest point of my life. But I came out of it.' A decade later, he started experiencing increased difficulty breathing on the rocking bed at night – 'I just felt like somebody heavy was sitting on my chest' – and suffered what is called sleep apnoea: 'I mean, every time I had a great night's sleep, I was really going into a coma – there's no other word for it.' Eventually, he was forced to seek advice.

> That's when I called – what's her name? – Gini. I'd met Gini at a couple of conventions I'd gone to, and talked to her on the phone. I asked her, 'Who shall I go and see?' She said, 'What are you using for breathing?' 'I'm using a rocking bed.' She said, 'You've got to get off that right away. You come down here and see Oscar' – Dr Oscar Schwartz. So I met him about a week later and he sent me home with this thing called a Lifecare PVEO, or something, and this old machine does wonders! It is so good – when I get that pressure in I feel like a king. I still sleep on the rocking bed, because it makes nursing care easier – being able to tip it up – but I don't rock at all.

It was at one of the GINI conferences that someone suggested to Tom that he get a bus to facilitate his travel. At first, it seemed too expensive an option. But then he sold one of his companies at an unexpectedly high profit, and thought he might just be able to afford a used bus.

> I went down and looked at this bus, and they had a brand new bus sitting right next to it. Well, I think they'd taken the old bus out and thrown mud

> at it for three hours, because it was *grungy* – with this beautiful new bus sitting next to it. So I decided, 'Oh, what the heck . . .'

The feeling that time might be running out prompted the question, 'What am I saving up for?' Tom bought the bus, though it stretched his resources, and the man who was responsible for adapting it came and video-taped his routine, so that he thoroughly understood what he had to do. After that, it was just a question of getting the team to learn how to operate it . . .

> I realise I don't have a lot of time left. I've been saying that for years and everybody says, 'Rogers, you're still around . . .' I remember when I was young, I thought if I could live to fifty that would be remarkable – because most of the boys and girls I knew of my age who'd had it badly were gone by then . . . I made it to fifty, and they all had a party for me and I said, 'Well, I've asked for ten more.' Unfortunately, that's only a year and a half away. So I'm going to have to go on back and ask for another extension, and He's probably going to say, 'Rogers, how long do you think you can carry this on? There's a limit to my patience.'

At the end of this, the final story here, I must drop the authorial mask of anonymity and add a personal postscript. For my visit to Tom Rogers coincided with an event which he described to me as 'the best thing in the world that ever happened to me' – his marriage to Kera, a widow with several children and grandchildren and a friend from the card-playing circle, to whom he had given financial advice when her architect husband died. 'I didn't think', he told me, 'that marriage was on the books for me . . .' Indeed, the possibility of a requited love seems to have been so far from his expectations that he apparently had to be prompted into an awareness of Kera's increasingly warm feelings towards him.

When we arrived (my wife Jenny was with me), Tom was seated in a wheelchair in his cramped office – the gently solicitous Millie at his side – wearing the kind of headset that telephone operators use in order to have their hands free to connect callers, only his hands lay lifeless on the arms of the chair. Clenched between his teeth like a jumbo-sized hookah was the mouthpiece of the positive-pressure respirator – which itself was neatly slotted on to a shelf under the seat of the wheelchair – and his surprisingly strong speech was punctuated with stertorous breaths. And on one of his legs there was what looked like a huge, yellow moonboot.

It was like entering a stage set, and that was probably the impression Tom intended, since work for him – even more than for most people – is intimately bound up with self-respect, and he wanted us to see him first in a work setting. He was certainly the centre of attention, cracking jokes between puffs on the respirator, organising the ordering of the Vietnamese take-away that we would all eat together later in the temporarily non-functional kitchen, and telling us within five minutes of meeting that we must, we absolutely must *attend his wedding the following week. There would be somewhere in the region of 500 guests, and he had hired a Mississippi gambling-boat for the reception. If we came back for the wedding, he added slyly, we would also get to see the bus. That clinched it: we would return to Moline for the wedding.*

In the large basement of the bungalow is Tom's shower-room, where he is taken to be 'hosed down', as he puts it. He gets down there from his bedroom in an open elevator, which his engineer father designed and made. The main feature of the bedroom is the rocking bed on which Tom sleeps, and shortly after ten that night, when he was ready for bed, he invited us to witness his bedtime ritual, an example of his practice of keeping open house – 'People just come on in, when I'm in the shower, when I'm on the toilet, when I'm getting dressed, and I love it; I encourage it; it's a zoo.' He talked all the time while his attendant, a medical student called Jeff, wheeled him in, attached to a hoist the two or three slings that went under his inert body, disconnected the respirator – at which point Tom had to stop talking and concentrate on frog-breathing – and hoisted Tom on to the slightly tilted rocking bed.

Once on the bed and undressed, his ablutions performed, Tom could be reconnected to the respirator. This was quite an elaborate procedure. Since the mouthpiece that he used during the day would be ineffectual at night – he would simply lose it when his mouth fell open in his sleep – and Tom was uncomfortable with the kind of oxygen mask that covers both nose and mouth, he had a French-made nasal attachment which he used. But there were problems with this, too. Jeff still had to bind up his jaw so that his mouth would not fall open while he slept and short-circuit the breathing process by expelling through his mouth the air coming in through his nose.

On our return to Moline the following week, Jeff took us to where 'The Ark' – as its license plate proclaimed – was parked. It was a shiny red, white, blue and grey monster the size of a Greyhound bus, with a map of the United States – white and pristine except for a solitary red star marking Moline – along its side, next to a list of names under 'Owned and operated by: Tom Rogers, Tim Garrity . . .' On the back were the three 'Home Ports': 'Moline, Il. – Ludington, Mi. – Sarasota, Fl.' The bus had its own generator and a hoist to lift Tom and his wheelchair. Inside, it was like a luxury caravan with a spacious living area behind the driver's seat at the front, in addition to the customised conversion of the back to Tom's domestic and office requirements. One garish detail, which Jeff gleefully demonstrated, was the horn which played any one of a hundred tunes at ghetto-blaster volume – 'The Ark' it might be, a stealth bomber it most certainly was not . . .

Earlier that wedding morning one of Tom's relatives had described the bus's arrival in San Francisco on one of its very first trips. 'When we heard the sirens and saw the number of cop cars surrounding it', he said, 'we knew Tom had arrived . . .' The Ark had failed to negotiate the summit of one of San Francisco's famous hills and was suspended over the peak – its wheels hanging helpless on either side. Tom, far from being put out at this turn of events, was apparently relishing all the attention. (I was told by one of his cousins that Tom used to send 900 Christmas cards every year and, when asked how he could possibly know so many people, had said it was simple: he sent them to everyone he had ever met – including one or two state troopers.)

Tom and Kera had chosen the music for the wedding and, in among the Bach and Mozart Preludes, Tom had insisted on having a medley of such numbers as 'Get Me to the Church on Time' and 'Those Wedding Bells Are Breaking Up That Old Gang of Mine'. Ten minutes before the start of the service – in time to catch the strains of 'Get Me to the Church on Time' – Tom (who was determined to get through the day without using his respirator) was wheeled in, frog-breathing away. Soon after, Kera, in a splendid white dress and hat, accompanied by her elderly and tiny Russian mother and her four grown-up children, walked up the aisle to join him. The minister's opening remarks set just the right tone, mixing informality with a sense of occasion. 'I have officiated at any number of marriages', he said, 'but I have never had so much fun *preparing a wedding . . .' (He disclaimed all responsibility for the choice of music.) The service was short and moving. Tom's responses came over loud and clear, Kera's rather more muted. By special request of the couple, Kera's children were formally invited to be part of this new family. The wording of the service was adapted in other ways, too. When it came to the placing of the ring, for instance, the minister, taking into account Tom's disability, used the phrase, 'that the ring be placed', and himself put it on Kera's finger. When it was over, the entire audience broke into spontaneous applause; it lasted several minutes and I swear there wasn't a dry eye in the church.*

We all trooped out, blinking in the summer sunshine, and made our way to the river and the berth of The Diamond Lady, *'an authentic 19th-century replica sternwheeler'. On board there were waitresses dressed in scarlet and black, with plunging necklines and short, frilly skirts; a little old man with a boater, also in scarlet and black, was strumming a banjo; and every guest had been provided with a $10 voucher for gambling. We hardly saw Tom, so surrounded was he with well-wishers. We did notice, however, that by the end of the day he'd been obliged to have recourse to the respirator. His doctor, Oscar Schwartz, agreed that the multiple-attendant system Tom had evolved for his own care was an insurance both against abuse – Dr Schwartz spoke of a lawyer he knew in Washington who was constantly being ripped off by his attendants – and against becoming too dependent on a single, devoted carer, a wife or a lover. In Tom's case, marriage at this late hour – whatever other changes it might effect in his lifestyle – was not going to alter his care arrangements.*

On the programme of 'The Marriage Celebration of Kera Bertrand Toline and Thomas Wallace Rogers, June 12, 1992', the following verse was engraved:

Hold fast to dreams
For if dreams die
Life is a broken-winged bird
That cannot fly.

PPS. September 1994: On his sixtieth birthday in February Tom toasted his family and friends and said: 'I am on such a roll with Kera . . . [our] bus . . . great health . . . I hope the Lord will stretch my contract to include another ten [years] . . .' Alas, it was not to be. Tom died in hospital in St Louis on Wednesday 24 August. But he had never expected to reach sixty, and his request for an 'extension' was partly tongue-in-cheek. If he was not granted another ten years, he had said at his party, 'I would not have changed my life for anything. I have been truly blessed . . . Let's enjoy the evening.' That was the authentic voice of Tom Rogers.

A Civil Wound: an autobiographical coda

And what about you, some of the people I spoke to would ask, what was your experience of polio?

Well, I would say, I got it in 1959, when I was twenty. It was in Hong Kong, during my National Service; I was an officer in the Gurkhas at the time; and it was an isolated case, not part of an epidemic. I was in an iron lung for about a month, and when I was weaned off that I was flown first to Singapore and then, a fortnight later, back to England. After months of intensive physiotherapy, I learned to walk again with a caliper on my weaker leg, initially with crutches and then with sticks and finally with just one stick. Twenty years later, I abandoned the caliper and reverted to two sticks, which seemed to provide a better balance, and that's how I still get around. My right side was generally more affected than my left, and though my right arm is as powerful as my left there is some residual weakness in the shoulder . . .

These are some of the facts. Another is that it changed, not just the direction of my life, but the whole tenor of it. On the rare occasions when I meet someone from Before Polio (BP), someone who has not been a constant in my life, I experience a sense of disorientation, which has nothing to do with whether or not I like that person. I look at him – and it's invariably a him – and listen to what he has to say about what we did and how we were, and I think, 'Yes, that must have been so . . .' Perhaps I even remember it, or some of it, but it is part of another life which might have been someone else's for all the connection I have with it. Sometimes I feel a fraud, as though I were laying claim to experiences I never had; I feel embarrassed, as much for the other person as myself, because I can't help thinking that if I'm having such difficulty in connecting with this past, he must sense my detachment and mistake it for criticism.

It happens in all lives, of course, the meeting with friends who once meant so much and now seem merely alien. But the traumatic experience of a serious, life-threatening illness at the age of twenty makes the dissociation that much more extreme. At first it bothered me, and I clung to the old life with a desperation born of impossibility; then I adapted and it ceased to matter; and now I regard the experience as more enriching than otherwise. But I still have difficulty in fitting the two halves of my life together.

I have written about the impact of polio on my life at various times and in various ways over the past thirty or so years – and here I shall be my own exegete. When I first tackled the subject, four years after the event, I was still too close to the experience to do anything more than attempt to convey how it felt to lie helpless in an iron lung, unable to move. But this account has an immediacy which – I hope – to some extent makes up for its lack of a context:

> I had been dreaming about the sea. When I woke up I was conscious only of darkness and a rhythmic noise, as of waves clawing at the sand. Slowly a painful awareness of my surroundings returned. My eyes probed the darkness for hints of the reality I knew. It was all there, walls, window, door, and the iron lung in which I was lying. I could now make out the figure of the night-sister, sitting under a dim reading-lamp, which was thoughtfully turned away from me. This lamp poured a concentrated circle of yellow light, a dull beam in which the dust swirled slowly, over the book she was reading. I must have moved my head slightly, because she stirred and, lifting the lamp so that the beam reached where I lay, peered myopically in my direction.
>
> 'Are you awake?' she asked softly, fearful of disturbing me.
>
> 'Yes. What time is it?' I spoke as the iron lung filled with air, causing me to breathe out.
>
> 'Ten past two,' she said, looking at her watch and stifling a yawn. 'Try and go to sleep again.'
>
> I watched her pick up her book, adjust the beam of light, and start to read again. For a while I lay still, content to observe her seated figure bent over the book attentively. Ten past two; another two hours or more until dawn. I began to feel again the onset of emptiness, an inability to control my thoughts or indeed to think about anything at all. I became aware only of my body, the numb patch on my right thigh where they must have pumped injection after injection into me, the heel of my left foot which was just touching the sheet – a pad under the ankle kept most of the weight off the heel, one of the 'pressure points' – and aching as if the sheet had burned a hole in it, and the base of my spine under which the sheet felt more like hot metal than cotton. I wanted to be moved again but I knew it was too soon. They could not be forever adjusting the pads and pillows. I'd have to sweat it out, telling myself that the pain was imaginary. Try to think about something else, talk to the sister . . .

'Sister,' I began, as soon as the idea had struck me.

She looked up almost guiltily, as if she shouldn't have been reading when the patient demanded her attention. Perhaps he needed a bottle, in which case she'd call one of the male orderlies or maybe even get it herself.

'What is it?'

'Sister . . . ,' I started again. But what was there to talk about, what could I say except that I would not sleep again until dawn? And she knew that already, but had to resist the assumption and encourage me to relax, forget, and to sleep. But I had to say something now. 'Sister, what do you do during the daytime, when you're off-duty, I mean?'

'I sleep, of course,' she replied, closing her book resignedly. 'How else d'you think I could stay awake all night looking after you?'

'But all the time?' I insisted. 'Can you sleep *all* day?'

'Well, no. I usually get up in the afternoon, when it's too hot to sleep, and go for a swim. But come on, it's you who must sleep now. You'll only tire yourself out talking.'

'I wish I could, if it meant that I'd sleep.' Nevertheless, I stopped talking. So she swam in the afternoons. I hung on to that like a drowning man to a piece of wood, trying to visualise the crowded swimming pool, the splashing and laughter, children in the shallow end and parents sitting at tables, anxiously watching . . . But it was no good, none of this was real to me any longer. It was part of the world outside, a brilliant phantasmagoria from which I was excluded.

Now the other heel was beginning to ache, too. I felt as though both my legs were clamped to the sheet and the iron lung, firmly clasping my neck, were a hideous form of straitjacket. Only my hands had any freedom of movement inside that box. There were moments when I felt so completely vulnerable that I broke into a cold sweat, when I was so overcome with claustrophobia that my head seemed to burst with a silent scream of panic. Then I would concentrate on the heavy, monotonous breathing of the machine which caused my chest to heave and subside, and stare at the engrossed form of the sister, and wonder if either of us was real.

I couldn't say how long I'd been in the hospital, but it seemed as if I'd never been anywhere else. It had started one night after I'd played a game of football in which, miraculously, I'd scored two goals. I had played in defiance of doctor's orders. That morning I'd told the battalion medical officer about the pains in my stomach, the feeling that I'd strained my stomach muscles, and he had said: 'Lay off the games for a bit and see how you feel.' But I'd promised to play that afternoon . . . I'd eaten heartily, talked with the other officers in the mess and gone to bed in the normal way. Only when I lay down did it really begin, the fever which set my brain racing and my body tossing in extremes of hot and cold, alternating so swiftly that I became confused and lost all consciousness of time and place . . .

It was dark, then it was light. Somebody came and went. Another person came in and talked about some parade which had been cancelled – because of rain, he said. Was it raining outside? Hundreds and hundreds of soldiers, tramping through the rain, their boots all hitting the ground in unison. Such banging inside my head – would it never end?

There was an ambulance outside. They brought a stretcher up the few steps to my room. 'I can walk,' I told them. 'It's only a few yards.' But did I actually say that? I struggled out from under the mosquito-net, and slumped to the floor. How stupid I felt, lying there like a sheep on its back. They carried me out to the waiting vehicle. The dark faces of the Gurkha soldiers inside the ambulance were closed. In their olive-green uniforms they already seemed far removed from me, and I felt naked under their indifferent gaze, naked and helpless, a stranger among men I knew. As the ambulance jerked forward, I was overcome with irritation at my state. Things were passing out of my control.

We were driving through the streets of Kowloon, but all I could see from my prone position were the tops of the taller buildings and patches of cloudy sky. When we reached the military hospital, I was lifted out of the ambulance by anonymous, white-coated orderlies who gave each other instructions on how to move me as though I didn't exist. My irritation had passed now and I just felt ill. There was even some comfort in being ignored, not having to make decisions any longer.

Once inside the hospital, we climbed interminable stairs and I lay in mortal terror of being dropped, as the orderlies negotiated tight corners and struggled to keep the stretcher on an even keel. We reached the landing of what I calculated to be the second floor and I was carried into a large, white and empty room – the officers' ward, as I was told. Another orderly appeared, carrying a pair of regulation hospital pyjamas. He asked if I could manage to change into them without help, and left me to it when I said yes. For a long time I remained motionless, the effort required to sit up being more than I could muster. I told myself I must make a move before someone came in and found me lying on top of the bed, having done nothing. I sat up, feeling terribly weak, my head spinning; but in reaching for the pyjamas I lost my balance and fell on the floor. I found I couldn't get up and, wondering if I might have hurt myself, feebly called for help. But no one heard.

On his return the orderly, not seeing me in the bed, ran over to find out what had become of me. He lifted me back into bed, hardly listening to my apologies for being so stupid.

'That's all right,' he said. 'You just lie there.' And he was gone again.

I didn't have long to wait before some other people arrived.

'This is the one, doctor,' said a large woman with a starched white cap. She was followed by a grey-haired man in the uniform of a colonel. They came over to my bed, the sister impassive and the doctor smiling warmly.

He asked me some question which I answered without thinking, because I was too absorbed in watching him. I liked his kindly face, the reassuring creases of laughter round his eyes, and I was impressed by the row of campaign ribbons on his chest, headed by the Military Cross. I trusted him, immediately, implicitly, and I felt an immense relief.

He examined me carefully, asking many questions. Then he straightened up, no longer smiling, but apparently satisfied with whatever conclusions he'd reached.

'You've got a temporary form of paralysis,' he said in a matter-of-fact tone of voice. 'You may find that it gets increasingly difficult for you to breathe – and you know what that'll mean.'

I had no idea, but I said yes. I was content to hand over all responsibility to this man whom I trusted. I was past caring for explanations of why I was ill and, for the moment, totally without curiosity about the future. The present was all that existed, and now I wanted to be left alone.

'I'm going now,' the kindly colonel put a hand on my shoulder, apparently understanding how I felt. 'But don't worry, I'll look in on you later.'

Alone again, I was conscious of my laboured breathing, frightened now at the possibility of not being able to breathe, of suffocating in silence. It would be like drowning; and I had a sudden recollection of struggling desperately under water in a crashing silence, swallowing great mouthfuls of the stuff and seeing all the colours of the rainbow swirling in front of my eyes, before breaking the surface, spitting and gasping for air, while everyone round the edge of the swimming pool was laughing, including the schoolmaster who'd thrown me in. Every time I breathed, I moaned and took comfort in the sound. It was almost an activity, this regular moaning, and it seemed to help me manage the pain.

They came for me with a stretcher again. I was to be moved upstairs to a 'more comfortable ward' – into isolation, in fact, though I didn't know it then. There was only one bed in this narrow room; there was also a primitive air-conditioning plant, which, it turned out, hardly ever worked.

The room grew dark, and my thoughts whirled in delirious confusion. People came and went, masked and robed in white, and I saw other people, other faces, familiar yet forgotten since childhood, touching off half-forgotten dreams . . .

I was struggling for breath, my head going from side to side like a metronome. Someone came over, looked at me for a second or two in silence and went away again. Then there were several people, all leaning over me – so that was why I couldn't breathe, they were blocking out my air. I wanted to push them out of the way; they were smothering me and I couldn't move. When they put their hands on me and lifted me, I screamed with the pain. From what I could see, their faces above their masks were taut and strained, full of an ugly determination. They lay me down between the jaws of a yawning box which had appeared from nowhere. For a moment I thought they were going to operate on me without an anaesthetic and I swore at them – in English and in Gurkhali, too, apparently. Yet a part of me remained detached, watching myself shout and struggle, shocked at my lack of control. I couldn't believe this was happening to me. Now my head was sinking backwards towards the floor; the whole box was being tilted. Suddenly it came to me that they had made a terrible mistake, that this was a coffin and they were burying me alive.

'I'm not dead!' I wanted to shout, but all I could do was pull my head away from their clutching hands. One of those hands held a tube, and they were trying to force it down my throat, which made me retch. I vomited over and over, until there seemed to be nothing left inside me. But they kept pushing that tube down my throat regardless – as if they would choke me to death, then pull down the lid of the coffin.

Gradually I became aware of somebody talking to me, urgently, angrily, but I could hear nothing. It was as if there were panel of glass between us. With a supreme effort I focused my attention on the speaker, whose face was contorted with the effort of trying to reach me, and I recognised the grey-haired doctor. His face was close to mine and it was his hand that held the tube. Something – his tone of voice perhaps, or the expression on his face – broke my resistance and I gave up the struggle. After that I sank into oblivion . . .

I am running, running over a field with the wind at my back – it fills my shirt like a sail and carries me along so that my feet scarcely touch the ground. I move across the earth with the speed of a cloud-shadow: one minute I am in the open, the next deep in a wood carpeted with bluebells and strewn with mossy logs. Then I am in a meadow and as I cross the stream I remember I'm thirsty and I lean over and cup my hands and scoop up the cool, clear water . . .

Now I am lying on a beach, deserted except for a solitary figure in the distance. The sun warms my bones and I feel a delicious lassitude. As the figure approaches through a shimmering heat haze, I can see it's a girl. I am struck by her incandescent beauty, yet I feel no desire. She comes closer still and I see that she is carrying, or rather cradling, something in her hands: a gull with a broken wing. I know the wing is broken because it hangs awkwardly at its side, yet the gull looks so peaceful, almost asleep . . .

The sister raised her hand to conceal a yawn. She was still reading, but seemed to be growing restless after the prolonged session under the dim lamp. She cast a routine glance in my direction, then looked away into the blankness in front of her before turning to me again as if she'd just realised something. Perhaps she could tell that my eyes were still open.

'It must soon be dawn,' I suggested.

'Yes,' she said, but without looking at her watch. 'Are you uncomfortable? Would you like to be moved?'

'Please,' I replied.

She went to the door and opened it a fraction to summon one of the orderlies. The machine was switched off, and together they opened up the top half of the box – I could breathe unaided for a few minutes while the box was open. The sister lifted one heel and rubbed it with surgical spirit, which stung the raw flesh but eased away the ache. She turned her attention to the other heel, then to my elbows and finally, lifting me gently with the help of the orderly, she rubbed my back. Before they clamped the iron lung shut again, they adjusted all the supporting pads, and replaced the wads of cottonwool around my neck to keep it airtight. This was the moment I both longed for and dreaded; comfortable again, I would have to stay in this position for hours, since I could hardly ask the sister to open up the lung again for anything so trivial as an itch that I couldn't reach, or an incipient ache in my heel . . .

The orderly went out.

'Please leave the door open,' I called after him. 'Then I can see when it's dawn.'

The door, or rather double doors, directly behind my head, were kept open during the day. There was a mirror above my head, which I could use like the rear-view mirror in a car to watch people go by on the balcony outside and look over the rooftops of Kowloon and see the island of Hong Kong itself rising steeply out of the sea, its waterfront a conglomeration of gleaming banks and office-blocks. But for the moment it was still dark and I could see nothing.

There was nothing dramatic about the dawn – just a gradual infusion of light, diluting the darkness to a murky grey in which familiar objects appeared as shadows of themselves. But it brought the promise of another day and, for me, a peace which the darkness withheld. Soon I would be able to submit myself to the train of routine and trivial events which made up my day and kept me floating on the surface.

I fell into a light and dreamless sleep in which I never really seemed to lose consciousness. Before my closed eyes everything was bright and still. To be unable to move was no longer pain but comfort. It was as though I were suspended in time and space, floating in the air on a cloud-cushion of forgetfulness – without memory or desire, past or future – living in the present, permanent moment of negation. 'And the peace of God, which passeth all understanding . . . ,' words from *The Book of Common Prayer* familiar from childhood, echoed in my sleeping-waking mind, an intimation of mortality, perhaps, a hint of the ultimate silence beyond the range of human voice and contact . . .

I heard the sister say, 'He's sleeping now, so don't disturb him. There's plenty of time.'

I kept my eyes closed, trying to retain the sense of floating on air, the ecstasy of the knowledge of peace experienced. But I heard the machine sucking and blowing, and could feel my chest rising and falling inside the lung. I also wanted to urinate. I would have to wake up soon . . .

It was broad daylight. The mirror above my head reflected the glint of the sun on the rooftops. The sister stood over me smiling – relieved, no doubt, that her vigil was over.

'It's a lovely day,' she said. 'I didn't wake you earlier because I thought the sleep would do you good. How are you feeling?'

'Fine,' I said. 'Have I slept long?'

'Over an hour. You look better this morning anyway.'

An orderly appeared with a bowl and a mug of water. He put them down on a table nearby and picked up my shaving brush.

'We'll get you cleaned up now, before the day staff arrive,' the sister remarked, all brisk efficiency. 'You give him a shave now, Smith, then we'll open the lung and give him a proper wash.'

The day had begun . . .[1]

I wrote this account of the onset of polio while I was an undergraduate at Cambridge, a place where – BP – I never imagined I would be going. I had left school at the age of sixteen and thought that was the end of my education. Reading English at Cambridge simply didn't enter into my calculations; the life I had envisaged for myself was an outdoor, quasi-colonial style existence, rather on the model of my father, who had been in teak forestry in Thailand between the wars before he came home and took up farming. The switch to a cerebral life was directly attributable to polio, which – I now see – opened more doors than it closed.

The next time I wrote about polio, it was with a very different focus. I was filling in time after Cambridge, teaching English part-time to GCE A-level students at the Dartington College of Arts near my home in Devon. I was provided with a room in the restored mediaeval courtyard there in return for my none-too-arduous labours; and the rest of the time I devoted to writing a novel, which stubbornly refused to transcend its autobiographical origins and take on an imaginative life of its own. In contrast to that 'fictional' writing, which was nothing if not *willed*, the following account of a visit to the Dartington estate of a group of spastics – as sufferers from cerebral palsy were then known – seemed to force itself upon me, demanding to be written . . .

Last night, during their party, one of the spastic girls broke down and cried; and went on crying although – or perhaps because – one of our students with a kind heart and strong maternal instincts tried to comfort her. They had to take her away, still crying, when all the spastics left by coach early this morning.

They'd been at the Devon Centre for a fortnight, on a course to test their aptitude for work; for they had just finished their schooling. They did practical work, they painted pictures; they visited a castle, and a factory; they wrote mock-application letters for jobs, and had a discussion on the subject, 'Why Work?' – 'It's very simple,' said Ian. 'There's only one reason, and that's money.' They passed what was probably the happiest fortnight of their lives.

The students of the art college gave a party for them and sang songs with guitar accompaniment. The girls, some barefoot with long hair and paint-spattered jeans, and the boys (who are smart this year and wear ties) moved among the drably or brightly – but always unfashionably – dressed spastics, bringing them cups of coffee and encouraging them to join in, which they readily did.

But the students only met them at parties; we, the staff of the college, had all our meals with them. We were their 'friends', and they greeted us on our entry into the dining-room and more or less told us where to sit. Sport and the world of television were the most common topics of conversation. They organised their own session of *Juke Box Jury* and borrowed a bell to register 'hits.'

It was Richard, our drama tutor and producer, who lent them the bell and did not complain when it came back cracked. Being popular, he was subjected to friendly physical assaults from both Ian and Philip. Philip could not speak coherently, but made noises when he wanted anything and smiled frequently; he threw his left arm back when he was excited and was usually the first to greet our entry into the room.

Judith often sat next to him and brought him his food; one of us would cut it up for him. Judith was a big Scandinavian-looking blonde and herself severely handicapped. Both she and Philip wore deaf-aids and, though she could speak, her speech was very difficult to understand. But she liked to do things for others and would go and collect our food and coffee. She was so shaky, however, that most of the coffee would be in the saucer by the time it reached our table.

'I didn't go to school until I was eleven,' she told me. 'I'm backward, you see. I would like to have more education, learn more.'

I sympathised. What had happened to my leg, she wanted to know. Polio. Had I had an iron machine on my chest? Yes, the iron lung. When was that? Seven years ago. How old was I then? Twenty.

At the next meal she told me I was twenty-seven. She was delighted when I appeared surprised that she should know my age. She explained how she had worked it out, and asked how far I could walk.

The night before last, as we entered the dining-room for our evening meal, Richard said: 'I can't face Ian tonight. Let's tell them we're having a business discussion and sit at another table.'

We collected our soup and there were the usual shouts: 'Over here, Richard,' and 'We've saved your seat, Tony.' Richard made his excuses and went to another table. I followed, but Judith pulled out a chair as I was about to go past her.

'Sit here,' she said.

'I can't tonight, Judith,' I explained. 'I have to talk with Richard – business, you know. You must excuse me tonight.' She looked at me and then her head dropped with a sudden spastic movement.

Yesterday morning at breakfast she ignored me, walking past as though I were a stranger. I said hello, but she didn't reply. I sat at Ian's table and was relieved to find I still had friends there. At lunch, Judith sat alone and I watched her from the next table; soon others joined her. By the evening excitement was running high: their farewell party was about to begin. Ian was the comedian, his only prop a battered trilby. He ordered Richard, who had shown them round the theatre earlier in the day, to be there at the beginning, at 8 p.m. sharp.

I arrived late. Everyone was seated around the four walls of the room and they were playing 'Pass the Parcel'. The one holding it when the music stopped stripped a layer of paper and had to perform a forfeit. 'Kiss the girl with the most beautiful nose,' read out the master of ceremonies, and there was general laughter – and embarrassment. The deed done, the package was passed on. The music stopped again and the girl who held it threw it on to her neighbour as if it contained a bomb; her neighbour threw it back. It passed to and fro between them until order was restored. The first girl opened the parcel: the prize was a packet of sweet cigarettes.

'I hope she smokes,' someone joked.

For the next game, a variation of 'Musical Chairs', all the boys were called to form a line in the centre of the room. The girls had to circle round, and grab a boy's arm when the music ceased. Boys were removed from either end of the line, until only one remained – and two girls, one of them badly deformed. Chairs were placed at each end of the room for the contestants to go round. But the more crippled girl was so flustered that she began by circling in the wrong direction; then she caught hold of one of the chairs and slithered about, unable to control her movements.

'I can't do it, I can't,' she squealed in a frenzy of fear and excitement. Prizes were found for both girls.

Ian did his comedy act, but the others had seen it before and knew all his jokes. He had his revenge by calling out half-a-dozen of them to kneel on the floor as part of a shaggy dog story involving a man in the desert looking for a camel.

Richard contributed impromptu – or almost impromptu – limericks. They were much appreciated: 'There was an old man called Dobby (Mr Dobson was in charge of the course), who had an extraordinary hobby . . .'

Then Dobby himself – for he was master of ceremonies – announced: 'There's been a request for another performance from a young lady whom some of us saw dance the other night. For those who didn't it will be something of an experience. Unfortunately we haven't got any ballet music here, so we're putting on the record *Tears*. Judith!'

There was a round of applause as Judith walked to the centre of the room. I couldn't believe it. She began to dance. In all their terrible distortion the movements were recognisably those of ballet: her hands met above her head, she kicked out her legs, and she bent over and swept the ground with her finger-tips – all this without a trace of self-consciousness; she even paused to adjust her right shoe. She was smiling, but she saw no one. I looked and listened, entranced and revolted. The grotesque sentimentality of the Ken Dodd song might have been specifically designed to emphasise the macabre element of the performance. The record would not end. I watched with horror the final shaky bow of the dancer and waited for her to keel over. But she didn't. Instead, she came over and sat on the sofa beside me.

'But that was beautiful, Judith,' I said. I hoped we might still be friends, though I was glad she was going the next day. She turned and stared at me; I was relieved when she smiled. She struggled for words. 'You see, it comes from my soul,' she spluttered.

I preferred facts of a different order: 'How long have you been dancing?'

'A long time,' she told me. 'I have dancing lessons . . .' I could not understand the rest of the sentence, so I nodded sagely.

Some of the students arrived with a guitar. They sang *We Shall Overcome* brightly and earnestly. The girl who had to be removed was weeping loudly. Carey sat with his arm around the young student as she held the head of the weeping girl against her breast. I moved to another seat, but Judith followed me over with a bowl of crisps which she presented to me as an offering, bowing in front of me and making elaborate gestures. Then, as if she suddenly recalled the prayers to Allah of Ian's comic act in the desert, she got down on her knees, raised her hands and prostrated herself before me. I pushed my chair back hard against the wall.

'Come on, Judith, get up – please. You must have some crisps, too.' She shook her head, but she got up and sat beside me.

'God will make you better,' she said. 'Tomorrow morning you'll be better. You can bend your leg more than you could before, can't you?'

'Let's hope we all get better,' I said weakly, aware of the pressure of her arm and thigh against me.

'Why don't you go and dance with the others, Judith?' I asked. But this was not her sort of dancing. She did get up, however, and when I saw someone stop to congratulate her on her ballet, I made my escape to the bar.[2]

This encounter helped me define for myself the parameters of disability. What were a flaccid leg and a few other muscular weaknesses in comparison with the disabilities of Judith and her fellow sufferers from cerebral palsy – their speech defects, their hearing-aids, their shaking limbs and general physical undesirability? They were the truly disadvantaged. All their lives they would be pitied, smothered with kindness, mocked for their infirmities, imitated by schoolchildren who thought them funny – or simply avoided. Yet these boys and girls were without self-pity; like the party of Spanish dwarfs who come chattering along the hotel corridor in a memorable sequence in Ingmar Bergman's *The Silence* and seem to epitomise healthy normality in a film otherwise devoid of it, the spastics were magnificently unselfconscious, uncomplicatedly themselves.

Their appearance at Dartington reminded me how fortunate I was. Among the disabled. I scarcely counted as disabled: my lower limbs might be wasted, but my face was not disfigured and I was not even greatly inconvenienced by my disability. Short of playing football, or going for a run, I could lead a more or less normal life. I no longer needed anyone to plump my pillows, cut up my food, or take me to the bathroom. I was not dependent on others for any essential function; I could live on my own, travel without too much difficulty and have unproblematic physical relationships. I hardly even thought of myself as disabled.

For this reason perhaps, I did not return to the subject of polio in particular, or disability in general, for many years – years in which I married (twice), fathered three children and pursued a career in broadcasting (radio) and in magazine journalism. When I did return to it, I concerned myself with the immediate after-effects of the trauma:

> One of the most shocking things about a serious illness is the loss of status it involves. At least, that is what I found. I was a young National Service army officer when I had my first experience of severe illness. I lost the use of all my limbs, yet that was somehow less shattering than the loss of rank that went with it.
>
> As a very junior officer of nineteen or twenty my power, of course, was more apparent than real. But strutting around with a pip on each shoulder and a swagger-stick under my arm, being saluted and addressed as 'sir' or 'sahib' by men twice my age, no doubt gave me an exaggerated sense of my own importance. In reality I had but two assets: a public school accent (it could scarcely be called an education since I'd left school at sixteen) and physical fitness – and I was about to lose one of them.
>
> The first inkling I had of a change in my status was when I was delivered by ambulance to the military hospital in Kowloon and the orderly who brought me a pair of hospital pyjamas casually asked if I needed help to change into them. My rejection of his offer, I'm afraid, derived more from

the want of deference I detected in his manner than from any certainty that I could manage on my own. My humiliation, when he came back and found me helpless on the floor, was no more than poetic justice.

For the next six weeks or so I had other things to worry about. I had anyway reacquired status through the gravity of my condition, the fact that I was in an iron lung and on the 'danger list', isolated from all other patients and with no less than three QARANC sisters whose sole function was to minister to my (medical) needs. On the principle, I presume, of allowing the condemned man a hearty breakfast, I could order whatever food took my fancy – the only drawback being that nothing did. Likewise with drink: I had my own bottle of Scotch which was not so much appetising as morale-boosting.

When I was flown back (cas. evac.) to England after two months in hospital and two years in the Far East, I came down to earth in more senses than one. Perhaps my reluctance to return, even after I got polio, was due to an instinctive desire to prolong the moment of extraordinariness in which, like The Man Who Would Be King, I had status, if not beyond my wildest dreams, at least such as I had only dreamed of.

I exaggerate, of course. National Service subalterns were two a penny. But at the age of nineteen or twenty all experience seems unique, and the army's hierarchical structure, which combines the vertiginous pleasures of rank with homely over-protectiveness, was both familiar (from boarding school) and seductive – in the sense that it seemed to justify extended adolescence and give it a spurious sort of manliness.

The military hospital at Wheatley, outside Oxford, which was to be my home for several months, consisted of a number of Nissen huts linked by long corridors. I shared a side ward in one of these huts with a paraplegic Pakistani ex-naval officer, who'd had an abscess under his spine which had gone too long undetected. His valiant attempts to walk – with calipers and crutches – were hampered by the fact that his legs could only be straightened with difficulty and tended to recoil into a flexed position. It could take two men, using all their strength, to straighten just one of his legs. In view of his single-minded determination to walk again, it is sad to have to report that the last time I saw him, some years later, he was still wheelchair-bound.

At Wheatley, normality reasserted itself. After the high drama of the iron lung, the daily routine of mornings and afternoons spent in the physiotherapy department – with arms and legs suspended, puppet-like, by springs from an overhead frame – was frustrating in its lack of perceptible progress. Yet there was progress. The day came when I stood up for the first time in months: holding on to the parallel bars for dear life, I was almost overcome with vertigo. Not long afterwards, when I took my first steps on my own, with a caliper on one leg and elbow crutches for support, I was as uncertain as a novice on stilts.

It was as well I had a simple goal – to walk again without props, to make an improbable 100 per cent recovery – because in other respects, though I was hardly aware of it, I was floundering. I had lost the protection of my pips and found no other; out of uniform I had no rank. In the hierarchy of the hospital it is the doctors and sisters who are the officers, not the patients, no matter what their rank. An elderly major might be allowed his rank as a kind of courtesy, but a 2nd lieutenant who needed help to dress and undress and to go to the lavatory did not merit a 'sir' in the eyes of his fellow National Servicemen, such as the ward orderlies, or from the regular NCOs who were the physios.

I did not resent this – rather the reverse: I was flattered by the intimacy of first-name terms and found it liberating to cut across barriers of rank and class. But psychologically, emotionally, I was still living in the past. On those evenings when I wasn't watching television, or at weekends when the physiotherapy department was closed, I stuck into albums the photographs I had taken while I was in the Far East (the only time in my life I have ever taken photos). I was clinging to the past because the future was a blank, an undiscovered country for which I had no qualifications.

The next stage of the normalising process was called 'rehabilitation'. The RAF appeared to be better equipped for this than the army; at least I was sent to the RAF rehabilitation unit for officers at Headley Court, an estate tucked away in a leafy bit of country between Epsom and Leatherhead, just south of London. It had been donated to the RAF by some grateful banker after the Battle of Britain. It had a gymnasium and a heated swimming pool, and there were remedial gymnasts to put one through one's paces just as the physios at Wheatley had done; only Headley Court was an officers' mess, not a hospital, and the atmosphere was quite different.

The NCOs in charge of remedial classes believed that all officers who came their way had been thoroughly spoiled in hospital, and would greet each limping newcomer with the words, 'Got a bad leg, sir?' This apparently ingenuous question was in reality a signal to all the old-stagers, who eagerly awaited the victim's reply. 'Oh yes' – he would invariably fall into the trap – 'broke it skiing', or, 'damn cartilege went when I was playing rugby'. The NCO would then turn to the grinning class and say, 'Sympathy for the squadron-leader', and the whole class would respond with a long-drawn-out, mock-sympathetic, 'Aaaaah'. Limping or not, the newcomer soon made himself scarce.

As this anecdote suggests, I was back in the schoolboy world of the services. My morale improved as a result of my release from hospital – in hospital, as in prison, one can develop recidivist tendencies – and had I been returning to the army, no doubt, I would have been speedily 'rehabilitated'. But as my National Service was over and I was in the process of being invalided out of the army, an officer's mess was the last place where I should have been. In desperation I clung ever more tenaciously to my officer status and became something of an old lag, haunting the bar every lunchtime and evening, unable to face the coming void.

I thought I was safe, at least for a while. I still believed it was only a matter of time before I made a complete recovery, and I knew I had some way to go yet. But that illusion was rudely shattered when the chief medical officer sent for me one day and told me straight out that he didn't think there was much more they could do for me. I was dumbfounded. What, were they going to evict me now, when I was so far along the road to recovery? Every doctor, physiotherapist and remedial gymnast I'd ever worked with had encouraged me in the belief that I would, eventually, walk without aids; and now he was saying that I could not realistically expect to make much more progress. It was like kicking away the last plank of my self-esteem.

Slowly the truth, which for so long I had refused to countenance, was forcing itself upon me: not only was I no longer an officer, I was not even a whole man; I was, in fact, a cripple. As cripples went, I was not a particularly unsightly object. Indeed, a squadron-leader who had lost both legs *à la* Douglas Bader endeared himself to me forever when – after staring at me appraisingly for an uncomfortably long time – he remarked, as though to himself, 'Hmmm, tall, dark and with an interesting limp . . .'

He was wiser than I realised; it seems to me now that he may have been giving me a warning rather than paying me a compliment. Certainly I had to find a new way of dealing with my disability. It was no longer something which was *there* – like Mount Everest – for me to conquer; it was, more prosaically, something I would have to accommodate, something which would be with me for the rest of my life.

After more than a year of daily exercising with the single aim of recovering full fitness, I had to adjust my sights. Instead of regarding anything short of total recovery as failure, I must recognise that the ability to walk at all, even with a caliper and a stick, was an almost priceless achievement – in that it enabled me to lead an independent existence.

With this recognition came a new determination – to use the opportunity provided by my disability to make up for lost time and acquire an education. I would go to university; if my body wouldn't function properly, at least my brain could be made to work. By the time I left Headley Court and returned to Civvie Street, I had already moved a long way in spirit. I no longer gave much thought to status or identity. But of course I didn't change overnight; that's not how things happen in the real world. I have an abiding memory of trying to learn to fly an Auster, though my legs were too weak to control the rudder bar. The impossible I might perform at once; maturity would take a little longer.[3]

This article, it now seems to me, ends on a note of bathos; the last sentence, in particular, is embarrassingly flippant. On the advice of a friend I changed what I'd originally written, which had something to do with 'crossing the shadow-line into maturity'. My friend's objection – that this was a bit presumptuous on my part – is unexceptionable. Nevertheless I'm sorry I changed the ending – and not just for the loss of a favourite Conradian image. Perhaps 'maturity' was the wrong word, but what I was getting at was that there *was* some kind of a sea-change in my life at this juncture – the passage, if you like, from one sort of life to another, not necessarily better but qualitatively different, from BP – to put it neutrally – to AP (After Polio).

In his lightly fictionalised autobiography, *Lions and Shadows*, subtitled *An Education in the Twenties*, Christopher Isherwood writes:

> Like most of my generation, I was obsessed by a complex of terrors and longings connected with the idea 'War'. 'War', in this purely neurotic sense, meant The Test. The test of your courage, of your maturity, of your sexual prowess: 'Are you really a Man?' Subconsciously, I believe, I longed to be subjected to this test; but I also dreaded failure. I dreaded failure so much – indeed, I was so certain that I *should* fail – that, consciously, I denied my longing to be tested, altogether. I denied my all-consuming morbid interest in the idea of 'war'. I pretended indifference. The War, I said, was obscene, not even thrilling, a nuisance, a bore.[4]

For anyone who grew up in the Fifties this will strike a chord, for that decade bore much the same relationship to the Second World War as the Twenties had to the First. I certainly regarded my National Service as a kind of test. This was partly why, when given a choice as to which regiment I wanted to be commissioned into, I opted for one that was engaged in active service – a sort of 'make or break', 'do or die' approach. It was one of life's little ironies that I damn near did die, but not on any battlefield.

In that – 'purely neurotic' – sense, of course, polio itself became The Test; and I crossed the shadow-line from BP to AP when I came to recognise that it was a very different sort of test from what I'd imagined it would be, indeed not really a test at all. With polio, the only possible failure is not to survive (though with the more seriously disabled – respirator-users in particular – that option isn't confined to the acute phase of the disease). Once the will to live has asserted itself, recovery becomes a matter of hard graft; bravery doesn't come into it. (I was the kind of child who, if I so much as grazed my knee, thought I was dying – ask my mother.) In the rehabilitatory phase, you go into training for life as an athlete trains for the games; and though individual athletes may run brilliant – even brave – races, nobody will say how brave they are, doing all that training. They would be stupid if they didn't.

But dreams – illusions – may help us through difficult and challenging times. I probably *needed* the illusion that I would make a 100 per cent recovery in order to make as much of a recovery as I did make; the concentration of purpose that produces results would otherwise have been lacking. Equally, though, I needed to know when what, up till then, might have constituted laudable effort tipped over into futility; and that was what the chief medical officer at RAF Headley Court was telling me in so many words when he said that he thought they had done all they could for me. My initial response was childish, even petulant, because I was so disappointed. I'd thought the old life was within my grasp; I had only to persevere and I would be rewarded. Instead, I was to be cast out of the ranks of the physically fit, just as I was being invalided out of the army.

That moment, a year or so after the onset of polio, was the nadir of my illness and, far from marking the end of my rehabilitation, was its true beginning.[5]

Appendix: Monkey Business

Polio research was from the beginning a voracious consumer of monkeys; for over half a century these beasts were sacrificed in their thousands. Jonas Salk and the other scientists involved in the poliovirus typing programme, for instance, 'used 17,500 monkeys in their tests'.[1] Only a fraction of this number would have been necessary if the technique of growing poliovirus in non-nervous tissue culture discovered by Enders, Weller and Robbins had been available a year or two earlier. That breakthrough – according to polio historian John Paul – 'marked the end, at least partially, of the monkey era'.[2]

In the late 1940s, when researchers complained to the National Foundation of the difficulty of getting hold of monkeys, Basil O'Connor went into 'the monkey business', just as he had gone into the respirator business when there was a shortage of iron lungs. He obtained clearances to catch and export Rhesus monkeys from India and Cynomulgus monkeys from the Philippines, and acquired a place called Okatie Farm near Hardeeville, in the southernmost corner of South Carolina, which combined the functions of a sanatorium and a clearing house, pending the monkeys' despatch to laboratories around the country.[3]

Monkeys are not naturally liable to polio – certainly not Rhesus monkeys, the kind most widely used in polio research. In experiments, Cynomulgus monkeys were given polio orally, but with Rhesus monkeys it was injected directly into the brain.[4] Chimpanzees, however, are prone to the disease in the same way as humans are, as a German researcher discovered in 1935, when he noticed that two chimpanzees in a children's zoo in Cologne had suddenly developed a flaccid paralysis during an outbreak of polio in the town.[5] The celebrated chimpanzee-watcher, Jane Goodall, observed an epidemic among wild chimps in 1966. She describes the few months it lasted as 'the worst I have ever lived through'.[6]

Goodall did not immediately recognise the sickness that ravaged the chimpanzee community. But there had been an outbreak of polio in the nearest African town and 'since chimpanzees are susceptible to almost every human infectious disease, and are known to get polio', there was not much doubt about the nature of the paralysis affecting as many as fifteen of the chimps.

Mr McGregor, so called because 'for some reason' he reminded Goodall of Beatrix Potter's gardener in *The Tale of Peter Rabbit*, was an elderly male chimpanzee, easily recognisable by the 'monk's tonsure' of hair round his otherwise hairless head. One evening, Jane and her (first) husband Hugo, alerted by the worried calls of a chimp family group, went to see what was happening and found a cloud of flies buzzing angrily over what they assumed must be a dead body. But Mr McGregor was not dead and did not even look discomforted; he was reaching for berries in a bush just above his head and stuffing them into his mouth. It was only when he tried to move that the 'horror' of what had happened became apparent: 'he put both hands behind him on the ground and inched his body backwards in a sitting position'. His legs were paralysed. Even so, he heaved himself up into the lower branches of a tree, using only his powerful arms, and constructed a small nest. But with every exertion he involuntarily released a spurt of urine. This, along with his raw and bleeding rump, was what attracted the flies.[7]

The next ten days 'had a nightmare quality'. Mr McGregor stayed in his nest every morning till nearly midday, then dragged himself over the ground to gather what food he could reach before returning to bed in the late afternoon. When Jane and Hugo spotted him turning somersaults, their hopes were raised; but this painful parody of child's play was merely his attempt to get along using only the astounding strength in his arms.

The couple did what they could to alleviate the chimp's agony, bringing him food and water, cleaning his nest and driving away the ever-present flies. But they could do little about the reactions of the other chimps, which would either attack him or ignore him by keeping their distance as they could so easily do. Goodall watched 'old Gregor', as she sometimes calls him, make an heroic attempt to join his peers in their companionable grooming, an important social activity which he had been missing.

> When at last he reached the tree he rested for a while in the shade and then made the final effort and pulled himself up until he was close to two of the grooming males. With a loud grunt of pleasure he reached a hand towards them in greeting – but even before he made contact they both swung quickly away and, without a backward glance, started grooming on the far side of the tree. For full two minutes old Gregor sat motionless, staring after them. And then he laboriously lowered himself to the ground. As I watched him sitting there alone, my vision blurred, and when I looked up at the

> groomers in the tree I came nearer to hating a chimpanzee than I have ever done before or since.

One chimp, though he too refrained from grooming Gregor, was never far from him during these ten days. Goodall had long suspected that Humphrey was Gregor's younger brother and his behaviour now seemed to confirm this. Once when a mighty chimp she called Goliath drove Gregor out of his nest in a tree, Humphrey, who was generally submissive towards this 'alpha male', actually attacked him. On other occasions, when Gregor inevitably fell behind in his attempts to follow the rest of the chimps by turning head over heels or dragging himself along on his bottom, Humphrey would come back from time to time and make encouraging gestures, as he might have to a reluctant female. When that proved futile, Humphrey detached himself from the group and built a nest close to Gregor's.

One evening when Jane and Hugo took him his supper, Mr McGregor was not in his nest. They traced his movements and found him not far away; he had dislocated an arm and could no longer lift himself into a tree. They made him a nest on the ground, and visited him during the night. The next morning, after giving him a couple of eggs, which were his favourite food, they put him out of his misery.

> We did not allow any of the chimps to see his dead body, and it seemed that, for a long time, Humphrey did not realise that he would never meet his old friend again. For nearly six months he returned again and again to the place where Gregor had spent the last days of his life, and sat up some tree or other, staring around, waiting, listening. During this time he but seldom joined the other chimps when they left together for a distant valley; he sometimes went a short way with such a group but, within a few hours, he usually came back and sat, staring over the valley, waiting, surely, to see old Gregor again, listening for the deep almost braying voice, so similar to his own, that was silenced for ever.

A Note on Sources

My initial research in the library of the Wellcome Institute for the History of Medicine in London gave me some inkling of the size of the project I was undertaking. From wondering whether I had a subject worthy of a book I soon switched to worrying about my capacity to encompass all the strands – medical, scientific, social and political – of such a rich and complex history.

The second phase of my research took me to the United States. The Franklin D. Roosevelt Library in Hyde Park, New York, enabled me to extend my biographical knowledge of FDR and, at the same time, examine original documents in relation to Warm Springs, the National Foundation for Infantile Paralysis, the President's Birthday Ball Commission and so on.

The March of Dimes Birth Defects Foundation headquarters in White Plains, New York, is a rich source of visual – and audio-visual – material, but otherwise disappointing. Too many office moves, perhaps, have decimated the archive. At Warm Springs, Georgia, too, the historical archive is patchy, though attempts are being made to improve this facility. The university archives at Tuskegee, Alabama, contain valuable documents relating to the treatment of polio among black Americans at the institute there.

The Library of the American Philosophical Society in Philadelphia has interesting material on the early history of polio in the United States, including the Simon Flexner papers. The National Library of Medicine in Washington, DC, is superbly stocked. The office of the Gazette International Networking Institute in St Louis has a useful collection of polio and disability-related books gathered over the years by Gini Laurie, as well as back numbers of GINI's own splendid magazines, *Toomey j Gazette*, *Rehabilitation Gazette* and *Polio Network News*. And the archive of the Sister Kenny Institute in Minneapolis contains a wealth of material about Sister Kenny's American years and the doings – and later misdoings – of the foundation set up in her name.

Foremost among published works, there is the splendid scientific/ medical *A History of Poliomyelitis*, written by the late Dr John R. Paul of Yale University, one of the leading polio researchers and therefore an insider – but one who tells a story of considerable scientific and political infighting with commendable objectivity. This has been an invaluable source, but it is very much a specialist's book.

Greer Williams's *Virus Hunters* (modelled on Paul de Kruif's classic *Microbe Hunters*), John Rowan Wilson's *Margin of Safety*, Richard Carter's *Breakthrough* and, most recently, Jane S. Smith's *Patenting the Sun* all tell the story of the development of the Salk vaccine – and, in Wilson's case, of Sabin's, too – for a general readership. The medical historian Saul Benison provides the essential background to the vaccine story, in a succinct and trenchant essay on 'Polio Research in the United States', published in Gerald Holton (editor), *The Twentieth Century Sciences: Studies in the Biography of Ideas* and in his oral history memoir, *Tom Rivers: Reflections on a Life in Medicine and Science*; and from a younger generation of medical historians, Naomi Rogers's *Dirt and Disease: Polio Before FDR* examines the impact of the 1916 New York epidemic.

Among the innumerable biographies of FDR, several of which focus on his polio and the Warm Springs story, two fairly recent works – both by polio survivors – deserve special mention: Geoffrey Ward's magnificent and exhaustively researched *A First-Class Temperament* (the second volume of his study of the making of FDR), and Hugh Gregory Gallagher's richly contentious *FDR's Splendid Deception* – neither of which, sadly, has yet been published in Britain. Another polio, Turnley Walker, should be commended for his *Roosevelt and the Fight against Polio* (original American title: *Roosevelt and the Warm Springs Story*), which, even if it is – in the words of yet another polio, Lorenzo Milam – 'gooey caramel', is a memorable account of the early years at Warm Springs.

The best – indeed, to date the only serious – biography of Sister Kenny is by Victor Cohn: *Sister Kenny: The Woman Who Challenged the Doctors*. Victor Cohn is also the author of a brief and lively history of polio in America, the National Foundation and the Salk vaccine, called *Four Billion Dimes*. The Warm Springs and National Foundation supremo, Basil O'Connor (FDR's loyal lieutenant and one-time legal partner) still awaits his biographer, as does his protégé, Jonas Salk – not to mention Salk's great rival, Albert Sabin, who died in the spring of 1993 while this book was in preparation.

A number of medical textbooks, dating from early in the century, deal with clinical aspects of polio; and since the Second World War there has been a plethora of polio-related – and now post-polio related – conferences, all of which have spawned medico-scientific papers, published by various presses and journals. The *British Medical Journal* and the *Journal of the American Medical Association* both reported the polio years fully.

Finally, there are the memoirs of polio survivors – almost a subgenre in itself – and the treatment of it in fiction. The Anglo-Irish novelist J.G. Farrell, whose *The Siege of Krishnapur* won the Booker Prize for fiction, wrote an early, and largely forgotten, novel called *The Lung*, which deals fairly directly – perhaps too directly – with his polio experience. (I once wrote an article in which I argued that the novel in which Farrell tackles his experience of the iron lung imaginatively, rather than literally, is *Troubles*, at the end of which the protagonist comes round after being bashed on the head to find himself buried up to his neck in sand, facing an incoming tide.)

Of the polio memoirs I have read, I have already mentioned in the Introduction my favourite, *I Can Jump Puddles*, Alan Marshall's touching account of his childhood in the Australian bush. Another moving Australasian memoir is New Zealander June Opie's *Over My Dead Body*, which focuses on her time in a London hospital. Two of the three American memoirs that I enjoyed most, Bentz Plagemann's *My Place to Stand* and Lorenzo Milam's exotically entitled *The Cripple Liberation Front Marching Band Blues*, are extensively quoted in the sections of the later historical chapters here which deal with patients' experiences of Warm Springs; the third, Edward Le Comte's *The Long Walk Back*, is remarkable mainly for its reflectiveness and unpretentious honesty.

Notes

Chapter 1

1. *New York Times*, 1 July 1916.
2. *Ibid.*, 1 & 8 July 1916.
3. *Ibid.*, 9 July 1916.
4. Haven Emerson, *A Monograph on the Epidemic of Poliomyelitis (Infantile Paralysis) in New York City in 1916, Based on the Official Reports of the Bureaus of the Department of Health* (Department of Health, New York 1917), p. 17.
5. *NY Times*, 4 July 1916.
6. *Ibid.*, 5 July 1916.
7. *Ibid.*, 29 June 1916.
8. Emerson, *op. cit.*, p. 17.
9. *NY Times*, 23 July 1916.
10. Roland H. Berg, *Polio and Its Problems* (J.B. Lippincott, Philadelphia 1948), pp. 4–5.
11. *NY Times*, 21 July 1916.
12. Cited in Al V. Burns, 'The Scourge of 1916 . . . America's First and Worst Polio Epidemic', *The American Legion Magazine*, Sept. 1966, p. 45.
13. Emerson, *op. cit.*, p. 63.
14. *NY Times*, 26 July 1916.
15. *Ibid.*, 5, 7, 14 & 18 July 1916.
16. Emerson, *op. cit.*, p. 29.
17. *NY Times*, 24 July 1916.
18. *Ibid.*, 5 July and 14 July 1916.
19. Haven Emerson to Simon Flexner, 26 Aug. 1916. Flexner Papers, American Philosophical Society, Philadelphia.
20. *NY Times*, 4 & 23 July 1916.
21. C.H. Lavinder, A.W. Freeman, & W.H. Frost, *Epidemiological Studies of Poliomyelitis in New York City and the North-Eastern United States during the Year 1916* (Washington 1918), p. 208
22. *NY Times*, 23 July 1916.
23. Charles S. Caverly, 'Notes of an Epidemic of Acute Anterior Poliomyelitis', *Journal of the American Medical Association*, 1896, xxvi, 1; from *Infantile Paralysis in Vermont 1894–1922; A Memorial to Charles S. Caverly, MD* (Burlington, Vermont 1924), p. 23.
24. *NY Times*, 26 July 1916.
25. John Ruhrah & Erwin E. Mayer, *Poliomyelitis in All Its Aspects* (Philadephia & New York 1917), p. 17.
26. *Ibid.*, p. 73; John R. Paul, *A History of Poliomyelitis* (Yale University Press, New Haven & London 1971), pp. 4–7.
27. J.G. Lockhart, *Memoirs of the Life of Sir Walter Scott* (London 1900), pp. 11 & 16.
28. Ruhrah & Mayer, *op. cit.*, p. 18.
29. *Ibid.*, p. 19.
30. Cited in Elisabeth F. Hutchin, 'Historical Summary', in *Poliomyelitis: A survey made possible by a grant from*

the International Committee for the Study of Infantile Paralysis organized by Jeremiah Milbank (Williams and Wilkins, Baltimore 1932), pp. 11–12.

31. Cited in Paul, *op. cit.*, p. 32.
32. *Ibid.*, p. 32.
33. *Ibid.*
34. *Ibid.*, p. 76.
35. Greer Williams, *Virus Hunters* (Hutchinson, London 1960), p. 55.
36. *Ibid.*
37. John Rowan Wilson, *Margin of Safety. The Story of Poliomyelitis Vaccine* (Collins, London 1963), p. 33.
38. *Ibid.*, p. 32.
39. Paul, *op. cit.*, pp. 98–100; Williams, *op. cit.*, p. 200.
40. Williams, *op. cit.*, p. 97.
41. Paul, *op. cit.*, pp. 116–17.
42. James Thomas Flexner, *An American Saga: The Story of Helen Thomas and Simon Flexner* (Little, Brown, Boston & Toronto 1984), pp. 21–2.
43. Saul Benison, 'The History of Polio Research in the United States: Appraisal and Lessons', in Gerald Holton (ed.); *The Twentieth Century Sciences: Studies in the Biography of Ideas* (W.W. Norton, New York 1972), p. 39.
44. Flexner, *op. cit.*, p. 40.
45. *Ibid.*, p. 103.
46. S. Flexner, 'Autobiographical Notes', cited in Benison, *op. cit.*, p. 310.
47. Abraham Flexner, *Medical Education in the United States and Canada* (New York 1910).
48. Benison, *op. cit.*, p. 311.
49. Paul, *op. cit.*, p. 108.
50. *Ibid.*, pp. 126–36.
51. *New York Times*, 9 March 1911, cited in Paul, *op. cit.*, p. 116.
52. *British Medical Journal*, 4 Nov. 1911.
53. *British Medical Journal*, 18 Nov. 1911.
54. Simon Flexner to Helen Flexner, 31 July 1911, Flexner Papers, APS, Philadelphia.
55. Simon Flexner to Helen Flexner, 29 June 1916, Flexner Papers, APS, Philadelphia.
56. F.W. Peabody, G. Draper, A.R. Dochez, *A Clinical Study of Acute Poliomyelitis* (New York 1912), pp. 71–2; this passage is reprinted, unattributed, in George Draper, *Acute Poliomyelitis* (New York 1917), and again, after Peabody's death – this time duly acknowledged – in George Draper, *Infantile Paralysis* (New York 1935).
57. Saul Benison, 'Poliomyelitis and the Rockefeller Institute: Social Effects and Institutional Response', *Journal of the History of Medicine and Allied Sciences*, Vol. xxix, 1974.
58. Guenter B. Risse, 'Revolt against Quarantine: Community Response to the 1916 Polio Epidemic, Oyster Bay, New York', an expanded version of the Samuel X. Radbill Lecture given at the College of Physicians of Philadelphia on 17 Oct. 1990, p. 12. I am indebted to this paper for the details of the Oyster Bay 'citizens' revolt'.
59. *Ibid.*
60. George Draper to Simon Flexner, 25 Aug. 1916, Flexner Papers, APS, Philadelphia.
61. Risse, *op. cit.*, p 26.
62. Kathleen Doherty to John D. Rockefeller, 29 Aug. 1916, Flexner Papers, APS, Philadelphia.
63. Emerson, *op. cit.*, p 75.
64. *Ibid.*
65. *Ibid.*, p 76.
66. *Ibid.*
67. K.P. Wilson to Simon Flexner, 21 Aug. 1916, Flexner Papers, APS, Philadelphia.
68. Paul, *op. cit.*, p. 132.
69. *NY Times*, 14 July 1916.
70. Lavinder, Freeman, and Frost: *op. cit.*, p. 208.
71. *Ibid.*, p. 210.
72. *NY Times*, 15 July 1916.
73. *Ibid.*, 22 July 1916.
74. *Ibid.*, 20 July 1916.
75. Paul, *op. cit.*, p. 190.
76. George Draper to Simon Flexner, 7 Aug. 1916, Flexner Papers, APS, Philadelphia.

77. *Ibid.*, 25 Aug. 1916.
78. F.W. Peabody, 'A report of the Harvard Infantile Paralysis Commission on the diagnosis and treatment of acute cases of the disease during the year 1916', *Boston Medical and Surgical Journal*, 176: 637–42 1917; cited in Paul, *op. cit.*, p. 193.
79. Paul, *op. cit.*, pp. 198–9.
80. *NY Times*, 23 July 1916.
81. *Ibid.*, 30 July 1916.
82. Emerson, *op. cit.*, p. 76.
83. Ibid., p. 77.
84. *NY Times*, 13 Aug. 1916.
85. *Ibid.*, 2 Aug. 1916.
86. 'Proposed method of organizing orthopedic treatment of Infantile Paralysis cases in New York City,' 9 Aug. 1916, p. 4, Flexner Papers, APS, Philadelphia.
87. Jerome D. Greene to Simon Flexner, 18 Aug. 1916, Flexner Papers, APS, Philadelphia.
88. *Ibid.*
89. *Ibid.*
90. Robert W. Lovett, *The Treatment of Infantile Paralysis* (Philadelphia 1916).
91. R.W. Lovett, 'The Treatment of Infantile Paralysis', *Journal of the American Medical Association*, 26 June 1915; from *Infantile Paralysis in Vermont 1894–1992; A Memorial to Charles S. Caverly, MD* (Burlington, Vermont 1924), p. 224.
92. Robert W. Lovett, 'The Treatment of Infantile Paralysis (with especial reference to the earlier stages)', *Health News* (monthly bulletin of New York State Department of Health, Albany, NY), Aug. 1916, pp. 189–90.
93. Lovett, *Memorial to . . . Caverly*, p. 225.
94. Lovett, Treatment of Infantile Paralysis, p. 39.
95. FDR to Eleanor Roosevelt, cited in Geoffrey C. Ward, *A First-Class Temperament: The Emergence of Franklin Roosevelt* (Harper & Row, New York 1989), p. 317.
96. *NY Times*, 10 Aug. 1916.
97. Paul, *op. cit.*, p. 152.
98. *NY Times*, 6 Aug. 1916.
99. Emerson, *op. cit.*, p. 80.
100. *Ibid.*
101. *Ibid.*, p. 292.
102. *Ibid.*
103. Lavinder, Freeman, & Frost, *op. cit.*, p. 214.
104. Paul *op. cit.*, p. 160.

Chapter 2

1. James Roosevelt (with Bill Libby), *My Parents: A Differing View* (Chicago 1976), pp. 72–3.
2. Geoffrey C. Ward, *A First-Class Temperament: The Emergence of Franklin D. Roosevelt* (Harper & Row, New York 1989), p. 416.
3. Joseph P. Lash, *Eleanor & Franklin* (André Deutsch, London 1972), p. 258.
4. Ward, *op. cit.*, footnote on p. 584.
5. FDR to William Eggleston, 11 Oct. 1924, cited in Vice-Admiral Ross T. McIntire, *Twelve Years with Roosevelt* (London 1948), p. 31, and in Hugh Gregory Gallagher, *FDR's Splendid Deception* (Dodd, Mead, New York 1985), p. 222.
6. Ward, *op. cit.*, p. 588 & footnote on p. 591.
7. *Ibid.*, p. 587.
8. Eleanor Roosevelt, *This Is My Story* (New York 1937), cited in Ward, *op. cit.*, p. 414.
9. Frank Friedel, *Franklin D. Roosevelt: Vol. 2, The Ordeal* (Boston 1954), p. 100.
10. Ibid., p. 101.
11. Ward, *op. cit.*, p. 594.
12. *Ibid.*
13. *Ibid.*, footnote on p. 620.
14. Eleanor Roosevelt, *The Autobiography of Eleanor Roosevelt* (Hutchinson, London 1962), p. 98.
15. *Ibid.*, p. 56.
16. Ward, *op. cit.*, p. 620; Anna Roosevelt, 'My Life with FDR, Part 3: How Polio Helped Father', *Woman*, July 1949.

17. Eleanor Roosevelt, *Autobiography*, p. 98.
18. Ward, *op. cit.*, p. 633.
19. *Ibid.*, p. 605.
20. FDR to William Eggleston, 11 Oct. 1924.
21. Ward, *op. cit.*, p. 616.
22. *Ibid.*, p. 617.
23. *Ibid.*, p. 622.
24. *Ibid.*, p. 626.
25. *Ibid.*, pp. 627–8.
26. Gallagher, *op. cit.*, p. 21.
27. Eleanor Roosevelt, *Autobiography*, p. 216.
28. Bernard Asbell, *The FDR Memoirs* (New York 1973), p. 246.
29. James Roosevelt, *op. cit.*, p. 104.
30. Lash, *op. cit.*, p. 507.
31. Asbell, *op. cit.*, p. 263.
32. James Roosevelt, *op. cit.*, p. 108.
33. Frances Perkins, *The Roosevelt I Knew* (London 1947), pp. 13–14.
34. *Ibid.*, pp. 28–9.
35. Elliott Roosevelt & James Brough, *An Untold Story: The Roosevelts of Hyde Park* (Dell, New York 1974), p. 213.
36. *Ibid.*
37. James Roosevelt, *op. cit.*, pp. 94–5.

Chapter 3

1. Etta Blanchard Worsley, 'Warm Springs', *Georgia Review*, Vol. III No. 2, Summer 1949, pp. 234 & 239.
2. Adam M. Lunoe, 'Notes on Georgia Warm Springs Foundation,' National Foundation for Infantile Paralysis (New York, undated), p. 6.
3. Charles Dudley Warner, *On Horseback* (Boston & New York 1888), p. 135 (extract sent to FDR with letter from Paul T. Pechstein of Roanoke, Virginia, 11 March 1937).
4. FDR to Mrs K. Lake, 23 July. 1923 (FDR Library).
5. FDR to Mr Paul D. Hasbrouck, 17 Oct. 1923 (FDR Library).
6. FDR to Mr Paul D. Hasbrouck, 12 Jan. 1925 (FDR Library).
7. Frank Friedel, *Franklin D. Roosevelt: Vol. 2, The Ordeal* (Boston 1954), p. 191.
8. Alden Whitman, 'Basil O'Connor, Polio Crusader, Dies', *New York Times*, 10 March 1972.
9. Turnley Walker, *Roosevelt and the Fight against Polio* (Rider, London 1954), p. 17.
10. Basil O'Connor to FDR, undated but probably Oct. 1926 (FDR Library).
11. *Ibid.*
12. Richard Carter, *Breakthrough: The Saga of Jonas Salk* (Trident Press, New York 1966), p. 12.
13. Walker, *op. cit.*, p. 102.
14. Theo Lippman, Jr, *The Squire of Warm Springs: FDR in Georgia, 1924–1945* (Playboy Press, Chicago 1977), pp. 42–3.
15. Fred Botts, 'Warm Springs – the Roosevelt Era and Beyond', *Wheelchair Review*, Vol. VI No. 9, 12 Aug. 1951, p. 3.
16. Fred Botts, 'Where Infantile Paralysis Gets Its "Walking Papers" ', typescript in the Roosevelt Warm Springs Institute for Rehabilitation Library, Georgia, p. 1.
17. *Ibid.*, pp. 2–3.
18. *Ibid.*, pp. 4–5.
19. Cited in Lippman, *op. cit.*, p. 39.
20. *Ibid.*
21. LeRoy W. Hubbard, 'The Treatment of Infantile Paralysis; By Exercises in a Natural Warm Water', *The Crippled Child*, Vol. V No. 2, July/Aug. 1927, p. 38
22. Lippman, *op. cit.*, p. 14.
23. Author interview with Mary Frances 'Pidge' Cole, St Louis, 3 June 1992.
24. Hugh Gregory Gallagher, *FDR's Splendid Deception* (Dodd, Mead, New York 1985), p. 29.
25. *Ibid.*, p. 30.
26. Cited in Geoffrey C. Ward: *A First-Class Temperament: The Emergence of Franklin D. Roosevelt* (Harper & Row, New York 1989), p. 745.
27. Author Interview with Mary Hudson Veeder, Warm Springs, Georgia, 26 May 1992.

28. *Ibid.*
29. *Ibid.*
30. Friedel, *op. cit.*, p. 255.
31. Cited in Ward, *op. cit.*, p. 798.
32. *Ibid.*, p. 794.
33. William Shakespeare, *Julius Caesar*, IV, iii, 217.
34. Ward, *op. cit.*, p. 755.
35. *Ibid.*, pp. 798–9.
36. Walker, *op. cit.*, p. 141
37. *Ibid.*, p. 144.

Chapter 4

1. Keith Morgan to Basil O'Connor, 1 July 1935 (FDR Library).
2. Turnley Walker, *Roosevelt and the Fight against Polio* (Rider, London 1954), p. 153.
3. Keith Morgan to Merle Crowell, 17 Sept. 1929 (FDR Library).
4. Richard Thayer Goldberg: *The Making of Franklin D. Roosevelt: Triumph over Disability* (Abt Books, Cambridge, Mass. 1981), p. 138.
5. FDR, 'Wanted: Enlistments for a Crusade', *Polio Chronicle*, July 1931.
6. Fred Botts, 'Announcing the New National Patients Committee', Polio Chronicle, July 1931.
7. Goldberg, *op. cit.*, p. 133.
8. *Ibid.*, p. 165.
9. Walker, *op. cit.*, pp. 162–3.
10. Arthur Carpenter to Basil O'Connor, 9 Sept. 1931 (FDR Library).
11. Keith Morgan to Basil O'Connor, 1 July 1935 (FDR Library).
12. Author interview with Mary Hudson Veeder, Warm Springs, Georgia, 26 May 1992.
13. Walker, *op. cit.*, p. 200.
14. *Polio Chronicle*, Nov. 1932.
15. *Ibid.*, Dec. 1932.
16. James Roosevelt (with Bill Libby): *My Parents: A Differing View* (Chicago 1976), p. 83.
17. Hugh Gregory Gallagher, *FDR's Splendid Deception* (Dodd, Mead, New York 1985) pp. xiii–xiv.
18. Author interview with Mary Frances 'Pidge' Cole, St Louis, 3 June 1992.
19. Goldberg, *op. cit.*, p. 145.
20. Martin Green, 'For Seven Years Banker and Politician has Waged Winning Battle', unidentified newspaper cutting in FDR Library, March 1928.
21. Fulton Oursler (ed. Fulton Oursler, Jr), *Behold This Dreamer!* (Boston 1964), pp. 369–70.
22. Richard Carter, *Breakthrough: The Saga of Jonas Salk* (Trident Press, New York 1966), p. 13.
23. Theo Lippman Jr, *The Squire of Warm Springs: FDR in Georgia, 1924–1945* (Chicago 1977), p. 203.
24. Walker, *op. cit.*, p. 180.
25. Goldberg, *op. cit.*, p. 154.
26. Walker, *op. cit.*, p. 182.
27. Keith Morgan to FDR, undated, but FDR acknowledges on 28 June 1934 (FDR Library).
28. Carter, *op. cit.*, p. 14.
29. *Ibid.*
30. Dr Michael Hoke to Marguerite LeHand, 7 Sept. 1935, and Marguerite LeHand to Dr Michael Hoke, 25 Sept. 1935 (FDR Library).
31. DNL to Eleanor Roosevelt, undated (FDR Library).
32. Paul H. Harmon, 'The Geographical and Racial Incidence of Poliomyelitis,' *International Bulletin for Economics, Medical Research & Public Hygiene*, Vol. A 40, Infantile Paralysis, 1939/40, p. 136.
33. Keith Morgan to FDR, undated – see note 27 above – (FDR Library).
34. John R. Paul, *A History of Poliomyelitis* (New Haven & London 1971), p. 212.
35. Harmon, *op. cit.*, pp. 137 & 143.
36. Eckford C. De Kay to FDR, 21 March 1934 (FDR Library).
37. FDR to Eckford C. De Kay, 18 April 1934.
38. Lippman, *op. cit.*, p. 196; Carter, *op. cit.*, p. 15; & Goldberg, *op. cit.*, p. 159.
39. Keith Morgan to Henry Kannee (assistant to Press Secretary Marvin McIntyre at the White House), 16 Sept. 1935 (FDR Library).
40. Typescript 'History of the National Foundation for Infantile Paralysis',

no author or date given, except for a handwritten 'appar. written 1953 or 1954' (March of Dimes Foundation), p. C–2.

41. Saul Benison (ed.), *Tom Rivers: Reflections on a Life in Medicine and Science* (MIT Press, Cambridge, Mass. & London 1967), footnote on p. 182.
42. Paul de Kruif, *The Sweeping Wind: A Memoir* (Hart-Davis, London 1962), pp. 197–8.
43. Jane S. Smith, *Patenting the Sun: Polio and the Salk Vaccine* (Anchor Books, New York 1990), p. 71.
44. Benison, *op. cit.*, pp. 181–2.
45. Saul Benison, 'The History of Polio Research in the United States: Appraisal and Lessons', in Gerald Holton (ed.), *The Twentieth Century Sciences: Studies in the Biography of Ideas* (New York 1972), p. 322.
46. Benison, *Tom Rivers*, p. 183.
47. Paul de Kruif, 'Infantile Paralysis: Progress Report No. 1 of Research in Immunization and Prevention', 12 Nov. 1935 (FDR Library), p. 27.
48. Carter, *op. cit.*, p. 19.
49. Paul, *op. cit.*, p. 117.
50. *Ibid.*, pp. 227–9.
51. *Ibid.*, p. 117.
52. Paul de Kruif, 'Progress Report: President's Birthday Ball Commission for Infantile Paralysis Research', 9 Nov. 1936 (FDR Library), p. 1.
53. De Kruif, Infantile Paralysis: Progress Report No. 1', pp. 3–4.
54. Paul de Kruif, 'The President's Birthday Ball Commission for Infantile Paralysis research: Progress Report', 28 March 1937 (FDR Library), p. 19.
55. Paul de Kruif to S.M. Greer, 8 Dec. 1936 (FDR Library).
56. Keith Morgan to Henry Kannee, 16 Sept. 1935 (FDR Library).
57. Benison, *Tom Rivers*, p. 184.
58. Carter, *op. cit.*, p. 20.
59. Paul, *op. cit.*, p. 256.
60. *Ibid.*
61. *Ibid.*, p. 258.
62. *Ibid.*
63. John Rowan Wilson, *Margin of Safety: The Story of Poliomyelitis Vaccine* (Collins, London 1963), p. 57.
64. Carter, *op. cit.*, p. 22.
65. De Kruif, 'Infantile Paralysis: Progress Report No. 1', p. 8.
66. *Ibid.*, p. 11.
67. De Kruif, *Sweeping Wind*, p. 199.
68. Benison, *op. cit.*, p. 189.
69. *Ibid.*, footnote on p. 190.
70. Carter, *op. cit.*, p. 23.
71. Benison, *op. cit.*, p. 188.
72. De Kruif, 'Infantile Paralysis: Progress Report No. 1', p. 9.
73. Benison, *op. cit.*, p. 189.
74. *Ibid.*, p. 190.
75. Paul, *op. cit.*, p. 261.
76. Greer Williams, *Virus Hunters* (Hutchinson, London 1960), p. 226.
77. Benison, *op. cit.*, p. 190.
78. Paul, *op. cit.*, pp. 247–8.
79. Benison, *op. cit.*, p. 193.
80. *Ibid.*, p. 262.
81. Paul, *op. cit.*, p. 307.
82. George Draper, *Infantile Paralysis* (New York 1935), pp. 58–68.
83. *Ibid.*
84. De Kruif, 'Infantile Paralysis: Progress Report No. 1', p. 15.
85. Paul, *op. cit.*, p. 162.
86. *Ibid.*
87. Cited in Benison, *op. cit.*, p. 230.
88. Benison, *op. cit.*, p. 306.
89. *Ibid.*, p. 307.
90. George Draper to FDR, 29 Aug. 1934 (FDR Library).
91. FDR to George W. Draper, 6 Sept. 1934 (FDR Library).
92. George Draper to the President (telegram), 6 March 1935 (FDR Library).
93. FDR to George Draper, 8 March 1935 (FDR Library).
94. Typescript 'History of the Nat. Found . . .', p. C–4.
95. Victor Cohn, *Four Billion Dimes* (Minneapolis 1955), p. 56.
96. Basil O'Connor to Stephen T. Early, 12 Nov. 1937 (FDR Library).
97. Keith Morgan to FDR, 4 Nov. 1937 (FDR Library).
98. Keith Morgan to Marvin H.

McIntyre, 4 Dec. 1937 (FDR Library).
99. Cohn, *op. cit.*, p. 52.
100. *Ibid.*, p. 53.
101. Benison, *op. cit.*, p. 229.
102. Cited in Carter, *op. cit.*, pp. 24–5.
103. Benison, *op. cit.*, p. 232.
104. Williams, *op. cit.*, p. 77.
105. *Ibid.*, p. 80.
106. *Ibid.*, pp. 77–82.
107. Rexford Tugwell interview with William D. Hassett (FDR Library).
108. Fred Botts to FDR, 16 Aug. 1936 (FDR Library).
109. Dr Michael Hoke to Keith Morgan, undated (FDR Library).
110. Basil O'Connor to FDR, 4 Feb. 1942; FDR to Arthur Carpenter, 1 June 1944 (FDR Library).
111. Oscar De Priest to Louis McH. Howe, 23 Jan. 1934 (FDR Library).
112. 'Rabbit' to Louis Howe, 3 Feb. 1934 (FDR Library).
113. M. Lipman to Stephen Early, 10 Feb. 1934; Stephen Early to M. Lipman, 13 Feb. 1934 (FDR Library).
114. Walter White to Eleanor Roosevelt, 20 Oct. 1936 (FDR Library).
115. C.H. Hamlin to FDR, 15 Sept. 1936; Stephen Early to C.H. Hamlin, 23 Sept. 1936 (FDR Library)
116. J.N. Peacock Jr to FDR, 29 June 1934; Robert G. Mack to J.N. Peacock JR, 28 June 1934; M.A. LeHand to J.N. Peacock Jr, 13 Aug. 1934 (FDR Library).
117. John F. Nolan to Stephen Early, 23 Sept. 1936 (FDR Library).
118. Stephen Early to Keith Morgan, 26 Sept. 1936 (FDR Library).
119. *New York Sun*, 14 Oct. 1936.
120. J.S. Brookens to H.L. Ickes, undated; Robert C. Weaver to Rev. J.S. Brookens, 15 July 1936 (FDR Library)
121. Robert C. Weaver to Walter F. White, 19 Oct. 1936 (FDR Library).
122. Eleanor Roosevelt to FDR, 21 Oct. 1936 (FDR Library).
123. M.A.L. to Mrs Roosevelt, 16 Nov. 1936 (FDR Library).
124. Basil O'Connor to Marguerite A. LeHand, 23 Dec. 1936 (FDR Library).
125. Basil O'Connor to FDR, 3 March 1937 (FDR Library).
126. Basil O'Connor to FDR, 3 March 1937 (FDR Library).
127. Georgia Warm Springs Foundation, Inc., Memorandum of replies re question of a cottage for Negroes at Warm Springs, undated (FDR Library).
128. Joseph P. Lash, *Eleanor & Franklin* (André Deutsch, London 1972), p. 346.
129. Lippman, *op. cit.*, p. 155.
130. Basil O'Connor to FDR, 3 March 1937 (FDR Library).
131. Dr LeRoy W. Hubbard to Basil O'Connor, 9 March 1937 (FDR Library).
132. Basil O'Connor to FDR, 10 March 1937 (FDR Library).
133. Henry N. Hooper to Basil O'Connor, 8 May 1937; Basil O'Connor to FDR, 11 May 1937 (FDR Library).
134. Basil O'Connor speech at the Tuskegee Institute, Alabama, 22 May 1939.
135. Fred Allhoff, 'Black Man's Miracles: The Story of a Great and Good American', *Liberty Magazine* (undated – 1937/8 – cutting in FDR Library), p. 21.
136. Cited in Linda O. McMurry, *George Washington Carver: Scientist and Symbol* (Oxford University Press, New York 1981), p. 253.
137. Programme for 'The Old Maid's Dilemma', Warm Springs, Georgia, 26 Nov. 1935 (FDR Library).
138. George Washington Carver to Dr Hardee Johnson, 28 Jan. 1936, cited in Edith P. Chappell and John F. Hume, *A Black Oasis: Tuskegee's Fight against Infantile Paralysis, 1941–1975* (unpublished thesis at Tuskegee University), p. 17.
139. McMurry, *op. cit.*, p. 255.
140. Chappell and Hume, *op. cit.*, p. 31.
141. *Ibid.*, pp. 73–4.
142. *Ibid.*, p. 75.
143. Allan M. Brandt, 'Racism and

Research: The Case of the Tuskegee Syphilis Study', *Hastings Center Report*, Dec. 1978, pp. 21–9.

Chapter 5

1. Author interview (by phone) with Victor Cohn, Washington DC, 15 May 1992.
2. *Ibid.*
3. Victor Cohn, *Sister Kenny: The Woman Who Challenged the Doctors* (University of Minnesota Press, Minneapolis 1975), pp. 41–2; I am deeply indebted to this excellent study throughout this chapter.
4. *Ibid.*, p. 45.
5. *Ibid.*, *passim.*
6. *Ibid.*, p. 85.
7. Desmond Zwar, *The Dame: The Life and Times of Dame Jean Macnamara, medical pioneer* (Macmillan, Australia 1984) p. 86.
8. Sir Henry Gauvain, 'Anterior Poliomyelitis in England', *International Bulletin for Economics, Medical Research & Public Hygiene*, Vol. A 40, Infantile Paralysis, 1939/40, p. 85.
9. *Ibid.*, p. 82; Norman M. Harry, 'The Recovery Period in Anterior Poliomyelitis', *British Medical Journal*, 22 Jan. 1938, p. 164.
10. *British Medical Journal*, 22 Jan. 1938, p. 179.
11. H.A.T. Fairbank *et al.*, 'Report on the Kenny Method of Treatment', *British Medical Journal*, 22 Oct. 1938, pp. 852–4.
12. Thelander *et al.*, 'Report of the Royal Commission', cited in Cohn, *op. cit.*, p. 107.
13. Zwar, *op. cit.*, p. 90.
14. Cohn, *op. cit.*, p. 120.
15. G.T. Spencer, 'Re-inventing the Wheel: or the History of Artificial Respiration'; Philip A. Drinker & Charles F. McKhann, 'The Iron Lung: First Practical Means of Respiratory Support', *Journal of the American Medical Association*, 21 March 1986, Vol. 255 No. 11, pp. 1476–80.
16. Drinker & McKhann, *op. cit.*
17. Catherine Drinker Bowen: *Family Portrait* (Little, Brown, Boston 1970), pp. 241–2.
18. Drinker & McKhann, *op. cit.*, p. 1477.
19. John R. Paul, *A History of Poliomyelitis* (Yale University Press, New Haven & London 1971), p. 328.
20. *Ibid.*
21. Leonard C. Hawkins & Milton Lomask: *The Man in the Iron Lung: The Story of Frederick B. Snite* (The World's Work, Kingswood, Surrey 1957), pp. 65–9.
22. Paul, *op. cit.*, p. 331.
23. *Ibid.*, p. 334.
24. Cohn, *op. cit.*, p. 117.
25. Arthur Carpenter to Basil O'Connor, 29 May 1936 (FDR Library).
26. LeRoy W. Hubbard to Keith Morgan, 18 June 1936 (FDR Library).
27. C.E. Irwin to Basil O'Connor, 22 Sept. 1936 (FDR Library).
28. K.G. Hansson, 'After-Treatment of Poliomyelitis', *Journal of the American Medical Association*, 1 July 1939, Vol. 113, p. 35.
29. *Ibid.*, p. 33.
30. Jessie L. Stevenson, *Care of Poliomyelitis* (New York 1940), pp. 61 & 94.
31. Robert Dwight Brown, 'Poliomyelitis: A Personal Experience', 11 Sept. 1933 (typescript in FDR Library).
32. *Infantile Paralysis: A Monthly Journal* (Milton H. Berry Institute, Van Nuys, California), Vol. 1 No. 2, June 1939, p. 7; Milton H. Berry, 'A Challenge in behalf of Crippled Children to the Universities of America', undated, p. 8 (FDR Library).
33. *Infantile Paralysis: A Monthly Journal*, Vol. 1 No. 2, June 1939, p. 7.
34. Saul Benison, *Tom Rivers: Reflections on a Life in Medicine and Science* (MIT Press, Cambridge, Mass, & London 1967) p. 282.
35. Cohn, *op. cit.*, p. 128.

36. Author interview with Henry Haverstock Jr, Minneapolis, 11 June 1992.
37. *Ibid.*
38. Henry Haverstock Sr, typescript of talk given at reception at Jerome Ruff Home, Robbinsdale, Minnesota, 6 Dec. 1940 (Sister Kenny Institute).
39. Cohn, *op. cit.*, p. 132.
40. Henry O. Kendall and Florence P. Kendall, 'Report to the National Foundation for Infantile Paralysis on the Sister Kenny Method of Treatment in Anterior Poliomyelitis', undated (Sister Kenny Institute).
41. *Ibid.*
42. Cohn, *op. cit.*, p. 142.
43. *Ibid.*, p. 143.
44. Robert Bennett to Miland Knapp, 26 May 1941 (Sister Kenny Institute).
45. Wallace H. Cole & Miland E. Knapp, 'The Kenny Treatment of Infantile Paralysis: A Preliminary Report', *Journal of the American Medical Association*, 7 June 1941, Vol. 116, pp. 2577–80.
46. Marvin L. Kline, 'The Most Unforgettable Character I've Met', *Reader's Digest*, Aug. 1959.
47. Lois Mattox Miller, 'Sister Kenny Wins Her Fight', *Reader's Digest*, Oct. 1942.
48. Paul, *op. cit.*, pp. 339–40.
49. Victor Cohn, *Four Billion Dimes* (Minneapolis 1955), p. 65.
50. Robert L. Bennett, 'Recent Developments in the Treatment of Poliomyelitis', *Southern Medical Journal*, Feb. 1943, Vol. 36 No. 2, pp. 152–6.
51. Roland H. Berg, *Polio and Its Problems* (J.B. Lippincott, Philadelphia 1948), p. 140.
52. Author interview with Margaret Ernest, Minneapolis, 9 June 1992; Margaret Ernest, 'She Was a Fighter', unpublished typescript (Sister Kenny Institute).
53. *Ibid.*
54. Victor Cohn, *Sister Kenny*, p. 172.
55. *Ibid.*, p. 173.
56. Basil O'Connor, 'The Story of the Kenny Method', *Archives of Physical Therapy*, April 1944, Vol. XXV, p. 234.
57. *Ibid.*
58. Cohn, *op. cit.*, p. 180.
59. Albert Deutsch, 'Sister Kenny Again Engaged In Battle With Medicos,' *PM*, 7 Feb. 1944.
60. *Ibid.*
61. Ralph K. Ghormley *et al.*, 'Evaluation of the Kenny Treatment of Infantile Paralysis', *Journal of the American Medical Association*, 17 June 1944, Vol. 125, pp. 466–9.
62. Cohn, *op. cit.*, p. 187.
63. Paul, *op. cit.*, p. 343.
64. Robert B. Lawson to John F. Pohl, 8 Dec. 1945 (Sister Kenny Institute).
65. J.E. Hulett Jr, 'The Kenny Healing Cult: Preliminary Analysis of Leadership and Patterns of Interaction', *American Sociological Review*, June 1945, Vol. 10, pp. 364–72.
66. *Ibid.*
67. *Ibid.*
68. Cohn, *op. cit.*, p. 135.
69. Sir Macfarlane Burnet: *Changing Patterns: an atypical autobiography* (Heinemann, London 1968), pp. 166–8.
70. Author interview with Richard Owen, Minneapolis, 10 June 1992.
71. *Ibid.*
72. Richard Owen, 'Sister Kenny and the Origins of Modern rehabilitation', typescript notes given to the author by Dr Owen.
73. Author interview with Richard Owen.
74. Bernhard J. Stern, *Social Factors in Medical Progress* (New York 1927), p. 25.

Chapter 6

1. Keith Morgan to FDR, 4 Dec. 1941 (FDR Library).
2. Victor Cohn, *Four Billion Dimes* (Minneapolis 1955), pp. 66–8.
3. Basil O'Connor to FDR, 23 March 1943 (FDR Library).

4. *The Miracle of Hickory* (National Foundation for Infantile Paralysis, Publication No. 53, undated).
5. FDR to Basil O'Connor, 1 Dec. 1944, *Archives of Physical Therapy*, Vol. XXV No. 12, Dec. 1944, pp. 743–4.
6. Basil O'Connor to Dr James E. Paullin, 15 June 1944 (FDR Library).
7. Bentz Plagemann, *My Place to Stand* (Farrar, Straus & Co., New York 1949), pp. 136–235, for all quotations in this section.
8. Adam M. Lunoe, 'Notes on Georgia Warm Springs Foundation', undated (March of Dimes).
9. Plagemann, *op. cit.*, pp. 136–235, for remainder of quotations in this section.
10. *Ibid.*, p. 208.
11. Cohn, *op. cit.*, p. 74.
12. *Ibid.*, pp. 74–5.
13. *Ibid.*, p. 75.
14. *Ibid.*, p. 76.
15. *Ibid.*
16. *Ibid.*, p. 77.
17. John R. Paul, *A History of Polio myelitis* (Yale University Press, New Haven & London 1971), p. 353.
18. *Ibid.*, pp. 364–5.
19. Elisabeth F. Hutchin, 'Historical Summary', in *Poliomyelitis: A survey made possible by a grant from the International Committee for the Study of Infantile Paralysis organized by Jeremiah Milbank* (Williams & Wilkins, Baltimore 1932), p. 18.
20. Saul Benison, *Tom Rivers: Reflections on a Life in Medicine and Science* (MIT Press, Cambridge, Mass., & London 1967), p. 266.
21. Paul, *op. cit.*, p. 282.
22. *Ibid.*
23. Paul de Kruif, 'Infantile Paralysis: Progress Report of Research in Immunization and Prevention No. 1', p. 25 (FDR Library).
24. Paul, *op. cit.*, p. 288.
25. *Ibid.*, p. 365.
26. Richard Carter, *Breakthrough: The Saga of Jonas Salk* (Trident Press, New York 1966), p. 26.
27. *Ibid.*, p. 57.
28. Jane S. Smith, *Patenting the Sun: Polio and the Salk Vaccine* (William Morrow & Co., New York 1990; Anchor Books, Doubleday, New York 1991), p. 115.
29. Carter, *op. cit.*, p. 69.
30. Benison, *op. cit.*, pp. 404–5.
31. Carter, *op. cit.*, p. 57.
32. *Ibid.*, p. 58.
33. *Ibid.*
34. *Ibid.*, p. 63.
35. Sir Macfarlane Burnet, *Changing Patterns: an atypical autobiography* (Heinemann, London 1968), p. 169.
36. Greer Williams, *Virus Hunters* (Hutchinson, London 1960), p. 211.
37. *Ibid.*
38. *Ibid.*, pp. 212–13.
39. Paul, *op. cit.*, p. 373.
40. *Ibid.*, p. 374.
41. *Ibid.*, pp. 412–13.
42. Carter, *op. cit.*, p. 110.
43. Benison, *op. cit.*, p. 467.
44. *Ibid.*
45. Cited in John Rowan Wilson, *Margin of Safety: The Story of the Poliomyelitis Vaccine* (Collins, London 1963), p. 149.
46. Benison, *op. cit.*, p. 498; Carter, *op. cit.*, p. 148.
47. Wilson, *op. cit.*, p. 152.
48. Paul, *op. cit.*, p. 410.
49. 'Second International Poliomyelitis Conference: II, The Virus and Its Fellow Travellers', *British Medical Journal*, 22 Sept. 1951, p. 731.
50. Carter, *op. cit.*, pp. 113–14.
51. Smith, *op. cit.*, pp. 147–8.
52. Cohn, *op. cit.*, pp. 87–8.
53. Carter, *op. cit.*, p. 121.
54. *Ibid.*, pp. 120–21.
55. *Ibid.*, p. 121.
56. Wayne Martin, *Medical Heroes and Heretics* (Devin-Adair, Connecticut, undated), p. 58.
57. Carter, *op. cit.*, pp. 133–4.
58. *Ibid.*, p. 134.
59. *Ibid.*, pp. 134–5.
60. Smith, *op. cit.*, pp. 140–41.
61. Carter, *op. cit.*, p. 137.
62. *Ibid.*, pp. 137–8.
63. Cohn, *op. cit.*, p. 101.
64. Carter, *op. cit.*, p. 142.

65. Benison, *op. cit.*, p. 496.
66. *Ibid.*
67. Paul, *op. cit.*, p. 419.
68. *Ibid.*
69. *Ibid.*
70. Carter, *op. cit.*, p. 144.
71. *Ibid.*, pp. 147–8; Benison, *op. cit.*, p. 497.
72. Benison, *op. cit.*, pp. 499–500.
73. Paul, *op. cit.*, p. 419.
74. Carter, *op. cit.*, p. 146.
75. *Ibid.*, p. 154.
76. Smith, *op. cit.*, pp. 183–4.
77. Carter, *op. cit.*, p. 156.
78. Charles L. Mee Jr, 'The Summer Before Salk', *Esquire*, Dec. 1983, p. 42.
79. Williams, *op. cit.*, p. 237.
80. Benison, *op. cit.*, p. 502.
81. *Ibid.*, pp. 503–4.
82. Carter, *op. cit.*, p. 176.
83. *Ibid.*, p. 182.
84. *Ibid.*
85. Benison, *op. cit.*, p. 425.
86. *Ibid.*, p. 533.
87. Smith, *op. cit.*, p. 250.
88. Carter, *op. cit.*, pp. 227–8.
89. *Ibid.*, p. 228.
90. David L. Sills, *The Volunteers: Means and Ends in a National Organization* (The Free Press, Illinois 1957), pp. 246–7.
91. *Ibid.*, p. 31.
92. Carter, *op. cit.*, pp. 228–9.
93. Paul, *op. cit.*, p. 411.
94. *Ibid.*, pp. 423–4.
95. *Ibid.*, p. 424.
96. *Ibid.*, p. 423.
97. Carter, *op. cit.*, pp. 179–80.
98. *Ibid.*, p. 180.
99. Smith, *op. cit.*, p. 245.
100. Carter, *op. cit.*, p. 219.
101. *Ibid.*, p. 181.
102. Aaron E. Klein, *Trial by Fury: The Polio Vaccine Controversy* (New York 1972), p. 100.
103. Smith, *op. cit.*, pp. 256–7.
104. Carter, *op. cit.*, pp. 231–2.
105. *Ibid.*, p. 232.
106. *New York World-Telegram and the Sun*, 6 April 1954.
107. *Washington Post*, 7 April 1954.
108. Walter Winchell, 9 p.m. broadcast over WABC (New York) and the ABC Network.
109. *Ibid.*
110. Carter, *op. cit.*, p. 233.
111. *Ibid.*, p. 234.
112. Smith, *op. cit.*, p. 266.
113. *Ibid.*, pp. 22–3.
114. *Ibid.*, p. 387.
115. *Ibid.*, p. 269.
116. Carter, *op. cit.*, p. 243.
117. Wilson, *op. cit.*, pp. 93–4.
118. Carter, *op. cit.*, p. 246.
119. *Poliomyelitis: Papers and Discussions Presented at the Third International Poliomyelitis Conference* (J.B. Lippincott, Philadelphia 1955), p. 206.
120. Carter, *op. cit.*, p. 249.
121. *Ibid.*, p. 266.
122. Williams, *op. cit.*, p. 254.
123. *Ibid.*, p. 255.
124. Carter, *op. cit.*, p. 269.
125. *Ibid.*, pp. 279–80.
126. Wilson, *op. cit.*, p. 100.
127. Williams, *op. cit.*, pp. 259–60.
128. *Ibid.*, p. 262.
129. Klein, *op. cit.*, pp. 110–11.
130. Carter, *op. cit.*, p. 300.
131. *Ibid.*, p. 122.
132. Williams, *op. cit.*, p. 288.
133. Klein, *op. cit.*, p. 115.
134. Williams, *op. cit.*, p. 265.
135. Klein, *op. cit.*, p. 118.
136. Carter, *op. cit.*, p. 323.
137. *Ibid.*, p. 329.
138. Williams, *op. cit.*, pp. 276–7.
139. Carter, *op. cit.*, p. 334.
140. *Ibid.*
141. *Ibid.*
142. Smith, *op. cit.*, pp. 354–5.
143. Speech by James Shannon, Director of the National Institutes of Health, Oklahoma City, 25 Oct. 1966 (March of Dimes).
144. Saul Benison to Dr Robert Greene, Executive Secretary, Committee on Science and Public Policy, National Academy of Sciences, Washington DC, 7 Dec. 1966 (March of Dimes).
145. *Ibid.*
146. Paul, *op. cit.*, p. 438.
147. Gerald Markle & James Peterson, *Politics, Science and Cancer* (Boulder 1980), p. 164, cited in James T.

Patterson: *The Dread Disease: Cancer and Modern American Culture* (Harvard University Press, Cambridge, Mass. & London 1987), p. 252.

148. Smith, *op. cit.*, p. 376.

Chapter 7

1. John Rowan Wilson, *Margin of Safety: The Story of the Poliomyelitis Vaccine* (Collins, London 1963), p. 120.
2. Saul Benison, *Tom Rivers: Reflections on a Life in Medicine and Science* (MIT Press, Cambridge, Mass., & London 1967), p. 562.
3. *Ibid.*, pp. 562–3.
4. Wilson, *op. cit.*, pp. 120–21.
5. *Ibid.*, p. 121.
6. Thomas Gucker of Boston, Mass., at a staff medical meeting of the Georgia Warm Springs Foundation, 9 March 1949 (GWSF).
7. *British Medical Journal*, 26 July 1947, p. 136.
8. *British Medical Journal*, 16 Aug. 1947, p. 259.
9. A.H. Gale, *Epidemic Diseases* (Penguin, London 1959), p. 115.
10. W.H. Bradley in discussion on 'The Importance of Poliomyelitis as a World Problem', in *Poliomyelitis: Papers and Discussions Presented at the First International Poliomyelitis Conference* (J.B. Lippincott, Philadelphia 1949), p. 54.
11. W.H. Bradley, 'Reports of Official Delegates – England and Wales', in *Poliomyelitis: Papers and Discussions Presented at the First International Poliomyelitis Conference*, pp. 335–6.
12. Author interview with Geoffrey Spencer, London, 24 Aug. 1992.
13. *Ibid.*
14. H.C.A. Lassen (ed.): *Management of Life-Threatening Poliomyelitis: Copenhagen 1952–1956* (E. & S. Livingstone, Edinburgh & London 1956), p. xi.
15. *Ibid.*
16. G.L. Wackers, 'Theaters of War, Truth and Competence: the rhetorics of innovative intermittent positive pressure ventilation during the 1952 polio epidemic in Copenhagen', unpublished draft 1991.
17. *Ibid.*
18. Author interview with Anne M. Isberg, London, 15 June 1993.
19. Author interview with Geoffrey Spencer.
20. Bradley, *op. cit.*, pp. 335–6.
21. British Polio Fellowship annual report 1979.
22. V.E. Leichty, 'Through American Eyes', IPF *Bulletin*, no date, p. 10.
23. Author interview with David Widgery, London, 6 Nov. 1991 (See Part II, pp. 229–75).
24. Author interview with Lancelot Walton, Alderney, 11 Aug. 1992.
25. Author interview with Mary Gould, Hampshire, 20 March 1992 (See Part II, pp. 229–75).
26. IPF *News* (undated, but on internal evidence 1956–7).
27. W.H. Bradley, 'Reports of Official Delegates – England and Wales', in *Poliomyelitis: Papers and Discussions Presented at the Fourth International Poliomyelitis Conference* (J.B. Lippincott, Philadelphia 1958), p. 39.
28. Author interview with Janet Keegan, Hertfordshire, 10 Feb. 1992.
29. *Ibid.*
30. Michael Frayn, 'Festival', in Michael Sissons and Philip French (eds): *Age of Austerity* (Oxford University Press paperback edn, Oxford 1986), p. 325.
31. *Ibid.*, pp. 307–8.
32. Author interview with Duncan Guthrie, Sussex, 27 July 1992.
33. 'The Problem of Infantile Paralysis', IPF (undated, but on internal evidence 1951–2).
34. Duncan Guthrie, 'The Energetic Pursuit of Research', IPF *Bulletin*, No. 130, July 1952.
35. Author interview with Duncan Guthrie.
36. *Ibid.*
37. Wilson, *op. cit.*, pp. 122 & 162.

38. Author interview with George Dick, Sussex, 27 July 1992.
39. *Ibid.*
40. Cited in Wilson, *op. cit.*, p. 165.
41. Author interview with George Dick.
42. Hilary Koprowski, 'Vaccination with Modified Active Viruses', in *Poliomyelitis: Papers and Discussions Presented at the Fourth International Poliomyelitis Conference*, p. 122.
43. John R. Paul, *A History of Polio myelitis* (Yale University Press, New Haven & London 1971), p. 452.
44. Wilson, *op. cit.*, pp. 137–8.
45. *Ibid.*, p. 190.
46. Valentin D. Soloviev, 'Reports of Official Delegates – Union of Soviet Socialist Republic', in *Poliomyelitis: Papers and Discussion Presented at the Fourth International Poliomyelitis Conference*, p. 60.
47. Richard Carter, *Breakthrough: The Saga of Jonas Salk* (Trident, Press, New York 1966), p. 359.
48. Wilson, *op. cit.*, p. 201.
49. Carter, *op. cit.*, pp. 359–60.
50. Wilson, *op. cit.*, p. 205.
51. *Daily Mail*, 8 April 1959.
52. *Daily Telegraph*, 20 April 1959.
53. *Ibid.*, 28 April 1959.
54. Author interview with Rosalind Murray, Hull, 11 March 1992.
55. P.J. Fisher, *The Polio Story* (Heinemann, London 1967), p. 115.
56. *Ibid.*, pp. 115–16.
57. Wilson, *op. cit.*, pp. 226–7.
58. Professor George Dick to the author, 7 June 1993.
59. Carter, *op. cit.*, pp. 363–4.
60. Wilson, *op. cit.*, p. 210.
61. Carter, *op. cit.*, pp. 366–7.
62. Wilson, *op. cit.*, p. 217.
63. Carter, *op. cit.*, p. 364.
64. Wilson, *op. cit.*, p. 216.
65. Sir Graham S. Wilson, *The Hazards of Immunization* (Athlone Press, London 1967), p. 50.
66. H.V. Wyatt, 'Did Thalidomide Promote Poliomyelitis Following Oral Poliovirus Vaccination in West Berlin in 1960?' *Perspectives in Biology and Medicine*, 27, 1, Autumn 1983, pp. 93–106.
67. A. Bradford Hill, 'Inoculation Procedures as a Provoking Factor in Poliomyelitis', in *Poliomyelitis: Papers and Discussions Presented at the Second International Poliomyelitis Conference* (J.B. Lippincott, Philadelphia 1952), p. 330.
68. John Rowan Wilson, *op. cit.*, p. 218.
69. G. Courtois, *et al.*, 'Preliminary Report on Mass Vaccination of Man with Live Attenuated Poliomyelitis Virus in the Belgian Congo and Ruanda-Urundi', *British Medical Journal*, 26 July 1958, p. 189.
70. John Rowan Wilson, *op. cit.*, p. 181.
71. Tom Curtis, 'The Origin of AIDS', *Rolling Stone*, 19 March 1992.
72. Phyllida Brown, 'Polio vaccine "did not cause AIDS epidemic"', *New Scientist*, 31 Oct. 1992.
73. Hilary Koprowski, 'AIDS and the Polio Vaccine', letter in *Science*, Vol. 257, 21 Aug. 1992.
74. Cited in John Rowan Wilson, *op. cit.*, p. 221.
75. Paul, *op. cit.*, p. 454.
76. Alexander D. Langmuir, 'Inactivated Virus Vaccines: Protective Efficacy', in *Poliomyelitis: Papers and Discussions Presented at the Fifth International Poliomyelitis Conference* (J.B. Lippincott, Philadelphia 1961), pp. 105–13.
77. *Ibid.*
78. Thomas Francis in panel discussion on 'Poliovirus Vaccines', in *Poliomyelitis: Papers and Discussions Presented at the Fifth International Poliomyelitis Conference*, p. 351.
79. Sir Macfarlane Burnet: *Changing Patterns: an atypical autobiography* (Heinemann, London 1968), p. 171.
80. Carter, *op. cit.*, p. 369.
81. Ruth and Edward Brecher, 'What's Delaying the New Polio Vaccine?', *Redbook*, April 1961.
82. *Ibid.*
83. *Ibid.*
84. Paul, *op. cit.*, p. 456.
85. John Rowan Wilson, *op. cit.*, p. 119.
86. *Ibid.*, p. 98.
87. Greer Williams: *Virus Hunters*

(Hutchinson, London 1960), p. 276.

88. Carter, *op. cit.*, p. 359.
89. David Bodian, 'Poliomyelitis: Pathogenesis, Policy, and Politics,' *Bulletin of the Johns Hopkins Hospital*, 1962.
90. Paul, *op. cit.*, p. 464.
91. *Ibid.*, p. 459.
92. Carter, *op. cit.*, p. 372.
93. Paul, *op. cit.*, pp. 465–6.
94. National Foundation press release, 26 Sept. 1962 (March of Dimes).
95. Bodian, *op. cit.*
96. *Ibid.*
97. Sir G.S. Wilson, *op. cit.*, p. 52.
98. Carter, *op. cit.*, p. 391.
99. Paul, *op. cit.*, p. 466.
100. Sir G.S. Wilson, *op. cit.*, p. 53.
101. Paul, *op. cit.*, p. 467.
102. Edward Edelson, 'Polio vaccine: now there's a choice,' *Daily News*, 20 Dec. 1976.
103. Jonas Salk, 'Polio: A Cure for the New Controversy', *New York Times*, 26 May 1974.
104. Daniel Jack Chasen, 'The Polio Paradox', *Science*, April 1986.
105. Basil O'Connor, 'A Time for Decision', National Foundation annual report 1957 (March of Dimes), p. 7.
106. John Rowan Wilson, *op. cit.*, p. 237.
107. *Ibid.*, p. 238.
108. Victor Cohn: *Four Billion Dimes* (Minneapolis 1955), pp. 128 & 132.
109. *Ibid.* (Introduction to 1984 reprint).
110. Peter Maas, 'Where Does Your Charity Dollar Go?' *Look*, 15 March 1960.
111. *Ibid.*
112. Roland H. Berg, 'How Healthy Are the Health Drives?' *Look*, Oct. 1960.
113. Cohn, *op. cit.*, p. 130.
114. 'Basil O'Connor: One Man's War Against Disease,' *Medical World News*, Vol. 5 No. 3, 31 Jan. 1964.
115. *Ibid.*
116. Alden Whitman, 'Basil O'Connor, Polio Crusader, Dies,' *New York Times*, 10 March 1972.

Chapter 8

1. R.L. Bennett, 'Warm Springs Looks to the Future', *Contact*, Vol. 1 No. 1, Jan. 1964.
2. Robert L. Bennett to Joseph F. Guyon, 2 March 1966 (Georgia Warm Springs Foundation).
3. Author interview with Emelie Simha, Warm Springs, 25 May 1992.
4. Author interview with Viva Erickson, Warm Springs, 28 May 1992.
5. Elaine Strauss, *In My Heart I'm Still Dancing* (New Rochelle, New York 1979), p. 91.
6. Author interview with Diane Rice Smith, Carmi, Illinois, 31 May 1992.
7. Dr Eugene H. Dibble, Jr, Medical Director, to Dr L.H. Foster, President, Tuskegee Institute, 2 Aug. 1957 (Tuskegee University).
8. Author interview with Emelie Simha.
9. *Ibid.*
10. Author interview with Lawrence Becker, Williamsburg, Virginia, 16 May 1992 (see Part II, pp. 289–94).
11. Turnley Walker, *Journey Together* (New York 1951), pp. 34–5.
12. Author interview with Viva Erickson.
13. Author interview with Hugh Gallagher, Washington DC, 14 May 1992.
14. Lorenzo Wilson Milam, *The Cripple Liberation Front Marching Band Blues* (Mho & Mho Works, San Diego, California 1984), p. 85.
15. *Ibid.*, pp. 53–4 & 83.
16. Author interview with Hugh Gallagher.
17. Author interview with Viva Erickson.
18. Author interview with Richard Owen, Sister Kenny Institute, Minneapolis, 8 June 1992.
19. Kenneth S. Landauer, 'A National Program of Respiratory and Rehabilitation Centers', in *Poliomyelitis: Papers and Discussions Presented at the Fourth International Poliomyelitis Conference* (J.B. Lippincott, Philadelphia 1958), p. 632.
20. Gini Laurie, 'Twenty Years in the

Gazette House,' *Rehabilitation Gazette*, Vol. XXI, 1978.

21. *Ibid.*
22. *Ibid.*
23. Mrs Joseph S. Laurie to *Toomey j Gazette* readers, Jan. 1966 (Gazette International Networking Institute, St Louis).
24. *Rancho Los Amigos Medical Center Centennial: 1888–1988* (Los Angeles 1993), p. 255.
25. Miss Ruth Locher, 'Vocational Aspects in the Rehabilitation of the Poliomyelitis Respirator Patient', in *Poliomyelitis: Papers and Discussion Presented at the Fourth International Poliomyelitis Conference*, p. 652.
26. Discussion in *Ibid.*, p. 647.
27. Landauer, *op. cit.*, p. 635.
28. *Toomey j Gazette*, Vol. II No. 3–4, Fall-Winter 1959.
29. *Ibid.*
30. Saad Z, Nagi *et al.*: *Report on a Survey of Respiratory and Severe Post-Polios* (Ohio Rehabilitation Center, Ohio State University, Columbus, Ohio, May 1962).
31. *Ibid.*
32. *Toomey j Gazette*, Vol. VII No. 1, Spring 1964.
33. Donna McGwinn, 'A Respired Life' (typescript at GINI, St Louis).
34. *Rancho Los Amigos Medical Center Centennial*, pp. 247 & 228.
35. Virginia Wilson Laurie, '*Toomey j Gazette*: Horizontally and Vertically Yours,' *Randolph-Macon Woman's College Alumnae Bulletin*, Summer 1963.
36. Peter Marshall, *Two Lives* (Hutchinson, London 1962), Preface & pp. 116 & 169.
37. Peter Marshall, *The Raging Moon* (Hutchinson, London 1964), p. 126.
38. *Ibid.*
39. Peter Marshall to Gini Laurie, 31 Dec. 1964 (GINI).
40. *Ibid.*
41. Michael Lee, *The Residue of Poliomyelitis* (Office of Health Economics, London 1965).
42. John Vaizey, *Scenes from Institutional Life and other writings* (Weidenfeld & Nicolson, London, 1986 – first published 1959), p. 63.
43. Lee, *op. cit.*, pp. 17–21.
44. Annual Report of the British Polio Fellowship (London 1965), p. 6.
45. Paul Bates and John Pellow, *Horizontal Man* (Longmans, London, 1964), pp. 119 & 121.
46. *Ibid.*, pp. 139–40.
47. Author interview with Geoffrey Spencer, London, 24 Aug. 1992.
48. *Ibid.*
49. Frank Kelly, *Private Kelly* (Evans Bros. London, 1954), p. 35.
50. *Ibid.*, *passim.*
51. *Daily Express*, 11 Dec. 1953.
52. Kelly, *op. cit.*, p. 223.
53. Author interview with Geoffrey Spencer.
54. *Ibid.*
55. *Ibid.*
56. *Ibid.*, Judy Raymond & Gini Laurie, 'The Polio Conference: A Blueprint of Creative Cooperation for All Who Are Disabled,' *Rehabilitation Gazette*, Vol. 24, 1981.
57. Gail Dibernado, 'Dear, Dear Editor', *St Louis Weekly*, 10 May 1985; Gini Laurie, 'Twenty Years in the Gazette House'.
58. Larry Schneider, 'Those Passing Years', *Rehabilitation Gazette*, Vol. XXII, 1979.
59. Gini Laurie, 'Respiratory Rehabilitation and Post-Polio Aging Problems', *Rehabilitation Gazette*, Vol. XXIII, 1980.
60. Larry Schneider, 'Those Passing Years – II', *Rehabilitation Gazette*, Vol. XXIII, 1980.
61. Alice Mailhot, 'Age and the Old Polio: Do the Virtuous Fade First?' *Rehabilitation Gazette*, Vol. XXIII, 1980.
62. Walker, *op. cit.*, pp. 102–7.
63. *Ibid.*
64. Daniel J. Wilson, 'Covenants of Work and Grace: Themes of Recovery and Redemption in Polio Narratives,' Talk given at the American Association for the History of Medicine convention, May 1991, published in *Medicine and*

Literature, Vol. 13, No. 1 (Spring 1994), pp. 22–41.

65. Fred Davis, *Passage Through Crisis: Polio Victims and Their Families* (Bobbs-Merrill, Indianapolis, 1963), pp. 71–2.
66. Minutes of Georgia Warm Springs Foundation staff medical meeting, 27 Jan. 1954 (GWSF).
67. R.L. Bennett & G.C. Knowlton, 'Overwork Weakness in Partially Denervated Skeletal Muscle', *Clinical Orthopaedics*, No. 12, 1958, p. 22.
68. *Ibid.*, p. 23.
69. *Contact*, Vol. II No. 2, May 1965.
70. Robert L. Bennett, 'Unusual Loss of Muscle Strength', *Contact*, Vol. IV No. 2, April 1967.
71. Albert D. Anderson *et al.*, 'Loss of Ambulatory Ability in Patients with Old Anterior Poliomyelitis', *Lancet*, 18 Nov. 1972, p. 1061.
72. Dibernado, *op. cit.*
73. Author interview with Lauro Halstead, Washington DC, 13 May 1992 (see Part II, pp. 285–8).
74. *Ibid.*
75. Author interview with Richard Owen.
76. British Polio Fellowship *Bulletin*, Nov. 1989.
77. David O. Wiechers & Susan L. Hubbell, 'Late Changes in the Motor Unit after Acute Poliomyelitis,' *Muscle & Nerve*, 4, pp. 524–8, Nov./Dec. 1981.
78. Lauro S. Halstead, 'Late complications of poliomyelitis', in Joseph Goodgold (ed.): *Rehabilitation Medicine* (C.V. Mosby, St Louis, 1988), pp. 330–31.
79. Author interview with Geoffrey Spencer.
80. *Ibid.*
81. *Rehabilitation Gazette*, Vol. 25, 1982.
82. Gini Laurie, 'The Role of Networking and Support Groups', in Theodore L. Munsat (ed.): *Post-Polio Syndrome* (Butterworth-Heinemann, 1991), p. 109.
83. *Ibid.*, p. 112.
84. *Ibid.*, p. 109.
85. *Ibid.*, p. 112.
86. Author interview with Todd Keepfer, Cherokee, North Carolina, 23 May 1992.
87. Author interview with Stanley Lipshultz, Washington DC, 12 May 1992.
88. Author interview with Adele Gorelick, Washington DC, 15 May 1992.
89. Davis, *op. cit.*, pp. 22 & 113.
90. Author interview with Jennine Speier, Sister Kenny Institute, Minneapolis, 8 June 1992.
91. Author interview with Adele Gorelick.
92. *Ibid.*
93. Jessica Scheer & Mark L. Luborsky, 'The Cultural Context of Polio Biographies', *Orthopedics*, Vol. 14 No. 11, Nov. 1991, p. 1180.
94. Author interview with Jack Genskow, Springfield, Illinois, 5 June 1992.
95. Halstead, *op. cit.*, p. 335.
96. Richard L. Bruno & Nancy M. Frick, 'Stress and "Type A" Behavior as Precipitants of Post-Polio Sequelae: The Felician/Columbia Survey', in Lauro S. Halstead & David O. Wiechers (eds), *Research and Clinical Aspects of the Late Effects of Poliomyelitis* (March of Dimes Birth Defects Foundation, White Plains, New York 1987), pp. 151–2.
97. Nancy M. Frick, 'Post-Polio Sequelae and the Psychology of Second Disability', *Orthopedics*, Vol. 8, 1985, p. 852.
98. Alice Munro, 'Open Secrets', *New Yorker*, 8 Feb. 1993.
99. *Ibid.*
100. See Ch. 7, and Bentz Plagemann, *My Place to Stand* (Farrar, Straus, New York 1949), p. 174.
101. Author interview with Leonard Kriegel, New York, 4 May 1992 (see Part II, pp. 279–83).
102. Private information.
103. Frick, *op. cit.*, p. 852.
104. Davis, *op. cit.*, p. 39.
105. *Ibid.*, *passim.*
106. Frederick M. Maynard & Sunny

Roller, 'Recognizing Typical Coping Styles of Polio Survivors Can Improve Re-Rehabilitation: A Commentary', *American Journal of Physical Medicine and Rehabilitation*, Vol. 70 No. 2, April 1991, p. 70.
107. *Ibid.*
108. Frick, *op. cit.*, p. 852.
109. Sandy Hughes Grinnell, 'A Post-Polio "Normal's" Reconciliation with the Ghost of Polio Past', *Polio Network News*, Vol. 5 No. 4, Fall 1989, p. 15.
110. *Ibid.*, Part II, *Polio Network News*, Vol. 6 No. 1, Winter 1990, p. 22.
111. Author interview with Todd Keepfer.
112. Maynard and Roller, *op. cit.*, p. 71.
113. *Ibid.*
114. Author interview with Richard Owen.
115. *Ibid.*
116. Scheer & Luborsky, *op. cit.*, p. 1180.
117. *Ibid.*
118. Author interview with Joan Headley, GINI, St Louis, 4 June 1992 (see Part II, pp. 294–7).
119. Max Starkloff, President of Paraquad, quoted in *Paraquad News*, Aug. 1989.
120. Author interview with Joan Headley.
121. US figures: National Center for Health statistics, National Health Interview Survey 1987, in *Polio Network News*, Vol. 7 No. 1, Winter 1991. India: Sunny Roller, 'Polio Eradication & Rehabilitation in India', *Polio Network News*, Vol. 6 No. 2, Spring 1990.
122. Dr G.T. Spencer to the author, 9 Sept. 1993.
123. Rachel Joce *et al.*, 'Paralytic Poliomyelitis in England and Wales, 1985–91', *British Medical Journal*, Vol. 305, No. 6845, 11 July 1992, pp. 79–82.
124. Albert Sabin, 'Conquest of Polio: Unfinished Business,' from Proceedings of GINI's Third International Poliomyelitis and Independent Living Conference, St Louis, 10–12 May 1985.
125. Peter F. Wright *et al.*, 'Strategies for the Global Eradication of Poliomyelitis by the Year 2000', *The New England Journal of Medicine*, 19 Dec. 1991, p. 1776.
126. Chris Anne Raymond, 'Worldwide assault on poliomyelitis gathering support, garnering results', *Journal of the American Medical Association*, Vol. 255 No. 12, 28 March 1986, pp. 1541–6.
127. *Ibid.*
128. *Ibid.*
129. Wright *et al.*, *op. cit.*
130. *Ibid.*
131. *Ibid.*
132. Sabin, *op. cit.*
133. 'PolioPlus: Poliomyelitis Eradication by Year 2005,' *Journal of the American Medical Association*, Vol. 269 No. 1, 6 Jan. 1993, p. 15.
134. *Ibid.*
135. Chris Anne Raymond, 'Polio rare in United States but experts counsel caution', *Journal of the American Medical Association*, Vol. 255 No. 12, 28 March 1986, p. 1547.
136. Wright *et al.*, *op. cit.*

A Civil Wound

1. Tony Gould, 'Night', *Granta*, 25 Jan. 1964, pp. 6–9 (this and other articles by the author have been subject to minor editing).
2. Tony Gould, 'The Visit', *New Society*, 20 Oct. 1966. p. 617.
3. Tony Gould, 'Down to Earth,' *New Society*, 29 March 1984, pp. 492–3.
4. Christopher Isherwood, *Lions and Shadows* (Methuen 1953, originally published by the Hogarth Press 1938), pp. 75–6.
5. For subsequent events in the author's life, see Tony Gould, *Death in Chile* (Picador 1992), pp. 3–89.

Monkey Business

1. Richard Carter, *Breakthrough: The*

Saga of Jonas Salk (Trident Press, New York 1966), p. 73.

2. John R. Paul: *A History of Poliomyelitis* (Yale University Press, New Haven & London 1971), p. 374.
3. Victor Cohn, *Four Billion Dimes* (Minneapolis 1955), p. 82; Carter: *op. cit.*, p. 73.
4. Paul, *op. cit.*, p. 387.
5. W. Müller, 'Spontane poliomyelitis beim Schimpansen', 1935, cited in Howard A. Howe and David Bodian, *Neural Mechanisms in Poliomyelitis* (The Commonwealth Fund, New York 1942), p. 101.
6. Jane van Lawick Goodall, *In the Shadow of Man* (Collins, London 1971), p. 198.
7. *Ibid.*, pp. 198–204, for all subsequent quotations here.

Select Bibliography

Anderson, Albert D., *et al.*, 'Loss of Ambulatory Ability in Patients with Old Anterior Poliomyelitis', *Lancet*, 18 Nov. 1972

Asbell, Bernard, *When FDR Died* (New York 1961)

Asbell, Bernard, *The FDR Memoirs* (New York 1973)

Bates, Paul, & Pellow, John, *Horizontal Man* (London 1964)

Beisser, Arnold R., *Flying without Wings: Personal Reflections on Being Disabled* (New York 1989)

Bell, Harriet, 'Polio Survivors: Their Quality of Life' (unpublished dissertation, Columbia Pacific University 1984)

Benison, Saul, *Tom Rivers: Reflections on a Life in Medicine and Science* (Cambridge, Mass., & London 1967)

Benison, Saul, 'The History of Polio Research in the United States: Appraisal and Lessons', in Holton, Gerald (ed.), *The Twentieth Century Sciences: Studies in the Biography of Ideas* (New York 1972)

Benison, Saul, 'Poliomyelitis and the Rockefeller Institute: Social Effects and Institutional Response', *Journal of the History of Medicine and Allied Sciences*, Vol. xxix, 1974

Bennett, Robert L., 'Recent Developments in the Treatment of Poliomyelitis', *Southern Medical Journal*, Feb. 1943

Bennett, R.L., & Knowlton, G.C., 'Overwork Weakness in Partially Denervated Skeletal Muscle', *Clinical Orthopedics*, No. 12, 1958

Bennett, R.L., 'Warm Springs Looks to the Future', *Contact*, Jan. 1964

Bennett, R.L., 'Unusual Loss of Muscle Strength', *Contact*, April 1967

Berg, Roland, *Polio and Its Problems* (Philadelphia 1948)

Berg, Roland H., 'How Healthy Are the Health Drives?' *Look*, Oct. 1960

Bigland, Eileen, *The True Book about Sister Kenny* (London 1956)

Bingham, Robert, 'The Kenny Treatment for Infantile Paralysis,' *Journal of Bone and Joint Surgery*, 3 July 1943

Bodian, David, 'Poliomyelitis: Pathogenesis, Policy, and Politics', *Bulletin of the Johns Hopkins Hospital*, 1962

Boettiger, John R., *A Love in Shadow* (New York 1978)

Botts, Fred, 'Warm Springs – the Roosevelt Era and Beyond', *Wheelchair Review*, 12 Aug. 1951

Bowen, Catherine Drinker, *Family Portrait* (Boston 1970)

Brandt, Allan M., 'Racism and Research: The Case of the Tuskegee Syphilis Study', Hastings Center Report, Dec. 1978

Brandt, Allan M., 'Polio, Politics, Publicity, and Duplicity: ethical aspects in the development of the Salk vaccine', *International Journal of Health Services*, Vol. 8, No. 2, 1978

Brecher, Ruth and Edward, 'What's Delaying the New Polio Vaccine?' *Redbook*, April 1961

Brown, Phyllida, 'Polio vaccine "did not cause AIDS epidemic" ', *New Scientist*, 31 Oct. 1992

Burnet, Sir Macfarlane, *Changing Patterns: an atypical autobiography* (London 1968)

Burns, Al V., 'The Scourge of 1916 . . . America's First and Worst Epidemic', *The American Legion Magazine*, Sept. 1966

Carter, Richard, *The Gentle Legions* (New York 1961)

Carter, Richard, *Breakthrough: The Saga of Jonas Salk* (New York 1966)

Caverly, Charles S., 'Notes of an Epidemic of Acute Anterior Poliomyelitis', *Journal of the American Medical Association*, 1896, xxvi, 1

Chappell, Edith P., & Hume, John F., 'A Black Oasis: Tuskegee's Fight against Infantile Paralysis, 1941–1975' (unpublished thesis at Tuskegee University)

Chasen, Daniel Jack, 'The Polio Paradox', *Science*, April 1986

Cohn, Victor, *Four Billion Dimes* (Minneapolis 1955)

Cohn, Victor, *Sister Kenny: The Woman Who Challenged the Doctors* (Minneapolis 1975)

Colbert, Judy, 'Polio Victims 30 Years Later', *American Way*, 19 March 1985

Cole, Wallace H., & Knapp, Miland E., 'The Kenny Treatment of Infantile Paralysis: A Preliminary Report', *Journal of the American Medical Association*, 7 June 1941

Cole, Wallace H., Pohl, John F., & Knapp, Miland E., 'The Kenny Method of Treatment for Infantile Paralysis', National Foundation for Infantile Paralysis (New York 1942)

Cook, Blanche Wiesen, *Eleanor Roosevelt: Vol 1, 1884–1933* (New York 1992)

Courtois, G., *et al.*, 'Preliminary Report on Mass Vaccination of Man with Live Attenuated Poliomyelitis Virus in the Belgian Congo and Ruanda-Urundi', *British Medical Journal*, 26 July 1958

Curtis, Tom, 'The Origin of AIDS', *Rolling Stone*, 19 March 1992

Daniels, Jonathan, *Washington Quadrille* (New York 1968)

Davis, Fred, *Passage Through Crisis: Polio Victims and Their Families* (Indianapolis 1963)

De Kruif, Paul, *Microbe Hunters* (New York 1926)

De Kruif, Paul, *The Sweeping Wind: A Memoir* (London 1962)

Deutsch, Albert, 'Sister Kenny Again Engaged in Battle with Medicos', *PM*, 7 Feb. 1944

Dibernado, Gail, 'Dear, Dear Editor', *St Louis Weekly*, 10 May 1985

Draper, George, *Acute Poliomyelitis* (New York 1917)

Draper, George, *Infantile Paralysis* (New York 1935)

Drinker, Philip A., & McKhann, Charles F., 'The Iron Lung: First Practical Means of Respiratory Support', *Journal of the American Medical Association*, 21 March 1986

Edelson, Edward, 'Polio vaccine: now there's a choice', *Daily News*, 20 Dec. 1976

Emerson, Haven, *A Monograph on the Epidemic of Poliomyelitis (Infantile Paralysis) in New York City in 1916. Based on the Official Reports of the Bureaus of the Department of Health* (New York 1917)

Fairbank, H.A.T., *et al.*, 'Report on the Kenny Method of Treatment', *British Medical Journal*, 20 Oct. 1938

Farrell, J.G., *The Lung* (London 1965)

Fisher, P.J., *The Polio Story* (London 1967)

Flexner, Abraham, *Medical Education in the United States and Canada* (New York 1910)

Flexner, James Thomas, *An American Saga: The Story of Helen Thomas and Simon* Flexner (Boston & Toronto 1984)

Frick, Nancy M., 'Post-Polio Sequelae and the Psychology of Second Disability', *Orthopedics*, Vol. 8, 1985

Friedel, Frank, *Franklin D. Roosevelt: Vol 2, The Ordeal* (Boston 1954)

Gale, A.H., *Epidemic Diseases* (London 1959)

Gallagher, Hugh Gregory, *FDR's Splendid Deception* (New York 1985)

Gauvain, Sir Henry, 'Anterior Poliomyelitis in England', *International Bulletin for Economics, Medical Research & Public Hygiene*, Vol. A 40, Infantile Paralysis, 1939/40

Ghormley, Ralph K., *et al.*, 'Evaluation of the Kenny Treatment of Infantile Paralysis', *Journal of the American Medical Association*, 17 June 1944

Goldberg, Richard Thayer, *The Making of Franklin D. Roosevelt: Triumph over Disability* (Cambridge, Mass. 1981)

Goodall, Jane van Lawick, *In the Shadow of Man* (London 1971)

Goodgold, Joseph (ed.), *Rehabilitation Medicine* (St Louis 1988)

Gould, Jean, *A Good Fight: The Story of FDR's Conquest of Polio* (New York 1960)

Gould, Tony, 'Night', *Granta*, 25 Jan. 1964

Gould, Tony, 'The Visit', *New Society*, 20 Oct. 1966

Gould, Tony, 'My Army Life', *New Society*, 25 Nov. 1976

Gould, Tony, 'Down to Earth', *New Society*, 29 March 1984

Grinnell, Sandy Hughes, 'A Post-Polio "Normal's" Reconciliation with the Ghost of Polio Past', *Polio Network News*, Fall 1989

Halstead, Lauro S., & Wiechers, David O. (eds), *Late Effects of Poliomyelitis* (Miami 1985)

Halstead, Lauro S., & Wiechers, David O. (eds), *Research and Clinical Aspects of the Late Effects of Poliomyelitis* (New York 1987)

Hansson, K.G., 'After-Treatment of Poliomyelitis', *Journal of the American Medical Association*, 1 July 1939

Harmon, Paul H., 'The Geographical and Racial Incidence of Poliomyelitis', *International Bulletin for Economics, Medical Research & Public Hygiene*, Vol. A 40, Infantile Paralysis, 1939/40

Harry, Norman S., 'The Recovery Period in Anterior Poliomyelitis', *British Medical Journal*, 22 Jan. 1938

Hawkins, Leonard C., & Lomask, Milton, *The Man in the Iron Lung: The Story of Frederick B. Snite* (Surrey 1957)

Hobson, Harold, *Indirect Journey: An Autobiography* (London 1978)
Horowitz, Joy, 'Polio's Painful Legacy', *New York Times Magazine*, 7 July 1985
Howe, Howard A., & Bodian, David, *Neural Mechanisms in Poliomyelitis* (New York 1942)
Hubbard, Leroy W., 'The Treatment of Infantile Paralysis: By Exercises in a Natural Warm Water', *The Crippled Child*, July/Aug. 1927
Hulett, J.E. Jr, 'The Kenny Healing Cult: Preliminary Analysis of Leadership and Patterns of Interaction', *American Sociological Review*, June 1945
Hutchin, Elisabeth F., 'Historical Summary', in *Poliomyelitis: A survey made possible by a grant from the International Committee for the Study of Infantile Paralysis organized by Jeremiah Milbank* (Baltimore 1932)
Infantile Paralysis in Vermont 1894–1922: A Memorial to Charles S. Caverly, MD (Vermont 1924)
James, Burnet, *Living Forwards* (London 1961)
Joce, Rachel, *et al.*, 'Paralytic Poliomyelitis in England and Wales, 1985–91', *British Medical Journal*, 11 July 1992
Kamenetz, Herman L., *The Wheelchair Book: Mobility for the Disabled* (Illinois 1969)
Kelly, Frank, *Private Kelly* (London 1954)
Kendall, Henry O. & Florence P., 'Care during the Recovery Period of Paralytic Poliomyelitis', *Public Health Bulletin No. 242* (Washington DC 1938)
Kenny, Sister Elizabeth (in collaboration with Martha Ostenso), *And They Shall Walk* (London 1951)
Kenny, Sister Elizabeth, *My Battle and Victory: History of the Discovery of Poliomyelitis as a Systemic Disease* (London 1955)
Klein, Aaron E., *Trial by Fury: The Polio Vaccine Controversy* (New York 1972)
Kline, Marvin L., 'The Most Unforgettable Character I've Met', *Reader's Digest*, Aug. 1959
Knapp, Miland E., 'The Kenny Treatment for Infantile Paralysis', *Archives of Physical Therapy*, Nov. 1942
Knapp, Miland E., 'The Contribution of Sister Elizabeth Kenny to the Treatment of Poliomyelitis', *Archives of Physical Medicine & Rehabilitation*, Aug. 1955
Knapp, Miland E., Sher, Lewis, & Smith, Theodore S., 'Results of Kenny Treatment of Acute Poliomyelitis', *Journal of the American Medical Association*, 10 Jan. 1953
Koprowski, Hilary, 'AIDS and the Polio Vaccine', letter in *Science*, 21 Aug. 1992
Kriegel, Leonard, *The Long Walk Home* (New York 1964)
Kriegel, Leonard, 'Uncle Tom and Tiny Tim: Some Reflections on the Cripple as Negro', *The American Scholar*, Summer 1969
Kriegel, Leonard, *Falling into life* (San Francisco 1991)
Lash, Joseph P., *Eleanor & Franklin* (London 1972)
Lassen, H.C.A. (ed.), *Management of Life-Threatening Poliomyelitis: Copenhagen 1952–1956* (Edinburgh & London 1956)
Laurie, Gini, 'Twenty Years in the Gazette House', *Rehabilitation Gazette*, Vol. xxi, 1978
Laurie, Gini, 'Respiratory Rehabilitation and Post-Polio Aging Problems', *Rehabilitation Gazette*, Vol. xxiii, 1980
Lavinder, C.H., Freeman, A.W., & Frost, W.H.: *Epidemiological Studies of Poliomyelitis in New York City and the North-Eastern United States during the Year 1916* (Washington 1918)

Le Comte, Edward, *The Long Road Back: The Story of My Encounter with Polio* (Boston 1957)
Lee, Michael, *The Residue of Poliomyelitis* (London 1965)
Levine, Herbert Jerome, *I Knew Sister Kenny: The Story of a Great Lady and Little People* (Boston 1954)
Lippman, Theo Jr, *The Squire of Warm Springs: FDR in Georgia, 1924–1945* (Chicago 1977)
Lockhart, J.G., *Memoirs of the Life of Sir Walter Scott, Bart.* (London 1900)
Lovett, R.W., 'The Treatment of Infantile Paralysis', *Journal of the American Medical Association*, 26 June 1915
Lovett, Robert W., *The Treatment of Infantile Paralysis* (Philadelphia 1916)
Lovett, Robert W., 'The Treatment of Infantile Paralysis (with special reference to the earlier stages)', *Health News* (New York), Aug. 1916
Low, Merritt B., 'Poliomyelitis with Residual Paralysis', in *When Doctors Are Patients* (New York 1952)
Lunoe, Adam M., 'Notes on Georgia Warm Springs Foundation', National Foundation for Infantile Paralysis (New York undated)
McIntire, Vice-Admiral Ross T., *Twelve Years with Roosevelt* (London 1948)
MacMahon, Edward B., and Curry, Leonard, *Medical Cover-Ups in the White House* (Washington DC 1987)
McMurry, Linda O., *George Washington Carver: Scientist and Symbol* (New York 1981)
Maas, Peter, 'Where Does Your Charity Dollar Go?' *Look*, 15 March 1960
Mailhot, Alice, 'Age and the Old Polio: Do the Virtuous Fade First?' *Rehabilitation Gazette*, Vol. xxiii, 1980
Marshall, Alan, *I Can Jump Puddles* (Australia 1955)
Marshall, Peter, *Two Lives* (London 1962)
Marshall, Peter, *The Raging Moon* (London 1964)
Martin, Wayne, *Medical Heroes and Heretics* (Connecticut undated)
Maynard, Frederick M., & Roller, Sunny, 'Recognizing Typical Coping Styles of Polio Survivors Can Improve Re-Rehabilitation: A Commentary', *American Journal of Physical Medicine and Rehabilitation*, April 1991
Mee, Charles L. Jr, *A Visit to Haldeman and Other States of Mind* (New York 1977)
Mee, Charles L. Jr, 'The Summer Before Salk', *Esquire*, Dec. 1983
Milam, Lorenzo Wilson, *The Cripple Liberation Front Marching Band Blues* (San Diego, Cal. 1984)
Miller, Lois Mattox, 'Sister Kenny Wins Her Fight', *Reader's Digest*, Oct. 1942
Miller, Nathan, *FDR: An Intimate History* (New York 1983)
Moley, Raymond, *After Seven Years* (New York 1939)
Morgan, Ted, *FDR: A Biography* (New York 1985)
Munro, Alice, 'Open Secrets', *New Yorker*, 8 Feb. 1993
Munsat, Theodore L. (ed.), *Post-Polio Syndrome* (London 1991)
Murphy, Robert F., *The Body Silent* (New York 1987)
Nagi, Saad Z., et al., *Report on a Survey of Respiratory and Severe Post-Polios*, Ohio Rehabilitation Center (Ohio, May 1962)
O'Connor, Basil, 'The Story of the Kenny Method', *Archives of Physical Therapy*, April 1944
O'Connor, Basil, 'A Time for Decision', National Foundation annual report 1957 (New York 1957)

Opie, June, *Over My Dead Body* (London 1957)
Oursler, Fulton, *Behold This Dreamer!* (Boston 1964)
Patterson, James T., *The Dread Disease: Cancer and Modern American Culture* (Cambridge, Mass., & London 1987)
Paul, John R., *A History of Poliomyelitis* (New Haven & London 1971)
Peabody, F.W., Draper, G., & Dochez, A.R., *A Clinical Study of Acute Poliomyelitis* (New York 1912)
Peabody, F.W., 'A report of the Harvard Infantile Paralysis Commission on the diagnosis and treatment of acute cases of the disease during the year 1916', *Boston Medical and Surgical Journal*, 176, 1917
Peabody, Francis Weld, *Doctor and Patient: Papers on the Relationship of the Physician to Men and Institutions* (New York 1930)
Perkins, Frances, *The Roosevelt I Knew* (London 1947)
Plagemann, Bentz, *My Place to Stand* (New York 1949)
Pohl, John F. (in collaboration with Sister Elizabeth Kenny), *The Kenny Concept of Infantile Paralysis and Its Treatment* (Minneapolis 1949)
Poliomyelitis: Papers and Discussions Presented at the First International Poliomyelitis Conference (Philadelphia 1949)
Poliomyelitis: Papers and Discussions Presented at the Second International Poliomyelitis Conference (Philadelphia 1952)
Poliomyelitis: Papers and Discussions Presented at the Third International Poliomyelitis Conference (Philadelphia 1955)
Poliomyelitis: Papers and Discussions Presented at the Fourth International Poliomyelitis Conference (Philadelphia 1958)
Poliomyelitis: Papers and Discussions Presented at the Fifth International Poliomyelitis Conference (Philadelphia 1961)
Raymond, Chris Anne, 'Worldwide assault on poliomyelitis gathering support, garnering results' and 'Polio rare in the United States but experts counsel caution', *Journal of the American Medical Association*, 28 March 1986
Raymond, Judy, & Laurie, Gini, 'The Polio Conference: A Blueprint of Creative Cooperation for All Who Are Disabled', *Rehabilitation Gazette*, Vol. xxiv, 1981
Risse, Guenter B., 'Epidemics and History: Ecological Perspectives and Social Responses', in Fee, Elizabeth, & Fox, Daniel M. (eds), *AIDS: The Burdens of History* (Berkeley, California 1988)
Risse, Guenter B., 'Revolt against Quarantine: Community Responses to the 1916 Polio Epidemic, Oyster Bay, New York', an expanded version of the Samuel X. Radbill Lecture given at the College of Physicians of Philadelphia on 17 Oct. 1990
Rogers, Naomi, 'Dirt, Flies, and Immigrants: Explaining the Epidemiology of Poliomyelitis, 1900–1916', *Journal of the History of Medicine and Allied Sciences*, Vol. 44, 1989
Rogers, Naomi, *Dirt and Disease: Polio Before FDR* (New Brunswick, NJ 1992)
Roller, Sunny, 'Polio Eradication & Rehabilitation in India', *Polio Network News*, Spring 1990
Roosevelt, Anna, 'My Life with FDR, Part 3: How Polio Helped Father', *Woman*, July 1949
Roosevelt, Eleanor, *This Is My Story* (New York 1937)
Roosevelt, Eleanor, *The Autobiography of Eleanor Roosevelt* (London 1962)

Roosevelt, Elliot, and Brough, James, *An Untold Story: The Roosevelts of Hyde Park* (New York 1974)
Roosevelt, James (with Bill Libby), *My Parents: A Differing View* (Chicago 1976)
Ruhrah, John, & Mayer, Erwin E., *Poliomyelitis in All Its Aspects* (Philadelphia & New York 1917)
Rusk, Howard A., *A World To Care For* (New York 1972)
Salk, Jonas, 'Polio: A Cure for the New Controversy', *New York Times*, 26 May 1974
Scheer, Jessica, & Luborsky, Mark L., 'The Cultural Context of Polio Biographies', *Orthopedics*, Nov. 1991
Schneider, Larry, 'Those Passing Years', *Rehabilitation Gazette*. Vol, xxii, 1979
Schneider, Larry, 'Those Passing Years – II', *Rehabilitation Gazette*, Vol. xxiii, 1980
Sills, David L., *The Volunteers: Means and Ends in a National Organization* (Illinois 1957)
Sissons, Michael, & French, Philip (eds), *Age of Austerity* (London 1963; Oxford 1968)
Smith, Jane S., *Patenting the Sun: Polio and the Salk Vaccine* (New York 1990)
Stern, Bernhard J., *Social Factors in Medical Progress* (New York 1927)
Stevenson, Jessie L., *Care of Poliomyelitis* (New York 1940)
Strauss, Elaine, *In My Heart I'm Still Dancing* (New York 1979)
Vaizey, John, *Scenes from Institutional Life and other Writings* (London 1959; 1986)
Wackers, G.L., 'Theaters of War, Truth and Competence: the rhetorics of innovative intermittent positive pressure ventilation during the 1952 polio epidemic in Copenhagen', unpublished draft 1991
Walker, Turnley, *Rise Up and Walk* (New York 1950)
Walker, Turnley, *Journey Together* (New York 1951)
Walker, Turnley, *Roosevelt and the Fight against Polio* (London 1954)
Ward, Geoffrey C., *Before the Trumpet: Young Franklin Roosevelt* (New York 1985)
Ward, Geoffrey C., *A First-Class Temperament: The Emergence of Franklin Roosevelt* (New York 1989)
Warner, Charles Dudley, *On Horseback* (Boston & New York 1888)
Watson, Frederick, *The Life of Sir Robert Jones* (London 1934)
Whitman, Alden, 'Basil O'Connor, Polio Crusader, Dies', *New York Times*, 10 March 1972
Wiechers, David O., & Hubbell, Susan L., 'Late Changes in the Motor Unit after Acute Poliomyelitis', *Muscle & Nerve*, Nov./Dec 1981.
Williams, Greer: *Virus Hunters* (London 1960)
Willis, Evan, 'Sister Elizabeth Kenny and the Evolution of the Occupational Division of Labour in Health Care', *Australian and New Zealand Journal of Sociology*, 3 Nov. 1979
Wilson, Daniel J., 'Covenants of Work and Grace: Themes of Recovery and Redemption in Polio Narratives', *Literature and Medicine* 13, no. 1 (Spring 1994)
Wilson, Sir Graham S., *The Hazards of Immunization* (London 1967)
Wilson, John Rowan, *Margin of Safety: The Story of Poliomyelitis Vaccine* (London 1963)
Worsley, Etta Blanchard, 'Warm Springs', *Georgia Review*, Summer 1949
Wright, Peter F., *et al.*, 'Strategies for the Global Eradication of Poliomyelitis by the Year 2000', *The New England Journal of Medicine*, 19 Dec. 1991

Wright, Wilhelmine G., 'Muscle Training in the Treatment of Infantile Paralysis', *Public Health Reports*, 23 Dec. 1927 (Washington DC 1928)
Wyatt, H.V., 'Did Thalidomide Promote Poliomyelitis Following Oral Poliovirus Vaccination in West Berlin in 1960?' *Perspectives in Biology and Medicine*, Autumn 1983
Zwar, Desmond, *The Dame: The Life and Times of Dame Jean Macnamara, medical pioneer* (Australia 1984)

Index

Action Research for the Crippled Child *see* National Fund
Adela Shaw Hospital, Kirkbymoorside (Yorkshire), 232, 238
ADL (activities of daily living), 240
adrenalin: used as treatment, 21
Ager, Dr Louis C., 21
AIDS, xi, 179–80
Alabama: 1936 polio epidemic, 69
Alleghany County, Pennsylvania: vaccinations in, 136
Alvarez, José, 76
American Cancer Society, 158
American Congress of Physical Therapy (1934), 95
American Cyanamid (company), 172, 178
American Epidemiological Society: Baltimore meeting (1938), 120
American Journal of Public Health, 68
American Magazine, 54
American Medical Association, 153, 183
American Orthopedic Association (AOA), 48, 110
American Public Health Association, 67–8
Amish people: 1979 polio outbreak, 226
anaesthetists: assist in respiration treatment, 163
Andrews, Eamonn: *This Is Your Life* (TV show), 201
Andrews, Julie, 166
animals: suspected as polio carriers, 8–9
Ann Arbor, Michigan: Salk test results announced from, 149–51; *see also* Poliomyelitis Vaccine Evaluation Center
Archives of Physical Therapy, 104
Argentina, 103, 195
Armstrong, Dr Charles, 69
Arthritis and Rheumatism Foundation, 187
Australia: Sister Kenny's work in, 87–9
Aycock, Dr Lloyd, 71, 120

bacteria: early investigation of, 12–13
Badham, Dr John, 10–11
Baltimore *see* Children's Hospital School
Barren Island, Brooklyn, New York, 26
Barrett, Dr (osteopath), 42
Bates, Paul, 201
Baylor College of Medicine, Houston, Texas, 210
Becker, Charlotte, 293–4
Becker, Lawrence: case history, 289–94
Bell, Dr Joseph, 139–40
Benison, Saul, 122, 138, 156–7
Bennett, Dr Robert, 100–101, 113, 189–90, 208–9
Berkeley, California, 199
Berlin: 1947 polio epidemic, 161; vaccination programme, 177–8
Bermingham, Dr E.J., 21

Berry, Milton, 95
bloodstream: poliovirus in, 130–31
Bodian, Dr David: on poliovirus types, 122; and poliovirus in bloodstream, 130–31; and use of monkeys, 131, 140; on Salk vaccine tests, 147; and Salk-Sabin vaccines conflict, 149, 183–4; and licensing of Salk vaccine, 151; George Dick works with, 170; attends 1983 post-polio conference, 212
Bradley, W.H.: quoted, 159; on polio in Britain, 161, 164
Brandt, Allan M.: quoted, 111
Brazil, 225
Brisbane, Australia, 88–92
Britain: polio made notifiable (1911), 16; polio incidence in, 88, 161, 174, 176, 224; temporises over Salk vaccine, 160, 164, 173; 1947 epidemic, 161; care and treatment of polio sufferers, 165–8; polio research in, 169; vaccination campaign, 173–6; surveys of polio survivors, 199–201
British Polio Fellowship (*formerly* Infantile Paralysis Fellowship), xiv, 165–9, 199, 201, 212, 250, 258
British Medical Journal, 16, 89, 127, 161
Brodie, Dr Maurice, 66–70, 125, 146
Brookens, Rev. J.S., 78–9
Brooklyn, New York, 4–5
Brooks, Donald, 243
Brown, Dr Robert D., 95
Brunhilde Type I poliovirus (Brunenders Type I), 160–61
Bruno, Dr Richard, 216
bulbar poliomyelitis, 92, 162, 236
Burnet, Dr (Sir) Frank Macfarlane, 65, 107, 122–3, 180–81
Bussey, Woodall, 189
Byoir, Carl, 60, 74

California: 1934 polio epidemic, 62
Calloway, Cason, 80
Canada: Salk vaccine distribution, 148; facilities for disabled, 271
cancer, 158
Cantor, Eddie, 74
Carey, Patricia, 165
Carpenter, Arthur, 54–5, 57, 60, 63, 74, 76–7, 93
Carshalton *see* Queen Mary's Hospital
Carter, Jimmy, 189
Carter, Richard, 149–50, 173, 176, 183, 185; *Breakthrough*, 65
Carver, George Washington, 81–3
Casement, Sir Roger, 8
Caverly, Dr Charles, 9, 11
Centers for Independent Living, 199
cerebral palsy, 316
Chailey Heritage Craft School, Sussex, 231, 243–6
Chenault, Dr John M., 83
Cheshire Foundation, 198
Cheshire, Group Captain Leonard (*later* Baron), 198
Chicago: 1981 symposium, 205–6, 209
children: vaccine trials on, 126, 132–3, 135–6, 140–41, 147–8; polio paralysis numbers reduced, 225–6
Children's Hospital, Boston, 124
Children's Hospital School, Baltimore, 93
Children's Hospital School, Los Angeles, 93
chimpanzees: susceptibility to polio, 322–4
China: polio in, xii, xv, 223
Chumakov, Mikhail, 173
Churchill, Sir Winston S., 115
Cincinnati, 182
Clarkson, Deryck, 201
class (social): and polio incidence, 119–20, 180–81, 196, 199–200
Cleveland, Ohio, 193, 195
Clinical Theology Association, 267
'clumping' (of vaccines), 156
Cobham (Surrey): 'Silverwood' (house), 166–7, 250–51, 267
Cochran, Lucile, 132
Cohn, Dr Edwin, 130
Cohn, Victor, 85–6, 186
Cole, Dr Rufus, 17
Cole, Dr Wallace, 98, 100
Cole, Mary Frances, 'Pidge', 48–9, 59
Colorado River War Relocation Project, Poston, Arizona, 112
coloured people *see* race (ethnic)
Congo: Koprowski experiments in, 179
Connaught Laboratories (Canada), 185
Consolidated Gas Company, New York, 90
Contact (magazine), 189, 209

Copenhagen: 1952 polio epidemic, xiv, 162–4, 195; *see also* International Poliomyelitis Conferences
Cox, Herald, 171–3, 176, 177–9, 182
Crossman, Richard H.S., 205
Culver, Bettyann, 129
Curtis, Egbert, 52
Cutter Laboratories, Berkeley, California: vaccine manufacturing errors, 153–4, 156–7, 160, 183

Dade County, Miami, Florida, 177–8, 182
Dail, Dr Clarence, 196
Daily Telegraph, 174
Dalrymple-Champney, Sir Weldon, 160
Daltry, Roger, 252
Dartington College of Arts, Devon, 312–16
Davis, Fred, 219–20
de Kruif, Paul, 63–7, 69, 71, 74–5, 120, 125, 146
Delano, Fred, 32
Denmark: manufactures Salk vaccine, 160, 164; *see also* Copenhagen
Des Moines, Iowa, 181
Deutsch, Albert, 104
Devonport (England): 1911 polio outbreak, 16
Dick, Professor George, 170–71, 176
disability and the disabled: attitudes to, 217–22
Division of Biologics Standard (USA), 182–3
Dochez, Dr Alphonse, 17–18
Doherty, Colonel Henry L., 60, 73
Doherty, Kathleen, 20
Draper, Dr George W.: quoted, 3; on 'dromedary form', 11; research, 17–18, 22, 25; on Roosevelt family tensions, 35; attends FDR, 35–6; correspondence with FDR, 72–3; theory of constitutional susceptibility, 70–71, 216; *Infantile Paralysis*, 70
Drinker, Philip: develops iron lung, 90–91
driving (motor cars), 241, 257–8, 272
D.T. Watson Home for Crippled Children, near Pittsburgh, 126, 132, 277
Dury, Ian, xiii, 229; case history, 230–31, 245–7, 250; career and personal life, 252–4

Early, Steve, 73–4, 77–8
East Germany, 177
education: for polios, 237–42, 246, 251–2, 284
Egypt: polio in, 119
Eiben, Dr Robert, 195
Eisenhower, Dwight D., 152, 172
electro-galvanic treatment, 247
Emerson, Dr Haven, 4–5, 7–8, 21, 23, 25–8, 91, 120
Emerson, Jack, 91
Enders, Dr John F.: character and style, 123; grows poliovirus, 123–4, 177, 322; at 1951 Copenhagen conference, 127; and O'Connor, 129; and Salk vaccine, 133–5, 154; not invited onto vaccine advisory committee, 138; declines invitation to Ann Arbor, 151; wins Nobel Prize, 152; British respect for, 160
England *see* Britain
Erickson, Viva, 189, 191–2
Ernest, Margaret, 101–3
Eve, Dr F.G., 196
exercise: value questioned, 208–9

Festival of Britain (1951), 168
Fishbein, Dr Morris, 96, 103, 105
Flanders, Michael, 274
Flexner, Morris, 14
Flexner, Dr Simon, 8, 13–15, 17, 19–22, 24, 26, 64–6
Flynn, Ed, 52
Ford, Edsel, 56
France: produces Salk-type vaccine, 160
Francis, Dr Thomas, Jr: and development of Salk's vaccine, 125, 136; supervises Salk vaccine field trials, 139–40, 143, 147; and announcement of Salk trial results, 149–51; and live-killed virus dispute, 149; quoted, 159; British respect for, 160; neutrality on vaccine types, 181
Frayn, Michael, 168
Freud, Sigmund, 222
Frick, Nancy, 216–17, 221

Friedel, Frank, 33
'frog-breathing' (glossopharyngeal breathing), 197, 298, 302
Frooks, Joseph, 23
Frost, Dr W.H., 8–9, 21, 27

Gallagher, Hugh Gregory, 49, 58, 191–2
gamma globulin, 22, 130–31, 133, 141
Garretson, Mr Justice (New York), 6
Garrity, Tim, 299, 302
Gazette International Networking Institute (GINI), 214, 223, 224, 276, 297, 300
Genskow, Professor Jack, 216
Georgia Hall, Warm Springs, 56
Germany: 1947 polio epidemic, 161
Ghormley, Dr Ralph K., 105
Gilot, Françoise: marries Salk, 158
GINI *see* Gazette International Networking Institute
Glasser, Melvin, 141–2
glossopharyngeal breathing *see* 'frog-breathing'
Goodall, Jane, 322–4
Goodpasture, Ernest, 177
Gordonstoun school, Scotland, 239
Gorrell, Dr John, 190
Gould, Mary: case history, 234–5, 250–52; career and personal life, 266–9
Gould, Tony: personal history and experience, 305–21
Governor's Island, New York, 26–7
Granger, Lester, 189
Greene, Jerome D., 24
Gregory, Cleburne, 43
Grinnell, Sandy Hughes, 220
Gudakunst, Dr Don W., 121
Guthrie, Duncan, 168–9, 201

Hall, Jeff, xiv, 173
Halstead, Dr Lauro, 209–13, 216; case history, 285–9
Hammon, Dr William McD., 130
Hansson, Dr Kristian G., 94
Hargreaves (Chailey Heritage orderly), 246
Hart, Herbert A.L., 294
Harvard Department of Bacteriology, 124
Harvard Infantile Paralysis Commission, 71
Hasbrouck, Paul, 43
Hassett, William, 76, 117–18
Haverstock, Henry, Sr, 98
Haverstock, Henry, Jr, 96–8
Headley Court rehabilitation unit, Surrey, 318–19, 321
Headley, Joan, 223, 276, 294–7
Heatherley (Cheshire Home), 198
Heatley, Mr Frederick, xv
Heine, Dr Jacob von, 11
Henderson, Dr Melvin, 96
Hendon: 1951 polio epidemic, 162
Henry, Anna, 5
Hershey, Pennsylvania, 125, 133, 135
Hickory, North Carolina: 1943 polio epidemic, 112
Higgins, Elizabeth Twistington, 201
HIV virus, 179–80
Hobby, Oveta Culp, 151–2, 155–6
Hoboken, New Jersey, 7
Hoke, Dr Michael ('Mike'), 57, 61, 76
Hooper, Henry, 81
Horstmann, Dr Dorothy, 130–31
Howe, Dr Howard, 134, 170
Howe, Louis, 31, 33–6, 40, 51–2, 77
Hubbard, Dr LeRoy W., 48, 54, 57, 80–81, 93
Hubbell, Susan L., 212
Hughes, Paul, 6–7
Hulett, J.E., Jr, 106
Hull: 1961 polio outbreak, 174–6
Hume, Dr J.F., 190
hydrotherapy, 48, 93–4, 233
hygiene: and causes of polio, 8–9, 16, 26–7, 119–20, 180

Ibsen, Dr Bjorn, 163–4
Ickes, Harold, 78
immunisation: discovery of, 12
India: polio in, xii, 119, 223
Infantile Paralysis Fellowship *see* British Polio Fellowship
insects: as supposed cause of polio, 21
International Poliomyelitis Conferences: First, New York (1948), 127, 161; Second, Copenhagen (1951), 127–8, 162; Third, Rome (1954) 148–9; Fourth, Geneva (1957), 167, 171–2; Fifth, Copenhagen (1960), 159, 177–8, 180, 185, 223

International Scientific Congress on Live Virus Vaccines, Washington (1959), 173
International Year of Disabled People (1981), 205, 253
invalid cars, 257
iron lung (respirator), xiv; developed, 90–91; Sister Kenny opposes, 90, 92; use of, 91–2; made obsolete, 164; in case histories, 232–5, 287–8, 290–91, 298, 305–7, 310–11
Irwin, Dr C.E., 80, 93, 190
Isberg, Anne, 164
Isherwood, Christopher: *Lions and Shadows*, 320
Izvestia (newspaper), 181–2

Jenner, Edward, 12
John A. Andrew Memorial Hospital, Alabama, 83
Johns Hopkins University: polio research at, 85
Johnson, Jean, 282; case history, 276–9
Johnstone, Mary (fictional figure), 217–19
Joseph, Louis, 42–3
Journal of the American Medical Association, 9, 25, 94, 105, 136, 149, 224
Julius Rosenwald Fund, 80

Keegan, Janet and John, 167–8
Keen, Dr W.W., 32–3
Kelly, Frank, 203–5
Kendall, Henry and Florence, 93–4, 98–100
Kenny, Sister Elizabeth, xii, xv; qualities and character, 85–6, 96, 98, 102–3, 106–8; background, 86–7; polio treatment methods, 86–9; 96–101, 110, 193, 235; moves to and works in USA, 89–90, 95–103; opposes use of iron lungs, 90, 92; attacks US after-care methods, 94–5; demonstrations, 98–9; disagreements over and opposition to, 99–101, 104–5; indifference to money, 102–3; honours and awards, 103; rift with O'Connor, 103–4; sets up Foundation, 104; snubbed by NFIP, 145; and Lancelot Walton, 166, 250
Kenny, Mary (Sister K's adopted daughter), 95–6, 103
Kenny Institute, Minneapolis *see* Sister Kenny Institute
Kerr, Randy, 147
Kessel, John, 122
Kingston Avenue Hospital, New York, 21
Kline, Marvin L., 100
Kling, Dr Carl, 15
Knapp, Dr Miland, 98–100
Knowlton, Dr Clint, 208
Koch, Robert, 12
Kolmer, Dr John, 67–8, 70, 125, 146
Koprowski, Dr Hilary, 125–7, 149, 170–73, 177, 179–80
Kriegel, Leonard, 276, 290; case history, 279–83
Kruif, Paul de *see* de Kruif, Paul

Laboratory of Biologics Control (LBC), USA, 140, 150–51, 155
Lake, Frank, 267
Lake, Kathleen, 35–6, 42
Lancet (journal), 126, 209
Landsteiner, Dr Karl, 13, 15
Lane Fox, Felicity, Baroness, 205
Lane-Fox unit *see* St Thomas's Hospital, London
Langmuir, Dr Alexander D., 180
Larooco (houseboat), 44
Larson, Mildred, 299, 301
Lash, Joseph, 31
Lasker, Mary, 158
Lassen, Dr H.C.A., 163–4
Laurie, Joe, 194, 206, 209, 214, 223
Laurie, Virginia Wilson ('Gini'): learns Kenny method, 193; works at Toomey Pavilion, Cleveland, 194–5; revives newsletter, 195; post-polio networking, 197–9, 206, 209, 214, 216, 296; organises 1981 Chicago symposium, 205; and 1983 St Louis conference, 211; and GINI organisation, 214, 223; death, 223; succeeded by Joan Headley, 223, 276, 297; character and style, 296; and Tom Rogers, 300
Lavinder, Dr C.H., 8
Lawick, Hugo van, 323–4
Lawson, Dr Robert, 105–6
Leake, Dr James, 68, 120
Lederle Laboratories, 125–6, 160,

170–72, 177–9, 182
Lee, Michael, 199–200
LeHand, Marguerite ('Missy'): relations with Roosevelt, 37, 39, 44, 52, 61–2; and Warm Springs racial policy, 77–9
Leiser, Dr Oscar M., 22
Lennon, John, 255
Lépine, Dr Pierre, 160
Levaditi, Dr C., 13
Levine, Dr Samuel, 33
Lewin, Dr Philip, 101
Lewis, Sinclair, 64
Lifecare PVEO, 300
Lilly, Eli (drug company), 149
London: polio epidemics, 162
Long Island, New York, 18
Lord, Cissy, 59
Los Angeles: 1934 epidemic, 62; *see also* Children's Hospital School
Lovett, Dr Robert W., 24–5, 33, 35–6, 42, 93, 208
Low, Merritt B.: quoted, 29
Lowman, Dr Charles, 93
Loyless, Thomas W., 42–3, 46–7, 52
Luborsky, Mark, 222
Lunoe, Adam L., 115

Macartney, Paul, 253
McCoy, Dr George, 66
McDonald, Dr William, 50
McDonnell, Dr Aeneas John, 86–8, 97
McDuffie, Irwin, 48
McGinley, Thomas, 7
McGwinn, Donna, 196
McIntyre, Marvin, 74
Macnamara, Dr (*later* Dame) Jean, 65, 88–9, 93, 107
Macon Country, Alabama, 84
Mahoney, Helena, 48–50, 56–7
Mahoney strain (of Type 1 virus), 143, 151, 154, 183
Mailhot, Alice: quoted, 188
Maitland, Drs Hugh and Mary, 123
Maling, Reg, 201
Malta: polio in, 119
'March of Dimes', xi, 74, 112, 135, 182, 186–7, 189, 196, 209, 291
March of Dimes Birth Defects Foundation (*formerly* National Foundation for Infantile Paralysis), 169
marriage: among polios, 196, 200
Marshall, Alan, xiii; quoted, 227
Marshall, Peter, 198–9
Marx, Karl, 121
Mauritius: polio epidemic, 170
Maynard, Dr Frederick, 220, 222
Medical Research Council (Britain), 169
Medical World News: quoted, 188
Medin, Dr Karl Oskar, 11
Mee, Charles L., Jr, 137
Melbourne, Australia, 88–9
Melnick, Dr Joseph, 127, 143, 177
Meltzer, Dr S.J., 21
Mercer, Lucy (*later* Mrs Winthrop Rutherfurd), 30–31, 38, 118
Meriwether Inn, Warm Springs, 42, 46–7, 50, 52; demolished, 56
merthiolate (antiseptic), 150–51
Metcalfe, Dr Richard, 232
Michigan State Medical Society, 146
Michigan, University of *see* Ann Arbor; Poliomyelitis Vaccine Evaluation Center
Middle East: polio in, 119
Milam, Lorenzo Wilson: quoted, 188; at Warm Springs, 192
Milton H. Berry Institute, Van Nuys, California, 95
Minneapolis: Sister Kenny in, 96–8, 100, 103
monkeys: experiments on, 21, 64, 67, 69, 122, 125, 131, 141, 322; and AIDS virus, 179
Montefiore Hospital, New York, 209
Moore, 'Big Jim', 86
Morena, Frederic, 165, 169
Morgan, Isabel, 85, 122, 125, 170
Morgan, Keith: as chief fund-raiser at Warm Springs, 54; and Roosevelt's insurance, 55; on role of Warm Springs, 57, 63; recruits Byoir and Doherty, 60; and Park-Brodie vaccine, 61; on 1934 California epidemic, 62; as Chairman of President's Birthday Committee, 73–4; and Warm Springs racial policy, 77–8, 81; writes to Roosevelt, 112
Morgan, T.H., 85
Mothers' March on Polio, 142
motor cars *see* driving
Munro, Alice, 217
Murray, Rosalind, 174–5

Murrow, Ed, 151

nasal passage: as 'portal of entry', 69
National Academy of Sciences (USA): attitude to Salk, 152; recommends live vaccine, 185
National Association for the Advancement of Colored People, 77–8
National Association for Mental Health, 187
National Cancer Institutes, 158
National Foundation for Infantile Paralysis (USA), xi, xiii, xiv; 20th anniversary, 12; virus research, 70; grant to Draper, 71; formed, 73, 77; de Kruif resigns from, 75; endorses Sister Kenny's methods, 100; non-cooperation with Sister Kenny, 103–4; and outbreak of World War II, 112; wartime fund-raising, 112; and death of FDR, 118–19; supports US Army virus research, 119; Weaver as Director of Research, 121–2, 125; and vaccine development, 135; and Salk vaccine, 138, 160, 181–2; Vaccine Advisory Committee, 138–9, 143–4; voluntary workers and funding, 142, 186; popular support for, 143, 150; and Sabin-Salk conflict, 145, 181; and Cutter incident, 153; status, 157–8; supports Sabin, 157, 182; British Polio Fellowship compared with, 165–6; becomes March of Dimes Birth Defects Foundation, 169; organises International Conferences, 185; criticisms of, 186–7; number of patients, 187–8; sets up respiratory centres, 193
National Fund (Britain, *formerly* Polio Research Fund; *later* Action Research for the Crippled Child), 169, 201
National Health Service (Britain), 255
National Institutes of Health (USA), 140–41, 144, 156
National Microbiological Institute (of NIH), 141
National Patients Committee (Warm Springs alumni), 56
Nebraska: 1952 polio epidemic, 189
negroes *see* race (ethnic)
Netherlands: polio in, 226
Netter, Dr A., 13
New Deal, 72
New England Journal of Medicine, 153
New York: 1916 epidemic, xi, 3–28
New York Academy of Medicine, 71
New York Herald Tribune, 153
New York Journal, 5
New York Sun, 78–9
New York Times, 4, 7, 9, 15, 23, 26, 187
Nixon, Richard M., 158
Northern Ireland: medical research in, 170
Nottingham University, 234–5, 251
Nuffield, William R. Morris, Viscount, 234
Nuremberg code, 127

occupational therapy, 241
O'Connor, Basil: co-founds NFIP, 12; partnership and friendship with FDR, 44–5, 156; and Mike Hoke, 57; and Draper's application for grant, 71; Presidency of NFIP, 73; and de Kruif, 75; management of Warm Springs, 76–7; and polio racial policy, 79–84; encourages Tuskegee Institute, 82–4; quoted on Sister Kenny, 85; in Warm Springs Hall of Fame, 85; and use of iron lungs, 92; and after-treatment, 93; meets Sister Kenny, 96; relations with Sister Kenny, 103–4, 106; honours, 111; on Salk vaccine, 111, 138; and outbreak of war, 112–13; at 1944 Warm Springs Founder's Day, 117–18; and FDR's health decline and death, 117–18; sustains NFIP after FDR's death, 118–19; criticises and activates researchers, 121–2; relations with Salk, 129–30, 137, 144, 152–3, 158; praises Rivers, 132; and Salk vaccine field trials, 133, 139–41; and Weaver-Van Riper conflict, 139; exploits popular support of NFIP, 143, 150; and live versus inactivated vaccine conflict, 143; on Bodian, 147; commits money to experimental vaccine doses, 148; and announcement of results of Salk field trials, 150–51; condemns National Academy of Sciences'

refusal to admit Salk, 152; and Cutter incident, 153–4; blamed for premature use of Salk vaccine, 155–6; circulates Benison's letter, 157; and popular demands for NFIP successes, 158; remarries after wife's death, 158; and conflict with Sabin, 173; disagreement with Rivers, 181; questions Sabin vaccine safety, 182; on infection risk from Salk vaccine, 184; commitment to polio eradication, 186; encourages volunteer principle, 186; salary, 186; character and personality, 187; death and obituary, 187; opposes federated fund-raising, 187; advocates total care, 189; and monkeys for research, 322
Okatie Farm, near Hardeeville, South Carolina, 322
Olitsky, Dr Peter K., 64, 67, 124
O'Sullivan, Maureen: case-history and life, 231, 243–5, 256–7; marriages and motherhood, 257–63
Oursler, Fulton, 60
Owen, Dr Richard, 108–10, 193, 212, 222
Oyster Bay, Long Island: 1916 polio outbreak, 18
Oz (magazine), 254

Pait, Charles, 122
Pan-American Health Organisation, 173, 225
Park, Dr William H., 66–9
Parritt, Simon: case history, 233–4, 235–8, 242; career and personal life, 263–6
Pasteur, Louis, 12–13, 119
Patterson, Colonel Dr, 286–8
Paul, Dr John R.: quoted, 3, 28, 159; researches, 11, 15; and 1934 California polio epidemic, 62; on Flexner's view of polio virus, 65; on vaccine research, 67–9; on Dr Brodie, 68; on physiological causes of polio, 71; on selective use of iron lungs, 91–2; on excessive immobilisation, 101; on discrediting of Kenny, 105; serves on US Army Epidemiological Board Commission, 119; on polio as enteric viral infection, 120, 122; and Enders, 124; and research at NFIP, 125; and ethics of experimentation on humans, 127; absent from Hershey meeting, 133; encourages Salk, 133–4; favours live-vaccine immunisation, 135; and Salk's vaccine, 136; and use of inactivated vaccine, 143–4; on Dick's denunciation of live-virus principle, 171; on Sabin and WHO committee, 180; and examination of Sabin vaccine, 182; in live-killed vaccine dispute, 183, 185; and infection risk from vaccination, 184; on experimental use of monkeys, 322
Pauling, Linus, 186
Peabody, Dr Francis, 17–18, 22
Peabody, George Foster, 42, 46
peanut oil: as nostrum for polio, 81–2
Pearl Harbor, 112
Peiping Union Medical College, China, 91
Perkins, Frances, xiii, 33, 35, 39, 44, 52
Perlman, Itzhak, xiii
Perry, Dr Jacqueline, 221
Peter Bent Brigham Hospital, Boston, Mass., 91
Pfizer Ltd (of Folkestone), 175
Philippines: polio in, 119
Phipps unit (St Thomas's Hospital, London), 205
Physiotherapy: in after-care, 93, 208; in individual case-histories, 237–41
Piel, Gerard, 187
Pius XII, Pope: on moral limits of medical experimentation, 126
Plagemann, Bentz, 113–17, 218
Plastridge, Alice Lou, 56, 98, 100, 113, 117
Plymouth (England): 1911 polio outbreak, 16–17
Pohl, Dr John F., 97–9, 101, 105–6
polio, xi–xvi; racial incidence, 8, 81, 83, 181; etymology and names, 9–10; history of, 9–10; symptoms, 10–11, 17–18; viral origins, 13, 15, 65–6, 70, 75, 120; becomes notifiable in Britain (1911), 16; popular beliefs about causes, 20–21, 215; early prevention and treatment, 21–5; after-care, 23–4, 92–5, 193–7; transmission and infection, 25–8;

and sexual activity, 37, 199, 218–19, 247, 250, 264; geographical distribution and incidence, 62; US research into prevention, 63–5; constitutional and genetic predisposition theories, 70–71, 75, 216–17, 222; and improved hygiene, 119–20; as intestinal disease, 120; natural immunities, 120; periodic epidemics, 120; US cases increase, 135; in Europe, 161; age susceptibility, 184–5; support groups, 214–16, 223, attitudes to sufferers, 217–19; world-wide incidence and propects, 223–6
Polio Chronicle, 55
Polio Network News, 295–6
'Polio Pioneer' buttons, 147
Polio Research Fund (Britain) *see* National Fund
Poliomyelitis Evaluation Center, University of Michigan, 140
Polk State School, Venango County, Pennsylvania, 126, 132
Poplar (London), 16
Popper, Dr Erwin, 13
Porteous, Jim: case-history, 231–2, 238–41; career and personal life, 273–5
'Possum' (Patient Operated Selector Mechanism), 201
post-polio deterioration, xii–xiii, xv, 210–15, 221; *see also* rehabilitation and after-care
President's Birthday Ball Commission for Infantile Paralysis Research (PBBC), 63–6, 69, 71, 74, 120
Prime Time Saturday (TV programme), 206
provocation poliomyelitis, 178
Public Health Bulletin No. 242 (USA): *Care During the Recovery Period in Paralytic Poliomyelitis*, 94
Public Health Service *see* United States Public Health Service

Queen Elizabeth's Training College for the Disabled, Leatherhead (Surrey), 245, 256
Queen Mary's Hospital, Carshalton, 88–9, 232, 235–9
Queensland, Australia: and Sister Kenny's methods, 87–9

'Rabbit, The' (Louis Howe's secretary), 77
race (ethnic): and polio incidence, 8, 81, 83, 181; and admission to Warm Springs, 77–81, 189–91, 291; and Tuskegee Institute, 82–4; and syphilis study, 84
Raleigh, North Carolina: 1935 polio outbreak, 66
Rancho Los Amigos Hospital, Los Angeles, 194, 199, 221
Raskob, John J., 52
Reader's Digest, 101
Redbook (magazine), 181–2
rehabilitation and after-care, 193–201, 205–12, 214–16, 219–20; *see also* post-polio deterioration
Rehabilitation Gazette (formerly *Toomey j Gazette*), 194, 206, 209, 211, 214, 218, 276; conducts survey, 196–7
religion, 267–9
respirators, positive-pressure, xiv, 164, 272, 301; and rocking, 196; *see also* iron lung
respiratory centres (USA), 193–5
Rivers, Dr Thomas: and funding of Rockefeller Institute, 64; serves on PBBC, 66, 69, 74–5; accuses Kolmer, Park and Brodie of vaccine haste, 68; questions Draper's views, 71; declines to meet Sister Kenny, 96; and US Army virus research, 119; praises Weaver, 122; resistance to Weaver, 125; and Koprowski's trials, 126; and Sabin, 128; disputes over monkeys with Bodian, 131; and Salk vaccine trials, 133, 136, 138, 140, 149; on British temporising over Salk vaccine, 160; dispute with O'Connor, 181
Robbins, Dr Frederick, 123–4, 127, 322
Rockefeller Foundation, 23
Rockefeller Hospital, New York, 17
Rockefeller Institute for Medical Research: early polio research, 13, 15, 17; de Kruif satirises, 64
Rockefeller, John D., 20
Rockey, Edna, 35–6
rocking: as respiratory aid, 196
Rogers, Kera, 301, 303–4
Rogers, Thomas: case-history, 297–301; marriage, 301–3; death, 304
Roller, Sunny, 220, 222

Rolling Stone (magazine), 179
Roosevelt, Anna (FDR's daughter), 34, 37
Roosevelt, Eleanor: marriage, 30–32, 38, 80; and FDR's polio, 33–6; conflict with mother-in-law, 34–7; momentary collapse, 36; on Warm Springs, 49; and FDR's political career, 51–2; receives begging letters, 61–2; and Draper's research grant, 71; and Warm Springs racial policy, 77, 79–80; in Warm Springs group photograph, 144
Roosevelt, Elliott (FDR's son), 36–7, 39
Roosevelt, Franklin Delano, xi, xiii, xv; co-founds NFIP, 12; on New York polio epidemic, 25; career, 29–30, 32; marriage relations, 30–31, 33, 38, 80; contracts polio, 32–5; and family tensions, 34–5; interests, 37; relations with Missy LeHand, 37–9; sex life after polio, 37–8; effect of polio on life, 39–40, 59, 116–17; sunshine treatments, 42–4; visits Warm Springs, 42–4, 58, 114–18; partnership and friendship with O'Connor, 44–5, 156; acquires and develops Warm Springs, 45–50, 53, 54–6, 76; attempts walking with cane, 50–51; runs for and wins governorship of New York, 51–2; quoted, 54; life insured in name of Warm Springs Foundation, 55; Presidencies, 56, 58, 72; photographed, 58–9; reaction to disability, 59–60; Birthday Balls, 60, 63, 72, 77–8, 103; declining popularity and support, 61, 72–3; correspondence with Draper, 72–3; and New Deal, 72; and polio crusade and appeal, 74, 112–13; and Carpenter, 77; and racial policy at Warm Springs, 77–81, 190–91; endorses use of peanut oil, 82; in Warm Springs Hall of Fame, 85; meets Sister Kenny, 103; and outbreak of war, 112; at Yalta, 117–18; health decline and exhaustion, 117–18; death, 118, 150
Roosevelt, Franklin, Jr, 38, 191
Roosevelt, James (FDR's son), 29, 37–9, 51, 58
Roosevelt, James Roosevelt 'Rosy' (FDR's half-brother), 34
Roosevelt, Sara (FDR's mother), 31, 34, 36–7, 52
Roosevelt, Theodore: holidays at Oyster Bay, 19; relationship to Eleanor, 30
Roosevelt Warm Springs Institute for Rehabilitation, 189; *see also* Warm Springs, Georgia
Rosenman, Sam, 38
Rotary International (Rotarians), 225, 250
Round Table (organisation), 239, 246
Royal Grammar School, High Wycombe, 246–7
Royal National Orthopaedic Hospital, Stanmore (Middlesex), 236–7
Russell, Dr Ritchie, 176
Russell, Rosalind, xii
Russia: polio research in, 172–3, 176–7, 183
Rutherfurd, Winthrop, 38

Sabin, Dr Albert, xii; changes from dentistry to medicine, 63; on US Army Board Epidemiological Commission, 119; virus research, 124, 177; supports live-virus vaccine, 125, 135, 143–4, 157, 173; relations with Salk, 127–9, 133–4, 135–7, 144–5; opposes field trials of Salk vaccine, 133–4; provokes NFIP, 145; and Salk trial results, 151; warns of Mahoney strain of virus, 154; vaccine delayed by Salk setback, 157; character and appearance, 159, 176; experimentation in Russia, 172–3, 183; action of oral vaccine, 175; oral vaccine accepted in Britain, 176; vaccine licensed in USA, 179, 183; and Koprowski, 180; effectiveness of vaccine, 181, 184; production and marketing of vaccine, 181–2; field trials in USA, 182; on moral issues of vaccine risk, 183, 185; infection risk from vaccine, 184–5; death, 185; at 1985 St Louis conference, 224; on world-wide incidence of polio, 224; on Brazil's vaccination programme, 225; vaccine success and popularity, 225
St Louis: post-polio conferences, 211,

214, 224, 296
St Thomas's Hospital, London, xv, 202–3, 205, 272–3
Salk, Dr Jonas Edward, xii, xv; develops vaccine, 22, 68, 111, 125, 130, 134–6, 140, 143–4; works on investigating virus types, 123–4; vaccine trials on children, 126, 132–3, 136–7, 139, 141, 149; relations with Sabin, 127–9, 133–7, 144–5, 173; friendship with O'Connor, 129–30, 144, 152–3, 158; premature publicity, 136–8; broadcast on vaccine, 137; Winchell attacks, 146; test results announced, 150–51; honours and fame, 152; refused election to National Academy of Sciences, 152–3; vaccine approved and licensed, 152; and vaccine production problems and dangers, 153–7; and risk of 'clumping', 156; effectiveness of vaccine, 157, 160, 180–81, 184; divorce and remarriage, 158; British caution over, 160; on vendettas and partisanship, 177; declares vaccine safe, 182; vaccine effectiveness and age susceptibility, 184; infection risks from vaccine, 185; opposes live vaccine, 185; vaccine combined with DPT vaccine, 225; uses monkeys in research, 322
Salk Institute, California, 158
salt tablets, 279
sanitation *see* hygiene
Scheele, Surgeon-General Leonard A., 151, 153–5
Scheer, Jessica, 222
Schlosser, Mrs (Warm Springs physical therapist), 114
Schneider, Larry, 206
Schulman, Sammy, 58
Schultz, Professor Edwin W., 67, 69
Schwartz, Dr Oscar, 300, 303
Schwartz, Dr Plato, 101
Science (magazine), 124
Scott, Sir Walter, 10, 113
scouts (Boy Scouts), 240
Sebrell, Dr W.H., Jr, 156
Seddon, Dr H.J., 170
Seidenfeld, Dr Morton, 191
serum therapy: early development, 21–2
Sexual and Personal Relationships of People with a Disability (SPOD), 264
Shannon, Dr James A., 156–7
Shaw, Louis, 90
Shepherd, Dr Gwendolyn, 195
Shoumatoff, Elizabeth, 118
'Silverwood' *see* Cobham (Surrey)
Simha, Emelie, 189–91
Singapore: 1958 polio epidemic, 175
Sister Elizabeth Kenny Foundation, 104, 173
Sister Kenny Institute, Minneapolis, 103, 110
Smadel, Dr Joseph E., 133, 135–6, 138
smallpox, 12
Smith, Al, 42, 51–2
Smith, Diane, 190
Smith, Jane S., 128, 132, 147–8; quoted, 188
Smorodintsev, Anatoli, 173, 183
Snite, Frederick B., 91
Soloviev, Dr Valentin, 172
Soltan, Dr A. Bertram, 16
South Devon and East Cornwall Hospital, 17
South Western Hospital, Brixton (London), 202
Southern Medical Association: annual meeting, Richmond, Virginia (1942), 101
Speier, Dr Jennine, 215
Spencer, Dr Geoffrey, xv, 162, 164, 202–5, 212–13, 223, 272
spinal fusion, 242
splinting, 92–3, 101
Spring, Dr William C., Jr, 183
Stalin, Josef V., 115
Stanley, Dr Wendell, 75, 155
Stanmore, Middlesex *see* Royal National Orthopaedic Hospital
Starr, Dr (osteopath), 42
State of the World's Children: 1993 report, 225
Sternberg, Dr George M., 13
Steuart, Dr (of South Africa), 90
Stevens, Ann: case-history, 232–3, 241–3; career and personal life, 269–73
Stevenson, Jesse L.: *Care of Poliomyelitis*, 94
Stimson, Dr Philip, 101
Stockholm: 1887 polio outbreak, 11

Stoke Mandeville Hospital, Buckinghamshire, 201, 240
Stonehouse, Devon: 1911 polio outbreak, 16
Strauss, Elaine, 189
stretching, 237–8
Sunny View (school, Pittsburgh), 277
surgery: as treatment, 248
survivors (polio) *see* rehabilitation and after-care
Sweden: 1911 polio epidemic, 15; *see also* Stockholm
syphilis: studied among US black males, 84

Talmadge, Eugene, 63
thalidomide, 178
Thayer, Dean (of Harvard), 79
Thomas, Dylan, xvi
Time magazine, 136
tobacco-mosaic virus (TMV), 75
Toluca, Mexico, 224
Toomey, Dr John A., 120, 193–4
Toomey j Gazette see *Rehabilitation Gazette*
Toomey Pavilion, Cleveland, Ohio, 193–5
Tower Hotel, London, 274
Townsville, Queensland, 87–8
tracheotomy, xiv, 164
Trask, Dr James D., 120, 122
Truman, Harry S., 117–18
Turner, Dr Thomas B., 133
Tuskegee Infantile Paralysis Center, 83–4
Tuskegee Institute, Alabama, 81–4, 190

Underwood, Dr Michael: *Diseases of Children*, 10
United Nations Children's Fund, 225
United States of America, xi, xii; ethical standards in medical experimentation, 126–7; numbers of polio cases, 135, 193; vaccination campaign, 177; licenses Sabin vaccine, 178–9; aims to eliminate polio, 224–5
United States Army Epidemiological Board Commission on Neurotropic Virus Diseases, 119
United Stated Public Health Service, 8, 21, 27, 141, 153–4, 180–82, 184

vaccination: practicality, 181; infection risks from, 183–4
vaccines: development of, xii, xiv, xv, 12–13, 66–9, 75, 111, 130–39; and virus types, 122–3; conflict between live/attenuated and killed/inactivated, 125–6, 134–5, 143–4, 149, 171, 181–5; field trials, 139, 141–3, 146–50
Vaizey, John, Baron, 200
Van Riper, Dr Hart, 137, 139, 143
Veeder, Mary Hudson, 50–51, 57
ventilators *see* respirators
Vermont: 1894 polio outbreak, 11, 208
viraemia, 130–31
viruses: early investigation of, 12–13, 15, 65–6, 70, 76; and periodic epidemics, 120; types, 122–4, 143, 160, 170–71, 175; cultivation, 123–4, 177, 322; and development of vaccine, 125, 130

Wackers, Dr G.L., 163
Walker, Flora, 207
Walker, Josephine: case history, 283–5
Walker, Turnley, 44, 54, 191–2, 207
Walker-Smith, Derek, 173
Walton, Lancelot, 166, 250, 266
Warm Springs Foundation: founded, 51; administration and fund-raising, 54, 56, 60–61; political connections, 60–61, 63; exercise policy, 208–9; support group function, 214
Warm Springs, Georgia, xi–xii, xv; Polio Hall of Fame, 12, 85, 144; character and atmosphere, 41–2, 76–7, 114, 190–93; FDR visits, 42–4, 58, 114–18; FDR purchases and develops, 45–6, 49–50, 53; attracts and treats polio sufferers, 46–50, 54, 57; hydrotherapy at, 47–9, 93–4, 188; benefactors, 52, 56; and FDR's political career, 52–3; administration, 54–6; constructions and improvements, 56; personnel changes, 56–7; charges and costs, 57; FDR's connection seen as detrimental, 73, 77; racial policy, 77–81, 189–91, 291; Negro School, 80; 'Poliopolitan Opera Company', 82; after-care methods, 92–4; wartime work, 113; swimming pools decay, 188–9; bought by state,

189; 1983 post-polio conference, 210, 212; and respiratory cases, 291
Warren, Earl: quoted, 111
Washington, Booker T., 81
Washington DC Polio Society, 215
Washington, Louis, 83
Wausau, Wisconsin, 142
Weaver, Dr Harry: as Director of Research at NFIP, 121–3, 125, 135, 157; and Copenhagen Conference, 127; praises Salk, 129; and vaccine trials on children, 133; Salk vaccinates, 138; conflict with Van Riper, 139; and Vaccine Advisory Committee, 139; and O'Connor, 144
Weaver, Robert C., 78
Weller, Dr Thomas, 123–4, 127, 322
Wheatley military hospital, Oxfordshire, 317
White, Walter, 77–9
Wickman, Dr Ivar, 11–12, 15
Widgery, Dr David: case-history, 166, 229–30, 247–9; political stance, 254–5; death, 256
Wiechers, Dr David O., 210–12
Willard Parker Hospital, New York, 21–3, 70
Williams, Greer: quoted, 3, 111, 159; on Brodie, 68; on approval of Salk vaccine, 152
Williams, Dr Stanley: quoted, 85
Wilson, Earl, 136–7, 145, 148, 182
Wilson, Sir Graham, 185
Wilson, John Rowan: quoted, 2, 159; on early vaccines, 13; on Koprowski's volunteers, 127; on official acceptance of Salk vaccine, 151; and British conservatism over vaccines 160; on National Fund, 169; on Herald Cox, 171; on Sabin and Russian research, 173; on British vaccination campaign, 176; on success of Sabin vaccine, 176; criticises NFIP, 186
Winchell, Walter, 145–7, 153
Wistar Research Institute, Philadelphia, 172, 179
Woillez, Dr (of Paris), 90
Workman, Dr William, 140
Worksop, Nottinghamshire, 10–11
World Health Organisation, xii, 157, 173, 180, 223
Worthing (Sussex): Lantern Hotel, 166
Wright, Dr Jessie, 132, 196
Wright, Dr Wilhelmine G., 93
Wrigley, Tom, 74
Wyatt, H.V., 178

Yale Poliomyelitis Unit, 69, 120
Yalta meeting (1945), 117–18

Zaire: AIDS in, 179
Zinsser, Hans, 123